D0875215

Rattlesnakes

Rattlesnakes

Their Habits, Life Histories,
and Influence on Mankind

Laurence M. Klauber

Abridged by Karen Harvey McClung

UNIVERSITY OF CALIFORNIA PRESS

Berkeley, Los Angeles, and London

Published for the Zoological Society of San Diego

University of California Press
Berkeley and Los Angeles, California
University of California Press, Ltd.
London, England
© 1982 by
The Regents of the University of California
Printed in the United States of America

1 2 3 4 5 6 7 8 9

Library of Congress Cataloging in Publication Data
Klauber, Laurence Monroe, 1883–1968.
Rattlesnakes, their habits, life histories, and influence on
mankind.
Bibliography: p.
Includes index.
1. Rattlesnakes.   2. Snake venom.   I. McClung,
Karen Harvey. II. Title.
QL666.069K52 1981   597.96   80-16660
ISBN 0-520-04038-4   AACR1

Contents

5. Behavior

6. Populations and Ecology 103

7. Food 123

8. Reproduction   142

9. Poison Apparatus   157

10. The Bite and Its Effects 167

11. Treatment and Prevention of the Bite 195

12. Control and Utilization 233

13. Enemies of Rattlesnakes 255

Publisher's Foreword

When Laurence M. Klauber died in 1968 at age eighty-four, one newspaper said he provided "an extraordinary example of what a man can do with his life." Rare is the person who achieves world eminence in a scientific field; rarer still, the person who does so when that field is merely a hobby. Such a person was Laurence Klauber, who did not begin a serious study of reptiles until he was in his forties.

An electrical engineer by training (BSEE Stanford University, 1908), Mr. Klauber joined San Diego Gas & Electric Company (SDG&E) in 1911 as a salesman. By 1949 he had become its chairman and chief executive officer, the position from which he retired in 1953. His success carried over into community service as well, for Klauber joined several scientific societies and was president of three; he was also president of two trade associations, the local Rotary Club, and the San Diego Zoo. During World War II he became the chief of civilian defense in San Diego and later was chairman of the city Library Commission.

Despite his professional involvement in SDG&E and his extensive community service, Klauber looked for still more ways to engage his curiosity. While on family picnics in the rugged San Diego back country, he began collecting snakes and lizards. But when he donated these specimens to the young and struggling San Diego Zoo, he found no one available to care for them and promptly volunteered to be the zoo's first Curator of

Reptiles. In time Klauber's fascination with the literature and taxonomy of the reptiles of the Southwest became more and more absorbing. As the collection grew and his knowledge broadened, he began publishing scientific papers that commanded immediate attention. These papers treated many species of snakes and lizards, but the rattlesnake eventually captured most of his attention.

In 1936, after having worked on the classification of rattlesnakes for fifteen years, Mr. Klauber published the results in the form of an identification key. He had previously noticed, he stated in the preface to the first edition of *Rattlesnakes*, "that these creatures commanded a sort of fearsome interest, even among people who certainly had no fondness for them. So I prefaced the key with a brief summary of rattlesnake ways of life, what they did and how, with the result that this pamphlet soon became an out-of-print rarity."

"Obviously," he continued, "it was not the key to the identification of the different kinds of rattlesnakes that gave the booklet its brief popularity; rather it was the short discussion of the lives and habits of the snakes. And I had observed, although I myself never brought up the subject, that rattlers, despite — or maybe because of — their sinister reputations, would always bridge a dull spot in a dinner conversation." So in 1945 he began work on what was to have been a revision of the 1936 pamphlet, except that the identification keys were to be subsidiary to the observations on life history. As to the latter, he expected to dwell largely on a single species — the western rattlesnake — which, in one of its several varieties, was the kind most often encountered by the covered-wagon emigrants westbound after crossing the Mississippi.

But the work grew vastly beyond the scope once contemplated, and much technical material of little interest to the casual seeker for the facts of rattlesnake life was included. Nevertheless, a consistent effort was made to supply the needs of "both the person who seeks some single fact to settle a bet, as well as the student with a broader purpose." In a way, the book became an encyclopedia of the rattlesnake; nearly all aspects of both the rattlesnake's environment and man's reactions to rattlesnakes were surveyed.

The result was a monumental 1500-page, two-volume book entitled *Rattlesnakes: Their Habits, Life Histories, and Influence on Mankind*, published in 1956 by the University of California Press. It received the highest critical acclaim and the author became internationally recognized as the outstanding authority on the subject. Over the next several years, Mr. Klauber continued to collect information for an eventual revised edition, but his work was interrupted by illness early in 1968 and terminated by his death on May 8, 1968. Because of the continuing high demand for the volumes, in 1972 a partially revised edition of *Rattlesnakes* was issued in the form in which Mr. Klauber left it. This second edition incorporated some of the changes and additions made possible by new knowledge of rattlesnakes obtained by herpetological students and field observers during the intervening years.

Several years later, the publisher decided to issue an abridged edition of *Rattlesnakes* that would concentrate on the less technical aspects of rattlesnakes' lives and their interaction with people, making this fascinating subject accessible to a much greater audience than it was in the original version. What remains is an eminently readable and inclusive text that can be enjoyed by anyone with an interest in snakes in particular or natural history in general.

Because several procedures in the treatment for rattlesnake bite have changed subsequent to Mr. Klauber's writing and in an emergency someone might look to this book for help, we asked Dr. Findlay E. Russell, Director of the Laboratory of Neurological Research, School of Medicine, University of Southern California, a leading authority in the field of snakebite treatment, to provide updated first aid information for this edition.

Dr. Russell has also kindly supplied revisions of the rattlesnake distribution maps, as the boundaries have changed considerably since Mr. Klauber prepared his maps. The publisher wishes to express appreciation to Dr. Russell for his contributions; and also to Dr. Nathan W. Cohen for reviewing the abridged edition before publication, and for providing a select bibliography and some of the photographs for this edition; to the Zoological Society of San Diego for supplying their photographs; and to Philip M. Klauber, Laurence Klauber's son for giving us the biographical information on his eminent father.

Abridger's Note

Because Mr. Klauber's writing style is clear and lively, I have tried to retain the wording of his original text wherever possible. Two chapters clearly intended for the specialist, "Classification and Identification" and "Paleontology and Zoogeography," have been entirely eliminated, as has most of the massive bibliography. With the rest of the book, because I felt its encyclopedic quality an attractive feature in itself, I have deleted the more technical parts of each section, rather than eliminating entire sections. Mr. Klauber's meticulous organization and high standard of presentation made this separation possible with very little rewriting. Other elements that have been deleted are the many references to the bibliography, some of the numerous quotations from Mr. Klauber's correspondents, and the repetitive material that is inevitable—sometimes intentional—in a work of this magnitude.

What I hope the reader will find in this abridged edition is not only the highlights of Mr. Klauber's surprisingly engaging account of rattlesnakes and people's reactions to them, but a heightened sense of Laurence Klauber as a scientist and a person—the way he conducted his experiments, the quality of his mind, his sense of humor, his life-time effort to study and present almost every detail of rattlesnake life and to disentangle the real rattlesnake from the creature people imagine it to be. Laurence Klauber was no rattlesnake booster—he advocated their destruction wherever they posed a danger to people—but how he must have loved his dangerous avocation, and the scientific method.

The decision to preserve the book as a kind of period-piece, with the charm of its original style, means that with the exception of the section on treatment for rattlesnake bite, most of the material has not been updated, though rattlesnake research has undoubtedly slithered forward in the past several decades. Nevertheless, for the most part Mr. Klauber's conclusions are still considered definitive, and the scientific aspects of the book retain their usefulness.

In characteristically wonderful spirit, the unabridged edition of this book was dedicated to Mr. Klauber's wife Grace, who "endured a basement full of rattlesnakes for more than thirty years."

Karen Harvey McClung

Introduction

Of all the books or pamphlets containing information about rattlesnakes, two that have probably most influenced or guided public ideas on the subject were written by men who certainly never saw a rattlesnake in its native habitat, and may not even have seen one alive. These authors were Oliver Goldsmith and the Rev. John G. Wood, neither a naturalist, both rather inaccurate and credulous compilers, but each with the gift of interesting popularization. So cherished were their natural histories—first published in 1774 and 1851, respectively—that they appeared in unnumbered editions; they were reissued, reprinted, revised, enlarged, pirated, and quoted without credit. They were read avidly by the children of successive generations and remained the standard natural-history reference works in many British and American homes down to the present century.

What made these and similar works so deservedly popular was not so much their vivid descriptions of the animals themselves as the information on their habits and the places where they live, often exemplified by stories of human encounters with them. It was not entirely the fault of these authors that many of their accounts were inaccurate, for naturally they were dependent for their information on travelers abroad who had had opportunity to make field observations. The compilers had no way of winnowing actual observations from myths and tales the travelers brought back. This difficulty of separating fact from fiction is particularly formidable in the case of rattlesnakes, creatures whose very nature invites exaggeration. Many rattlesnake stories still believed today date back to misunderstood incidents of colonial days, or to tales invented at the campfire to spoof a gullible traveler.

Purpose

This book is written to outline our present knowledge of rattlesnake habits and life histories. It includes numbers of field observations from varied sources in the hope that it may aid in the correction of some of the dubious accounts long current in popular natural histories.

I have no desire to exaggerate the importance of rattlesnakes in the scheme of nature, or in their influence on mankind. To people going afield they constitute a relatively minor danger, yet the hazard is sufficient, or is believed to be, to cause many persons to suffer almost continuously from fear of rattlers when in the woods or brush. An unexpected encounter with a rattler has broken up many a picnic party; the fear of meeting one has kept many another from ever leaving home. Those most familiar with these snakes learn to take them in their stride; they soon find that, compared with man-made hazards, rattlesnakes constitute a negligible danger.

Along with their danger, real or imagined, rattlers are of economic value to the farmer or stock raiser, for they are predators on injurious rodents. They are handsome yet sinister creatures, with curious ways of life. They are expert performers on a musical instrument they themselves cannot hear. I infer from the conversations of visitors at the San Diego Zoo, from letters of inquiry, and from the frequency of rattler items in the newspapers, that the general interest in them is great. In fact, throughout this book I have dwelt to a considerable extent on the relationships of rattlesnakes and men, for certainly one of the most remarkable aspects of rattlesnakes has been their effect on people. Quite apart, and often quite different, from their existence as reptiles in the forest or desert, rattlesnakes have had an existence in the minds of men—in unnatural natural history, in myth and folklore, in primitive medicine, and even in aboriginal religion. Certainly, the rattlesnake of song and story is a creature that quite surpasses nature. It is my hope that this book may lead to a better understanding of rattlesnakes, and this objective can be achieved only by including a survey of some of the less accurate ideas, their sources, and their deviations from field experience—as well as the factual information available. What people have thought about rattlesnakes and why is, in its way, as interesting as the snakes themselves are in another.

Sources of Information

In compiling these life histories of rattlesnakes, I have used four sources of information: (1) correspondence with field observers and naturalists; (2) published accounts; (3) studies of captive rattlesnakes made at the San Diego Zoo; and (4) the personal observations of the writer in the field and laboratory.

Data from Correspondents. There are certain people whose occupations keep them out-of-doors and in continuous contact with nature. They become naturalists in the best sense of the word. When they meet a rattler in the field it's a part of the day's work, not something to form the basis of a sensational story. Their observations are usually sound and accurate, in the aggregate comprising a volume of material far beyond the field notes of even the most fortunate and active herpetologist. What these observations may lack in coordination and continuity, they make up for in the corroborative evidence of their widespread sources.

Prior to the preparation of this book, I sent more than three thousand questionnaires on aspects of rattlesnake life to a variety of outdoor people, including National Forest rangers, U.S. Fish and Wildlife Service employees, Soil Conservation Service workers, National Park naturalists and rangers, and state game wardens and patrol officers. Other inquiries were sent to a list of field naturalists, hunters, trappers, stock and poultry raisers, county agricultural agents, and others likely to have firsthand information on the subject.

During the past years I have carried on an extended correspondence on the subject of rattlesnake life histories and habits with amateur and professional herpetologists who have had much experience with these snakes. From them I have also secured many useful observations.

Printed Material. My second source of information has been published material. I have examined a large number of books and articles, either about rattlesnakes or containing pertinent incidental statements on their habits. Some of these I have quoted verbatim; to many others I have made reference, summarizing the authors' observations or conclusions.

The literature on rattlesnakes is extensive, for not only are there many technical articles on the subject, but there are general natural histories, books of travel, medical journals, ethnological reports, nature magazines, hunting and fishing periodicals, and, finally, Sunday supplements, all containing material of interest, if not always of sound value.

Observations at the Zoo. Although artificial conditions under which captive specimens live may distort their behavior patterns, it is still possible to gain many facts of value by observing them. Of this we have taken advantage at the San Diego Zoo, where thousands of rattlesnakes of many species have passed through our hands since 1922. The exhibit series at the Zoo comprises only a part of the many specimens kept under observation. Feeding and mating habits, venom yields, shedding, and other activities have been recorded, some of which can only be studied in captive specimens. C. B. Perkins, who was in charge of the collection since 1932, and his successor, Charles E. Shaw, have been unusually successful in simulating natural conditions, and their efforts have been rewarded with new records of longevity and of breeding in captivity. Under their supervision many original or confirmatory data on rattler habits have been secured.

Personal Field and Laboratory Experiences. Occasionally, in judging the dependability of articles on rattlers, I have wondered about the extent of the experiences of the authors and the backgrounds of their statements. Since many of my readers may mentally raise the same question respecting the validity of this compilation, I trust I may be pardoned for summarizing my own experience.

I have been interested in rattlesnakes for more than sixty years and during the past forty have put in whatever spare time has been available in a study of snakes in general and rattlers in particular. Some of this work has been in the field—for I have collected extensively in the Southwest—but more in the laboratory. In connection with these studies, some of which have been published, I have accumulated scale counts, color notes, and measurements from about 12,000 rattlesnakes, of which some 7,500 were preserved in my own study collection. I have seen specimens of all the kinds of rattlesnakes known to exist today, most of them alive. In the course of a venom-gathering program, I extracted the venom from somewhat more than five thousand live rattlers.

My training, however, has been in engineering, rather than in biology or medicine, and my lack of technical training in these fields has placed certain obvious limitations on this work. Particularly, the reader will find little on the physiology of the rattlesnakes, a subject I should be ill-equipped to discuss. However, much material is available elsewhere for readers with a particular interest in this aspect.

CHAPTER ONE

Status

It is sometimes surprising to learn the extent to which the public may misunderstand basic terms. Thus, although the term "rattlesnake" is familiar to every American, it is astonishing to find what different ideas people have as to the kinds of creatures included by the term. To obviate this confusion, and to permit the nonherpetologist to orient himself with respect to the position of rattlesnakes in the snake world, the summary below is presented.

Classification and Nomenclature

Animals are classified by division into groups of successively narrowing scope. Thus the animal kingdom is divided into a few main groups, first into phyla, then the phyla into subphyla, and these, in turn, into classes. One of the classes of the subphylum Vertebrata, of the phylum Chordata, is the class Reptilia, which includes all the reptiles, living and extinct. Following down the line of increasingly restricted categories toward our ob-

The Components of the Family CROTALIDAE

Genus	Common name	Distinguishing characteristics	Habitat
Crotalus	Rattlesnakes	With rattles; small scales on crown	North and South America
Sistrurus	Massasaugas and pigmy rattlesnakes	With rattles; large plates on crown	North America
Lachesis	Bushmaster	Without rattles; small scales on crown; small scales under end of tail	Central and South America
Bothrops	New World pit vipers	Without rattles; small scales on crown; large scales under end of tail	Mexico to South America
Trimeresurus	Asiatic pit vipers	Without rattles; small scales on crown	Asia
Agkistrodon	Moccasins	Without rattles; large plates on crown	North American, southeastern Europe, and Asia

jective, the rattlesnakes, we find the subclass Diapsida, then the order Squamata, and, finally, to separate the lizards from the snakes—for both are included in the Squamata—the suborder Serpentes, to which all snakes, venomous or harmless, belong. Suborders, in turn, are divided into families, and among others, under Serpentes, is the family Crotalidae or pit vipers, so called because of their possession of a remarkable sense organ visible externally as a facial pit, placed below and back of the nostril. By this family designation, pit vipers are segregated from true vipers of the family Viperidae, which have no pits, and which, incidentally, do not occur in the New World.

The next category below the family level is that of genus (plural: genera). As our narrowing categories are now bringing us close to the two genera of rattlesnakes, their position with respect to their nearest relatives can best be clarified by recourse to the table above which presents a summary of all the genera belonging to the family Crotalidae.

Now we are in a position to define the term "rattlesnake" properly. Rattlesnakes are pit vipers—popular name for the whole family Crotalidae—belonging to the genera *Crotalus* and *Sistrurus*. They are found only in the Western Hemisphere. All possess rattles. All are venomous, although, by reason of differences in size and other characteristics, there is a wide difference in the degree of danger from their bites. All are rather heavy-bodied and have broad heads. They are of various colors and are marked by blotches or by cross bands along the back.

Above all, the crucial characteristic that distinguishes rattlesnakes from all other snakes—even from other pit vipers—is possession of the rattle. This is a loosely articulated, but interlocking, series of horny rings at the end of the tail, which, when vibrated, produces a hissing sound. All rattlesnakes have rattles, and no other kind of snake has them. No snake is a rattlesnake because it is shaped like a rattler, or because it has blotches like those of a rattler, or because it is venomous, or because it is found among rattlers, or because it will coil like a rattler, or because it will vibrate its tail as does a rattler. Many harmless and venomous snakes have some or all of these characteristics, but lacking rattles, they are not rattlesnakes. (The term "rattler," as used in this book, is a short and popular synonym for rattlesnake; the term "rattle" refers only to the noise-making device at the end of the tail.)

Even when born rattlesnakes have a blunt segment called a prebutton, which, although soundless, is quite different from the pointed tail end of other young snakes. It is true that rarely—maybe once in a thousand—a rattler is found that has lost the end of its tail (including the rattle) in some accident. But in such cases there is no difficulty of identification because of the short stubby tail that remains, provided, of course, it has the other characteristics of a rattlesnake.

The two genera of rattlesnakes, *Crotalus* and *Sistrurus,* differ in the nature of the scales that cover the crown—the forward half of the top of the head. In *Sistrurus* this area is covered with large plates, usually nine in number and regularly arranged in cross rows thus: two-two-three-two, from front to rear. In *Crotalus* the crown is covered by small scales, usually quite irregularly disposed, particularly from the eyes rearward; although each eye generally has a single large plate (the supraocular) above it.

The genus *Crotalus* is the more important of the two; it includes the most species, the largest and most dangerous snakes, and ranges over much the greater territory. But the members of the genus *Sistrurus* are

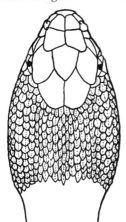

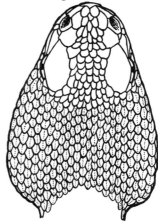

Fig. 1 Dorsal head plates of *Sistrurus* (*S. catenatus*)

Fig. 2 Dorsal head scales of *Crotalus* (*C. atrox*)

just as deserving of the name rattlesnakes, although, to distinguish them, they are generally referred to as massasaugas or pigmy rattlesnakes. Sometimes they are called ground rattlesnakes, not a particularly apt name as all rattlesnakes are ground dwellers.

The next lower category below the genus is the species, the fundamental unit of the divisional system. Each species is given a name composed of two parts, the first indicating the genus to which it belongs, the second the specific name applicable only to that species. (Examples: *Crotalus viridis* and *Sistrurus catenatus.*) Species, in turn, for the purpose of further segregation, may be divided into subspecies or races, in which case a third term, the subspecific name, is added. (Examples: *Crotalus viridis oreganus* and *Sistrurus catenatus tergeminus.*) Often in longer works, where the same subspecies is repeatedly mentioned, only the initials of the first terms may be used and even these are omitted if there is no sacrifice of clarity. (Example: *C. v. oreganus,* or, simply, *oreganus.*) There is a rule requiring that technical names below the family level be italicized. Generic names are always capitalized; specific and subspecific names—in zoölogy, but not in botany—are never capitalized, even though derived from proper names.

The question arises as to how species and subspecies are segregated: upon what bases are they differentiated? Species are populations that interbreed naturally; they cannot or will not interbreed with members of another species with which they may conjointly occupy a territory, and thus each preserves its separate identity and genetic integrity.* It is found that these groups of rattlers—these species—differ from each other in one or more of a variety of ways: in adult size, in bodily proportions, in male organs, in pattern and color (the most obvious but not always the most valid difference), in bone structure (especially of the skull), and, finally, in squamation—in the number of scales in certain series, and in their relative sizes, positions, and the contacts made with other scales. For rattlesnakes—in common with other snakes—are not clothed with scales haphazardly arranged like pebbles scattered on a beach. On the contrary, within each species the scales follow, with considerable consistency, certain patterns of size, number, and arrangement. Scale differences are the most practical for purposes of classification, and most identification schedules or keys are largely based on them, as well as on color and pattern.

A List of Rattlesnake Species and Subspecies

The list of rattlesnakes—that is, the number of species and subspecies—is continually growing, as new kinds are discovered in areas hitherto little explored, or as species are divided into more subspecies through the recognition of previously unnoted geographical divergences

---

*Rarely there may be crossbreeding or hybridization, especially under the unnatural conditions of captivity.

in characters. Nearly half of the subspecies now accepted have been described within the last thirty-five years. As of the present date, thirty-one existing species and seventy subspecies are recognized. Some are large dangerous snakes while others are small with a less serious bite. Some are found in considerable numbers near populated areas or are distributed over wide territories. Others are restricted to a single island. Still others although found on the mainland, occur in districts difficult of access to naturalists or collectors, so that only one or two specimens may ever have been seen by any herpetologist. There follows a list of the species or subspecies of rattlesnakes at present recognized as being valid:

Rattlesnakes of the Genus *Crotalus*

C. adamanteus. Eastern diamondback rattlesnake

C. atrox. Western diamondback rattlesnake

C. basilicus basilicus. Mexican west-coast rattlesnake

C. basilicus oaxacus. Oaxacan rattlesnake

C. catalinensis. Santa Catalina Island rattlesnake or rattleless rattlesnake

C. cerastes cerastes. Mojave Desert sidewinder

C. cerastes cercobombus. Sonoran Desert sidewinder

C. cerastes laterorepens. Colorado Desert sidewinder

C. durissus durissus. Central American rattlesnake

C. durissus culminatus. Northwestern Neotropical rattlesnake

C. durissus terrificus. South American rattlesnake

C. durissus totonacus. Totonacan rattlesnake

C. durissus tzabcan. Yucatan Neotropical rattlesnake

C. enyo enyo. Lower California rattlesnake

C. enyo cerralvensis. Cerralvo Island rattlesnake

C. enyo furvus. Rosario rattlesnake

C. exsul. Cedros Island diamond rattlesnake

C. horridus horridus. Timber rattlesnake

C. horridus atricaudatus. Canebrake rattlesnake

C. intermedius intermedius. Totalcan small-headed rattlesnake

C. intermedius gloydi. Oaxacan small-headed rattlesnake

C. intermedius omiltemanus. Omilteman small-headed rattles[

C. lannomi. Autlán rattlesnake

C. lepidus lepidus. Mottled rock rattlesnake

C. lepidus klauberi. Banded rock rattlesnake

C. lepidus morulus. Tamaulipan rock rattlesnake

C. mitchellii mitchellii. San Lucan speckled rattlesnake

C. mitchellii angelensis. Angel de la Guarda Island rattlesnake

C. mitchellii muertensis. El Muerto Island speckled rattlesna[

C. mitchellii pyrrhus. Southwestern speckled rattlesnake
C. mitchellii stephensi. Panamint rattlesnake
C. molossus molossus. Northern black-tailed rattlesnake
C. molossus estebanensis. San Esteban Island rattlesnake
C. molossus nigrescens. Mexican black-tailed rattlesnake
C. polystictus. Mexican lance-headed rattlesnake
C. pricei pricei. Western twin-spotted rattlesnake
C. pricei miquihuanus. Eastern twin-spotted rattlesnake
C. pusillus. Tancitaran dusky rattlesnake
C. ruber ruber. Red diamond rattlesnake
C. ruber lucasensis. San Lucan diamond rattlesnake
C. scutulatus scutulatus. Mojave rattlesnake
C. scutulatus salvini. Huamantlan rattlesnake
C. stejnegeri. Long-tailed rattlesnake
C. tigris. Tiger rattlesnake
C. tortugensis. Tortuga Island diamond rattlesnake
C. tansversus. Cross-banded mountain rattlesnake
C. triseriatus triseriatus. Central-plateau dusky rattlesnake
C. triseriatus aquilus. Queretaran dusky rattlesnake
C. unicolor. Aruba Island rattlesnake
C. vegrandis. Uracoan rattlesnake
C. viridis viridis. Prairie rattlesnake
C. viridis abyssus. Grand Canyon rattlesnake
C. viridis caliginis. Coronado Island rattlesnake
C. viridis cerberus. Arizona black rattlesnake
C. viridis concolor. Midget faded rattlesnake
C. viridis helleri. Southern Pacific rattlesnake
C. viridis lutosus. Great Basin rattlesnake
C. viridis nuntius. Hopi rattlesnake
C. viridis oreganus. Northern Pacific rattlesnake
C. willardi willardi. Arizona ridge-nosed rattlesnake
C. willardi amabilis. Del Nido ridge-nosed rattlesnake
C. willardi meridionalis. Southern ridge-nosed rattlesnake
C. willardi silus. West Chihuahua ridge-nosed rattlesnake

Rattlesnakes of the Genus *Sistrurus*

S. catenatus catenatus. Eastern massasauga
S. catenatus edwardsii. Desert massasauga
S. catenatus tergeminus. Western massasauga
S. miliarius miliarius. Carolina pigmy rattlesnake
S. miliarius barbouri. Dusky pigmy rattlesnake
S. miliarius streckeri. Western pigmy rattlesnake
S. ravus. Mexican pigmy rattlesnake

It will be observed that only two popular names do not use the term "rattlesnake"; these are "sidewinder" and "massasauga." The name sidewinder is derived from the snake's peculiar method of crawling. The term massasauga is said to have been derived from the Mississauga Indians and Mississauga (also spelled Mississagi) River, in Ontario, Canada.

Venomous Snakes in the United States Other than Rattlesnakes

Thus far I have pointed out the category in which the rattlesnakes fall as members of the family Crotalidae, and have listed the nonrattler genera in that family, as well as the rattlesnakes themselves. Another approach is justified—a glance at the position of the rattlesnakes in relation to the other venomous snakes found in the United States. For rattlesnakes are not the only poisonous snakes in the United States, although it is true that, as a group, they are the most widespread, the most prevalent in numbers, and, because of the size attained by some species, the most dangerous.

The distinction between the world's venomous and harmless snakes is neither as sharp nor as easy to ascertain as is often presumed. There is no ready means of telling, from a superficial examination, whether a snake is venomous. But, fortunately, in the United States there is a ready means of identification applying to all of our dangerously venomous snakes except one group—the coral snakes. The rest, including rattlers, are all pit vipers, and have the characteristic facial opening below and behind the nostril that gives them that name. In this country, as is the case everywhere except in Australia, the harmless snakes greatly outnumber the venomous, both in number of kinds and in number of individuals.

There are, in this country, as in many others, snakes that are, technically speaking, venomous, but which are quite harmless to man. They have venom glands and fangs, but the fangs are located in the back instead of in the front of the upper jaw. The fangs are short and grooved, rather than hollow. In consequence, they can only be imbedded in things of relatively small diameter, such as lizards and other small creatures upon which these snakes feed, and they must chew the victim, rather than merely biting it, to get the venom into the wound. The largest of these back-fanged species found in the United States are the lyre snake, the Texas cat-eyed snake, the black-banded snake, and the Mexican vine snake.

In returning to the really dangerous snakes of the United States, it may be observed that the most striking thing about venomous snakes, and, in a way, the most important, is not the venom itself but the means for injecting it—that is, the venom glands, ducts, and fangs, which, together, compose natural hypodermic syringes of great efficiency. True, the venom in some species is exceedingly powerful, and if injected into the blood stream of a victim produces disastrous results. But this is true of many protein substances, the white of an egg, for example, or the blood of an unrelated animal.

The front-fanged venomous snakes in our country that are not rattlers are the coral snakes, the cottonmouth moccasins, and the copperheads. The coral snakes, allied to the cobras of the Old World, have a powerful venom. However, they are usually mild-tempered creatures, and their fangs are relatively short; such accidents as have occurred have nearly always resulted from careless and even rough handling of them. The cottonmouths and copperheads have much the same shape as rattlesnakes, except that they are without rattles and have tapering tails. The bite of a cottonmouth is about as dangerous as that of a rattlesnake of similar size. The bite of a copperhead is painful and requires treatment, but human fatalities caused by this snake are extremely rare.

Confusion Regarding Rattlesnakes

I mentioned initially a rather widespread confusion that exists with regard to rattlesnakes, the inability of many persons to distinguish rattlers from quite harmless snakes, or from other kinds of venomous snakes. This confusion may have various important consequences. It leads to the killing of harmless snakes having considerable economic value, and whose destruction may actually increase the rattlesnake population because there will then be fewer harmless snakes competing with rattlers for food. It leads to serious complications in snake-bite cases, for the proper treatment of rattlesnake bite is, in itself, a painful procedure, and, if unnecessarily undertaken, puts the patient to great discomfort and some hazard. Further, the psychological effects of the bite of a harmless snake mistaken for a rattler may be quite serious. Such cases are well authenticated in medical literature. In one instance, a man bitten by a harmless gopher snake—which he thought was a rattler—almost died of shock.

Finally, these mistakes of identification are the bane of the snake specialist. He gets reports of the presence of rattlers in areas where they presumably never existed or have long since been exterminated, and when he runs these down the only evidence usually found is the smashed body of some harmless snake.

Many persons think that all venomous snakes are aggressive, or even vicious and vindictive creatures, whereas harmless snakes are thought to be kindly and timid. So, when they see a harmless garter or bull snake go through all the actions usually attributed to a rattler, such as coiling and striking, flattening the body and hissing, and even vibrating its tail—which, if done among dry leaves, makes a fair imitation of a rattle—they jump to the conclusion that the snake is a rattler, or at least a close relative, even though it lacks the telltale appendage on the tail. As a matter of fact, many harmless snakes, when cornered, will put on just as spectacular a posture of defense as any rattler—more so, in fact, since the threat is purely a bluff, not backed, as is the rattler's, by any really potent weapon.

Part of the confusion arises from a misunderstanding to the effect that the rattle is not always present. Some of the early naturalists advanced the theory that rattlesnakes do not get their first rattles until they attain their third year. Others had been told by Indians that only the males have rattles; and still others thought rattlers were without their rattles during part of each year. Whoever might believe any of these stories would have no trouble convincing himself that some threatening but harmless snake on the defensive was a rattlesnake without rattles.

Another cause of confusion is the failure of the supposed criteria for segregating venomous snakes—particularly rattlers—from the harmless: the broad head, thick body, and vertical pupil. Since these characteristics are also possessed by many harmless snakes, confusion is inevitable.

A further source of misunderstanding of what a rattlesnake really is lies in the misapplication or misunderstanding of popular terms. For example, in some areas certain harmless snakes are known as rattlesnake pilots and this is often followed by confusion with the rattlesnakes themselves. In other sections of the country, not only are all small rattlesnakes called sidewinders—although they are not the true sidewinder, or horned rattlesnake, of our southwestern deserts—but even some harmless snakes become known as sidewinders, and therefore are confused with rattlesnakes. In still other areas, all snakes that vibrate their tails when alarmed —a common habit of many harmless snakes, as well as venomous snakes other than rattlers—are popularly known as "ground rattlesnakes."

Distinguishing Venomous from Harmless Snakes

The difficulties of distinguishing venomous from harmless snakes are by no means confined to rattlesnakes; quite the contrary, for the rattlesnake, through possession of the rattle, is the most easily recognized of all venomous snakes. It is unfortunate that there is not some simple criterion by which the poisonous character of any venomous snake might be made readily apparent, but there is none. As a matter of fact, there is no sharp line dividing venomous and harmless forms, for snakes pass by degrees from the purely harmless through the back-fanged species, to the front-fanged and the folding-fanged snakes, to which groups nearly all of the really dangerous species belong. But even of these there are some that, because of small size, short fangs, mild venom, or like reasons, produce hardly as serious a result as a bee sting. I would like to repeat that none of the popular criteria such as a broad, triangular head, a heavy body, cat's eyes (vertical pupils), a flat body, or rough scales, are safe criteria, since there are both harmless and dangerous snakes known with any or all of these characteristics. The only unfailing method is an examination of the snake for hollow or grooved fangs and venom glands, and even this will not disclose the degree of danger from its bite.

The Use of Common Names

I shall, in discussing rattlesnakes in the succeeding chapters, use common names, for the most part, to indicate the several species, usually following this with the technical name as well. I do this in recognition of the fact that technical names are unfamiliar to most people, though preferred by scientists because their composite nature indicates certain relationships and because a code of rules and an international authority governing their promulgation and use make them more stable and precise.

Unfortunately, with respect to popular names, there is no order, agreement, or stability. At least as far as herpetology is concerned, no recognized authority has yet stepped in to recommend a coordinated set of names for adoption. Thus each author is free to adopt any he chooses; and although he may conscientiously try to follow the suggestions of his predecessors and colleagues, he is beset with duplications, uncertainties, and confusions, complicated by local usages. This conflict of names I shall try to surmount by adopting for the purposes of this book the set of popular names already presented in this chapter. But often I shall use both names with the idea of achieving definiteness and also to increase the reader's familiarity with the technical names.

This matter of rattlesnake names, and the use of methods whereby the same creature will be identified by the same name by everyone, is by no means solely of academic interest. Where several species of rattlesnakes occur together in a locality, it might be of great value to a physician treating a case to know which one had caused the bite. In the future, snake-bite treatment will increasingly recognize the great differences in venom quality among the several species of rattlesnakes, as well as other species differences affecting the gravity of the bite.

Of course, these confusions of names and identifications represent only a small part of the misunderstandings and exaggerations, often resulting from fantastic statements or stories, that have obscured our knowledge of the rattlesnakes from the earliest days. In subsequent chapters we shall review the real character of the rattlesnake, the reptile once referred to as "the snake without a friend."

CHAPTER TWO

Morphology

LENGTH

Few rattlesnake characteristics are of greater interest to the average person than size. When a rattlesnake story is told, the size of the snake is always an important feature. And, from a practical standpoint, size is of significance, for several elements of the relative danger from rattlesnake bite vary with size, such as the distance a snake can reach in its strike, the length of the fangs, and the quantity of venom discharged. Rattlesnake species differ greatly in size. Some of the smallest kinds rarely attain an adult length of two feet, whereas others may reach six feet or more. Even these figures give a rather inadequate idea of the size differences, for the larger snakes have a bulk or weight of from thirty to fifty times that of the smaller.

It must be mentioned that there are practical difficulties in determining length accurately. Rattlesnakes can be measured accurately only "in the round," for skins can stretch up to thirty-five percent. However, it is almost impossible to measure the lengths of live rattlesnakes accurately, especially the large specimens that would be of the greatest interest. The most accurate measurements are those made on snakes just after killing for preservation, but before they have stiffened; at this time they can be laid out along a ruler and measured exactly. Unfortunately, the difficulties of preserving large rattlesnakes, involving cumbersome containers and much preservative, are such that preserved collections seldom represent fair population samples of the larger species; and the record-breakers especially, of which it would be of decided interest to have the measurements, are never available. Mere estimates of the lengths of live rattlesnakes are quite useless; amateurs almost invariably overestimate the lengths to a marked degree; and even experienced observers can rarely guess the length of a snake within ten percent of actuality.

Also there is difficulty in drawing a line between obviously mythical or fantastic reports and those that have at least some semblance of truth or accuracy. An unreliable report may recur in print so often — usually without acknowledgment of the original source — that it acquires, through mere persistence, a quite undeserved authenticity. Many of the older records — of ten- to twenty-foot rattlers — represent only the repetitions by overcredulous travelers of campfire tales.

Finally, there is the normal variation in size within any population group. Just as we see exceptionally tall men in basketball teams and sideshows, so also there are rattlers much larger than their fellows. But I doubt whether a rattler exceeding its fellows by fifty percent is much more common that an eight-foot man. This transfers to the mythical any rattlesnake approaching a length of ten feet. Having mentioned some of the possible inaccuracies surrounding the reports of very large — but not mythical — rattlers, I shall report on the maximum length attained among the largest species.

The eastern diamondback *(Crotalus adamanteus)* is the largest of the rattlesnakes. Specimens exceeding seven feet in length are well authenticated, although I cannot claim to have measured one myself; and eight- and nine-foot snakes have been reported, possibly with some basis of truth.

Reviewing all the reports and statistics concerning the maximum size attained by this snake, I should guess that, while the average adult male measures five feet, very rarely the eastern diamondback does measure eight feet, give or take an inch or so.

The western diamond rattlesnake *(C. atrox)*, especially in some parts of Texas and Oklahoma, is a large rattlesnake, probably second only to the eastern diamondback *(C. adamanteus)* in size. I have seen some large specimens, certainly exceeding six feet in length, but as they were alive they could not be measured either safely or accurately.

The Mexican west-coast rattler *(C. b. basiliscus)* reaches six feet or slightly more. Several very large and heavy individuals of this species were born and raised in the San Diego Zoo. One, a female, attained a length of six feet eight-and-a-half inches and a weight of seventeen pounds at the age of five years. Another, a male, measured six feet seven-and-a-half inches and weighed fifteen pounds fourteen ounces at the age of ten years. Knowing the considerable effect that captivity has on the life and growth of rattlesnakes, we cannot be sure that *C. b. basiliscus* attains a length of six feet in the wild, although it probably does.

There seems little doubt that the canebrake rattlesnake *(C. h. atricaudatus)* occasionally attains a length slightly in excess of six feet; I have heard several reliable reports of members of this subspecies reaching this length.

Two extinct kinds of rattlesnakes have been described that are believed to have been larger than any existing species, this conclusion being based on a comparison of the size of their vertebrae with those of existing rattlesnakes. One may have reached a length of twelve feet; the other, a probable ancestor of the existing eastern diamondback, is presumed to have been nine or ten feet long. The fossils from which these forms were described were found in Florida, and are of the Pleistocene, or glacial, epoch. The evidence of these large rattlers of a bygone age does not in any way validate the stories of huge rattlers existing today.

BODY PROPORTIONS
Bulk

Rattlesnakes are heavy-bodied, that is, they are thick in proportion to length, when compared with most snakes. Yet they are by no means outstanding in this respect, for some African vipers are proportionately much thicker, to say nothing of some boas and pythons. The thickness of a snake's body, whether in terms of diameter or circumference, does not lend itself to accurate measurement. A live snake can swell its body with an intake of air, and can flatten it as well; and a dead snake is affected by the position and the diameter of the coil in which it has been allowed to set in preservative. For these reasons the length-weight relationship is a better criterion of bulk than the correlation of length with the diameter at midbody. By the length-weight relationship (useful only for rattlers in the wild) a seven-foot rattler would weigh about fifteen pounds, and an eight-footer about twenty-three pounds. The heaviest rattler of which I have heard was a seven foot five inch western diamond said to have weighed twenty-four pounds. Rattlesnakes, while by no means the long-est venomous snakes — the king cobra, attaining a length of at least eigh-teen feet, holds this distinction — probably attain a greater weight than any venomous snake, although no doubt closely approached by some of the thick-bodied vipers of Africa.

Head and Tail Dimensions

The head sizes of rattlesnakes, in proportion to the lengths of their bodies, differ considerably among the several species, some kinds having conspicuously large heads and others small, compared with the generic average. These differences are of some value in segregating species and determining their relationships. They are also of some practical impor-tance, since large-headed species tend to have longer fangs and more venom, and therefore are likely to be the more dangerous.

Rattlesnakes have relatively short tails. In most snakes, the tail, aside from being the repository of the scent glands and the retracted hemipenes in the male, is used primarily as an aid to locomotion, and in some snakes for holding and climbing. But in the rattlesnakes the tail is foreshortened by the attachment of the rattle, and concurrently serves principally as a muscular rattle-vibrator. Rattlesnakes exhibit some species difference in tail length when they are compared in terms of the ratio of the length of the tail to the length of the snake over-all.

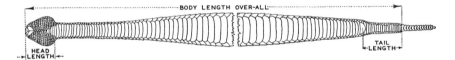

Fig. 3 Methods of measurement

SQUAMATION

Squamation, or scale arrangement, in rattlesnakes offers an interest-ing and important subject for study. As previously mentioned, the scales are arranged, on both the head and body, in fairly regular patterns, of which various series can be counted accurately. These series are found to have a considerable consistency within a subspecies, as well as constant differences between subspecies, so that squamation is of the greatest im-portance in the classification of snakes. Various kinds of snakes that are quite similar in size, form, and color may be readily distinguished by scale differences.

The scales of snakes are not separate, removable parts of an outer covering, as they are in most fishes; rather, they are formed by folds or creases in the skin. This type of formation is advantageous both in allowing flexibility of body movement and in enhancing strength and durability of the surface. The scales themselves comprise somewhat thickened and hardened sections of skin, which is thinner and more flexible between them. Yet the actual continuity of the skin surface—of the parts covering the tops of the scales, as well as the folds and creases between them—is shown by the unbroken character of the shed skin. In this, every scale top, every fold and crease, is represented in a single flattened sheet. It is by means of this flattening of the creases between scales that the skin is sufficiently enlarged so that, during the shedding process, the narrow neck-section can be slipped backward over the much thicker mid-body part of the snake.

Many species of rattlesnakes have certain eccentricities of head scalation quite apart from quantitative differences in the various scale series. These peculiarities are seldom invariant or completely consistent. But one of the few scale peculiarities of this type always present in one species is the raised supraoculars—the horns—of the sidewinder (C. cerastes). These are invariably evident in sidewinders and never in any other rattlesnake. What the purpose of the horn may be is not known. I once suggested that it might serve to leave the eyes exposed, yet protected from the drift, as the snake lies partly buried in the sand; but now I doubt that

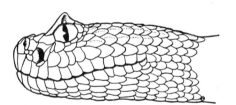

Fig. 4 C. cerastes, showing hornlike supraocular (lateral view)

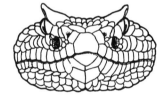

Fig. 5 C. cerastes (front view)

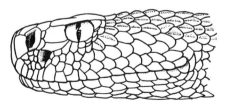

Fig. 6 Lateral head scales of Crotalus (C. atrox)

Fig. 7 Head scales of Crotalus (C. atrox) (front view)

the snake buries itself often enough to just the required depth to derive such a benefit from the horns. The theories that the horn comprises either a shade for the eye or a radiator of heat are doubtful in view of the sidewinder's nocturnality.*

COLOR AND PATTERN

"As touching Serpents, wee see it ordinarie that for the most part they are of the colour of the earth wherein they lie hidden: and an infinite number of sorts there be of them." So said Pliny many centuries ago.

Rattlesnakes are indeed of many colors and patterns. Although their colors can often be attributed to procrypsis or concealing coloration, this is by no means invariably the case. In general, rattlesnakes are blotched, as is usually true of heavy-bodied, slow-moving snakes. If any pattern can be considered typical of the rattlers, it comprises a primary series of dark dorsal hexagons, each bordered by a row of light scales. Posteriorly, the dorsal blotches merge with the laterals to form crossbands, a characteristic of many species. The ventrum is usually mottled with irregular dark marks on a lighter background. In most species, the patterns become less clear when the snakes become adult, and most individuals darken with age.

Rattlesnake heads are usually marked with a dark streak passing backward and downward through the eye toward the angle of the mouth. Rattlesnake tails are characteristically barred with dark cross-bands or rings on a lighter background. Sometimes the tail colors are sharply differentiated from those of the body. A close examination reveals that the rattlesnake's color is usually applied in the form of dots, stippling, or punctations. The blotches, marks, and ground colors vary from white to black, through many hues, shades, and tints. White, or almost white, is not at all uncommon ventrally, and many species have rows or clumps of white scales dorsally or laterally.

Jet-black rattlers are to be found as an adult color phase of several subspecies, including the eastern massasauga *(S. c. catenatus)*, the Mexican pigmy rattler *(S. ravus)*, and the timber rattler *(C. h. horridus)*.

Although most rattlesnakes run to grays and browns, bright colors are by no means absent from their patterns; but such colors are less conspicuous than in unicolored snakes. Certainly the old dictum that rattlers are never brightly colored, is far from accurate. Several kinds of rattlesnakes include bright yellow in their patterns. Some are almost unicolored orange or burnt-orange.

*Ed. note: since Klauber's writing, B. H. Brattstrom has shown that the horns fold down over the eyes when the snake is buried or travelling in an underground burrow — an effect which protects the eyes from sand abrasion.

No rattlesnake is a brilliant red, with the pure color occasionally to be seen in other kinds of snakes, but many are red-brown or even brick-red. Many rattlesnakes are pink ventrally; and the southwestern speckled rattlesnake *(C. m. pyrrhus),* from some mountain chains in central Arizona and southern California, has a beautiful coral-pink or salmon dorsal color, so little marred by darker blotches that it answers the requirements of any bibulous tale. Green is a common color among the rattlers, although it usually tends toward a dull olive rather than the brilliant green of some tree snakes of the tropics. No rattlesnakes are blue with the brilliance that characterizes the tails of many skinks, and few can be called blue in any sense, although some specimens of the banded rock rattler *(C. l. klauberi)* are slightly bluish. Some grayish rattlers have a lavender tint.

Snakes, rattlesnakes among others, have some power of color change, although only to a small degree compared with that possessed by many lizards. The change involves a lightening of some areas with increased temperatures, particularly the dorsal areas between blotches. At lower temperatures the pigment disperses, thus making the dark color on the surface more apparent.

As might be expected, the skin-shedding process results in a brightening of the color and pattern. Although a skin ready for shedding has little color, through surface wear and the accretion of foreign matter it loses some of its transparency, and the keratin of which it is composed has a natural straw color. Thus, upon its removal, the snake's pattern shows with renewed brilliancy.

Sexual differences in pattern and color are not important among the rattlesnakes.

FREAKS AND ABERRANTS

Various kinds of freakish or aberrant rattlesnakes are occasionally seen. Some are abnormal in form; of these, two-headed snakes are the most interesting to the public. Others have various defects, many of which are such as to prove lethal at birth or shortly thereafter. Hybrid rattlers also occur, more often in captivity than in nature. Abnormalities are commoner in pattern and color than in form. In color they range all the way from albinism to melanism. The most notable pattern aberrancy is a peculiar 90-degree rotation of the pattern, whereby the transverse blotches or rings that characterize most rattlers are transformed into longitudinal stripes.

Two-Headed Snakes

Two-headed snakes are freaks or monsters that occur as a very small percentage of normal births, just as they do among most other kinds of animals. As they are usually defective anatomically, they rarely survive

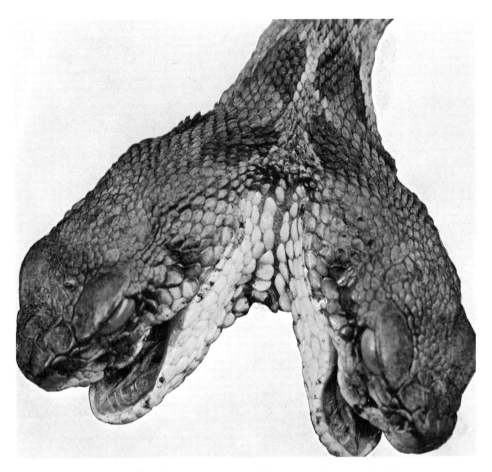

Fig. 8 Two-headed timber rattlesnake (*C. h. horridus*). (Photographed through the courtesy of Joseph Rimkus)

beyond a few days after birth. Although two-headed snakes have long been known—they were even discussed by Aristotle—the early accounts were elaborated into, or confused with, various mythical creatures described in the early natural histories, such as the seven-headed hydra.

Another confusion that persists to the present date, results from the application of the unfortunate term "two-headed" to various narrow-headed, blunt-tailed snakes, owing to a fancied similarity of their heads to their tails. One spectator at the San Diego Zoo was heard to decry the attention given a live two-headed king snake; he thought it unworthy of notice since both heads were at the same end. Needless to say there are no snakes with a head at each end.

I have few numerical data as to the frequency with which two-headed snakes occur, but the percentage must be quite small. One of the largest commercial snake dealers in the country informed me that no two-headed rattlesnakes had been brought in by his field collectors. During the past thirty years some twenty thousand snakes, of more than thirty subspecies, have been recorded in a species census of snakes from San Diego County, California, and there was a single two-headed individual (a king snake) among them. It should be stated, with regard to these two-headed freaks, that the popular interest in them exceeds the scientific. When people find such a creature, they are likely to have quite fantastic ideas concerning its value. If such a snake could be kept alive for awhile, it might be worth something to a roadside snake show, but to any scientific institution it would have only a passing interest and a purely nominal value.

Defective Young

I will elsewhere comment on the frequency with which defective young occur in broods of rattlesnakes born in captivity. I do not know whether they occur more often in the case of captive mothers than in the wild. Some are so deformed that they obviously would not survive, as, for example, snakes without eyes, with malformed heads, or twisted bodies. Adhesions between two parts of the ventral surfaces are sometimes noted. Occasionally these defects run in families.

Albinism

Albinism, a pigment deficiency, has been noted from time to time in practically all animals of which adequate series have been observed. It is not a particularly rare phenomenon. Once occurring, it is likely to be most persistent in inbred populations. Albino snakes probably suffer in competition with their normal fellows: their eyesight is perhaps defective, as is the case with most albinos, and they lack protective coloration, making them more readily subject to predation. Hence, most albino snakes found in the wild are juveniles.

A completely albinistic rattlesnake has pink eyes and lacks all of the darker colors usually evident in the normal pattern of the species to which it belongs. Yet the pattern may still be faintly discerned, because the pink of the tissue within is more evident through the patches of skin normally dark, than through those normally straw colored.

I have two albino rattlesnakes preserved in my collection, and have seen two others.

Melanism

Melanism — the occurrence of black individuals — seems to be of two kinds. The first involves the continued presence of many black, or almost black, individuals in a population otherwise not distinguished by being conspicuously darker than neighboring populations that do not produce melanos. This type of melanism is usually age-related, for the individuals destined to be black as adults are normally colored at birth. The second type involves only a very few specimens, and may, like albinism, be the result of some chance genetic disturbance that may appear in one or several individuals and then be eliminated from a population.

Among the rattlesnakes, the first kind of melanism occurs among the eastern massasaugas *(Sistrurus c. catenatus)*. These rattlesnakes ordinarily are quite dark-colored, but with a conspicuous pattern of squarish brown or black blotches on a gray-brown ground color. In some areas, however, many individuals are uniformly black.

Hybrids

There are various myths concerning the hybridization of rattlesnakes with other snakes. Throughout the West, but particularly prevalent in the upper Missouri Valley, there is a widespread idea that the rattlesnakes have crossed with the bull snakes, producing an especially

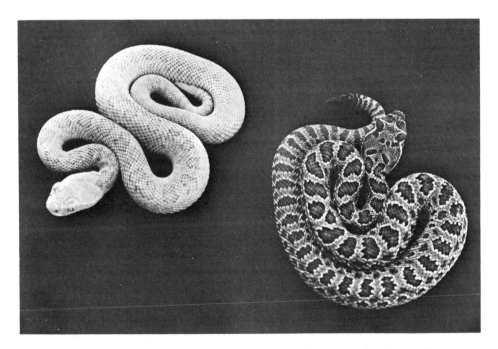

Fig. 9 Juvenile albino prairie rattlesnake *(C. v. viridis)* compared with one of normal coloration

Fig. 10 Melanistic eastern massasauga (*S. c. catenatus*)

vicious and venomous offspring; which further, being rattleless, is the more dangerous since it does not warn its victims. There is not the slightest reason for believing this story; it probably gained credence from the fact that a bull snake on the defensive will vibrate its tail, and will coil and strike at an intruder. Another myth, heard in the Southeast, credits the copperhead with being the female of the diamondback rattlesnake.

But hybridization between rattlesnakes of different species does occur. Three specimens have been collected in the wild, which, because of the intermediacy of their characters, give every evidence of being hybrids between two species known to occur in the areas where the anomalous individuals were found; and at the San Diego Zoo we have had two instances of interspecies matings resulting in live young. Another occurred at the University of California, Berkeley.

CHAPTER THREE

The Rattle

The rattle is the peculiar attribute of the group of New World snakes that form the subject of this study. This remarkable appendage, whose purpose has been a source of speculation and controversy ever since these snakes were first observed by European naturalists with an astonishment approaching incredulity, is a fruitful subject of research. For we have in this organ a mechanism unique in both development and function; and it is worthwhile not only to study it as a queer animal instrument, with its parts unchanged by growth and time, but to follow, also, the gradual emergence of accounts of the rattle from a plane of myth and folklore to one of understanding, although admittedly with problems still unsolved.

First, as to the development of European knowledge of the rattle and the origin of some of the peculiar beliefs as to its structure and use: the earliest reference to the rattle in print, as far as I have been able to determine, was made by Magalhães de Gandavo in 1578. In this short account two important ideas were presented: first the myth—persistent for two hundred years or more thereafter—that the rattle sounds whenever the snake moves; and, second, that passersby, when they hear the rattle, avoid the snake. In this second statement there is a suggestion both of the accepted theory of the function of the rattle and of its origin and development.

The first mention of the rattle in English referred only to its use by the Indians as an ornament; this is contained in an account of Virginia by Captain John Smith in 1612.

The indestructible myth that the rattlesnake adds one segment to its rattle every year (thus the age of the snake can be determined by counting the segments) was first stated in print by Francisco Hernández in 1615; he likewise observed (correctly) that the rattle was sounded furiously when a snake is angry or molested. Guilielmus Piso was the first (1648) to publish the belief that the rattle itself is a dangerous weapon—more deadly, in fact, than the snake's fangs and venom. This myth persisted for about a hundred years.

The first adequate description of the internal structure of the rattle was given in 1681. Its author clearly understood the multiplication of the sound produced by the clashing of the successive lobes against each other. He doubted that the rattles indicated accurately the age of the snake and found no verification of Piso's theory that the rattle itself is a dangerous instrument. In 1683 an author noted, correctly, that the flat side of the rattle is perpendicular to the ground, rather than parallel as was popularly believed.

In 1693 it was stated, quite erroneously, that young rattlers have no rattles for a year or so after birth. This author was probably confused by the young of some other kind of snake, but, at any rate, this notion still persists. He likened the rattle substance to that of a horn or fingernail. (The rattle is composed of keratin, the albuminoid substance that is also the basis of horn, hair, nails, and feathers.)

One of the first, if not the first, statements that the rattle is invariably sounded as a warning before the snake strikes was made in 1722. This supposed altruism of the rattler was questioned in 1752, but is a bit of folklore still widely accepted.

The exaggerated rattle string story originated in 1723. Its author tells of the killing of a rattlesnake with seventy to eighty rattles, a yarn often recurring in subsequent accounts; he also was the first to report that the Indians were afraid to travel through the woods in wet weather, since the dampness rendered the warning rattle inaudible. He further stated that when one rattler sounded off, all those in the vicinity would take up the chorus; it had previously been reported (1720) that the rattle was used by a snake as a call upon its fellows for assistance.

The early writers inferred a bell-like sound to the rattle, which probably few, if any, had ever actually heard as made by a live snake. The English no doubt got the bell idea from the translation of the Spanish *cascabel* ("small bell"), and Portuguese *cascavel,* words for the rattle — as well as the snake — used by the early explorers. This characterization of the noise was related to the erroneous idea that the rattle is sounded automatically whenever the snake moves. The earliest appropriate description of the sound was that of Megapolensis in 1644, who likened it to that of a cricket. Since crickets, locusts, and cicadas were often confused in those days, it is possible that he had a cicada in mind, for the sound produced by some cicada species does strongly resemble that of a rattle.

To Hernández belongs the credit for the first published picture of a rattlesnake and its rattle (1628). It shows a coiled rattler with the rattle pressed too close to the body for a clear view, although it is accurately drawn. Johann E. Nieremberg (1635), who plagiarized Hernández' work, improved the picture by showing the rattle somewhat separated from the body coil; however, the rattle is turned as if it were carried with the widest face parallel to the ground, instead of perpendicular as is actually the case.

FUNCTION AND UTILIZATION
The Function of the Rattle

The literature and folklore of the rattle are replete with speculations upon its function; these have been proposed in great variety since the earliest encounters between man and rattlesnake. Even today, when there is a practically unanimous agreement among those herpetologists who have had a wide field experience with rattlesnakes, that the rattle is a warning mechanism, the statement may still be seen that its function is in doubt. But, as one reviews the literature, one notes that most criticisms of the accepted warning theory have resulted from a failure to specify the circumstances under which the rattlesnake employs the warning, and the character of the creature warned. When these are defined, the usual criticisms of the warning theory are adequately answered. The problems that remain concern the evolution of so remarkable an instrument.

I shall first review some of the theories of the function of the rattle that have been advanced. These may be divided into four broad categories, involving: (1) purposes other than the conveyance of sound; (2) sounds intended for other rattlesnakes; (3) sounds intended for prey; (4) sounds aimed at other creatures.

Before discussing these categories, I shall amplify one fact that has a bearing on the credibility of several of the theories, namely, the infrequency with which rattlers are heard to rattle while undisturbed in the wild. Certainly it has been the experience of those of us who live in areas where rattlesnakes are quite common, that we almost never hear one unless we ourselves have disturbed it. An author in 1828 claimed that the love calls of the rattlers were the only sounds to strike the ear of the traveler at noon in the heat of the forest wilderness. I know of no one who has heard such a chorus, or has had an experience paralleling that of one recent writer who thought he had heard snakes rattling continuously all one night in British Columbia. I can only presume that the sounds came from other sources.

Returning to the discussion of theories of the purpose of the rattle, there is, in the first category — purposes other than the production of sound — only one that was ever taken seriously; this was Piso's claim made in 1648 that the rattle itself is a poisonous instrument, more dangerous to man than the fangs and venom. But this theory was denied as early as 1681, and has received little serious consideration since.

An echo of this idea, in modern folklore, is the myth that rattle dust is poisonous. Indeed, there is a myth that a snake rattles, not to make a sound, but to shake the poisonous dust into an intruder's eyes.

In 1889 the thought was advanced that tail vibration by snakes — a widely existing nervous reaction upon the part of a threatened or angered snake, which is by no means restricted to rattlers — might be designed to protect the tail from enemies, or to draw the attention of prey away from the threatening head. Since this custom of vibrating the tail was important in the genesis of the rattle itself, these hypotheses, if verified, would have a bearing on the function of the rattle. But I know of no facts supporting them.

The first of the functional theories involving the sound of the rattle assumes its purpose to be an exchange of signals between rattlesnakes, either (a) as a mating call, or (b) as a warning to, or call for help from, other rattlers.

Against the mating theory various arguments may be advanced. I have previously mentioned how rarely the sound of the rattle is heard in the wild. From observations of mating pairs found afield, the mating season — about four weeks in the spring — is well defined. It is inconceivable that, if the rattle were a mating call, it would not be frequently heard during this climactic period of rattlesnake activity; yet it is not. The courting pattern of male rattlesnakes is known from observations of large numbers of attempted and successful matings of captive snakes. The pattern is basically uniform; in no case has the use of the rattle been noted. Of course, if a rattler intent on mating is disturbed, then the rattle is used, but only as a warning to the intruder.

No advocate of the mating theory can claim that the mating call is the sole use of the rattle, for, as is well known, immature snakes, when disturbed, sound the rattle as readily as adults, and females as readily as males; indeed, the freshly born infant just out of its fetal sheath will vibrate its soundless button and threaten an intruder with all the spirit of its elders. Surely this instinctive reaction can have no relation to the sexual instinct. An advocate of the sex-call theory claimed that the rattle of the courting snake was used with less stridency than when sounded in anger. One enterprising writer suggested the use of an artificial rattle to decoy romantic rattlers to their destruction. But we must conclude that the extended observations of recent years lend no support to the mating-call theory.

The exchange of signals between rattlers for mutual warning, protection, or attack, has been suggested by a number of authors. It stems from the observation—which has a factual basis—that when a group of rattlesnakes is gathered together and one rattles, several or all may join the chorus. This suggestion of mutual alarm led to the related theory of mutual assistance.*

Observations of a great many captive rattlers, as well as field experiences, lead me to place no credence whatever in this theory of mutual-aid pacts. True, alarming one rattler of a group may cause others to rattle, but usually this can be attributed to the same disturbance—a visible movement of the intruder, or his foot-fall—that caused the first snake to be aroused, without any necessity for the assumption that the alarm was communicated by the rattle. Indeed, when the first snake sounds its rattle it almost always rises into the striking coil, and this movement may well alarm its fellows.

On several occasions in the field, I have come upon two or three rattlers together. One, alarmed at the intruder, would rear up and rattle violently, yet the others a few feet distant often remained quite unperturbed. It has been observed that a rattler in captivity will sometimes rattle when a second snake approaches food that the first has struck or is eating. This is a rather natural extension of the use of the rattle as a warning; it is like the growl of a dog, and shows that rattlers do not change this primary purpose of the rattle when using it on their fellows.

Finally, as a conclusive argument against either the mating-call or call-for-help theories, the fact should be cited that rattlers are quite deaf to the sound of the rattle. Elsewhere I discuss the hearing ability of rattlesnakes; and how an extraordinary sensitivity to vibrations of the substratum upon which a snake rests, would lead one to suppose that rattlesnakes can hear, whereas they are, in fact, quite deaf. Such experiments as

*The tall-story addicts eventually carried this call-to-arms propensity to its logical conclusion. In eastern Kansas, so they say, the tocsin is flashed from one den to the next, and soon all the 200 million rattlers in that section of the state are rattling in unison, and the human inhabitants quickly flee to their cyclone cellars.

I have conducted confirm the belief originally expressed in 1923, that no rattler ever heard another's rattle.

A number of theories regarding the purpose of the rattle have to do with its effect on the snake's prey. Before examining these theories, let me mention again the infrequency with which undisturbed rattlers are heard rattling in the field. Surely, in the spring in snake-infested areas, the hungry snakes would be heard everywhere broadcasting their lures or charms, and rattler-collecting would be appropriately expedited. No such condition exists.

Some of the theories involving the use of the rattle as a prey-securing mechanism hold that the prey is charmed or fascinated by the rattle. The advocates of these theories believe it effective in one of three ways: (a) the rattle itself charms the prey by its attractive sound; or (b) the rattle draws the attention of the prey to the snake, whereupon it is brought within the influence of the rattler's hypnotic eye; or, (c) the prey is so startled by the rattle as to become paralyzed and thus an easy catch.

I deal elsewhere with the myth of the rattlesnake's power of fascination, which, although still widely current, is entirely unsupported by modern studies. With regard to the part played by the rattle in the theories of fascination, the following comments are appropriate: the first and third of the proposals listed above are, of course, virtually antithetical; the rattle could not at once be both attractive and paralyzing. The second of these subtheories requires a belief in the mythical power of the rattler to charm with its eye. The writer who, in 1745, may have been the first to suggest this use of the rattle did not claim it was backed by field observation; on the contrary, he merely concluded that the rattle must be useful to the snake in some way and this was evidently its beneficial purpose. Others have believed in the power of fascination because they thought rattlers too lazy and slow-moving to capture prey by any other method than some form of remote influence, which overlooks their ability to ambush prey or to follow it down holes.

Finally we have the theory, still believed by some, that the sound of the rattle paralyzes the prey with fright or startles it into immobility. One naturalist stated that he had seen a rattler bring a bird to a standstill by sounding its rattle. One cannot state categorically that no bird or mammal was ever startled into immobility, but it is certainly extremely doubtful that the rattle is ever deliberately or instinctively sounded for this purpose. In the particular instance cited, assuming the observation to have been accurate, the snake may have rattled because the naturalist disturbed it, rather than in an endeavor to halt the bird. Observations of captive rattlers indicate that if the rattle does startle the prey it is more likely to cause it to run or fly away than to be paralyzed with fear.

The other theories assigning the rattle to the securing of prey are based on the premise that the rattle simulates the sound of the prey of some creature that the rattler seeks, such as the sound of the cicada that birds might hunt. That the rattle sounds much like some kinds of cicadas or locusts is unquestionably true, but that the snake actually buzzes the rattle to attract the birds is to be very seriously doubted—because the rattle is seldom heard in the wild, because rattlers seldom eat birds, and because the noise would tend to frighten rather than to attract the birds.

One theory has it that birds might be attracted to the sound of the rattle out of pure curiosity, but not because of a fancied resemblance to the sound of some insect. Another states that the rattle may decoy thirsty creatures to their destruction because of the resemblance of the sound to that of running water. This is quite fantastic since there is no such resemblance. In conclusion, it may be stated that no experience with rattlers in the field or in captivity tends to encourage credence in any of the various prey-securing theories.

Finally we come to the real purpose of the rattle, namely, its use as a warning—not a warning addressed to prey, nor the altruistic warning of the intruder for the intruder's protection, but a warning or threat intended to drive away creatures that might harm the rattler itself.

Before proceeding to a discussion of the warning theory, one belief that misled the early naturalists must be corrected, this being the entirely erroneous idea that the rattle sounds involuntarily whenever the snake crawls—clearly the presumption of one who had never seen a live snake in action. A typical expression of this belief, so contrary to every field observation or laboratory test, is that made in 1869: "Imagine these [the rattles] constantly clattering against each other, as the reptile moves, with a hoarse, dull, echoing sound, and you will be able to form some idea of the permanent warning of its approach which the *Crotalus* carries about with it."

But as early as 1634, it was said: ". . . at her taile is a rattle with which she makes a noyse *when she is molested*" (italics mine); and the theory that the rattle sounds involuntarily whenever the snake moves was specifically denied in 1745. Occasionally, the compromise statement is made that the rattle makes some noise as the snake travels, but much more when it is annoyed. There is an element of truth in this, for although a rattler crawling at a normal gait makes no sound with the rattle, if it be terrified into thrashing along as fast as possible through brush and rocks, the rattle may occasionally strike some obstruction and be audible as a click. But this is far from the strident hiss of the deliberately sounded rattle; and any thought that the rattle can be heard as the snake goes about its ordinary affairs is completely incorrect.

When we consider the many theories suggested to explain the fundamental purpose of the rattle, it is at once apparent that the usual rattlesnake reaction, which is almost universal, gives strong support to a single theory in preference to all others. For what does any rattlesnake do with its rattles? It sounds them in warning when disturbed or frightened, as by some movement or the approach of an intruder. Upon this there can be no argument; it is the common experience of everyone who has encountered a rattler in the field or startled one in captivity. I have no desire to assume an irritating attitude of assurance, but certainly I have seen this happen a thousand times; and there are others of more extensive experience who corroborate this observation.

The warning theory is very old, but from the first it has been, and still is, a matter of argument because of a confusion, often unrecognized, between three types of warning: that is, the warning of (1) intruders, possibly dangerous, warned for the protection of the snake; (2) intruders warned for the protection of the warnee, instead of the warner; or (3) prey, the last being a special case, already discussed, of (2). If we sharply distinguish between these three and point out the manner in which the rattle is beneficial to its possessor, by frightening away creatures that might otherwise injure the snake, we at once eliminate the objections that have been so often advanced against the warning theory. Certainly no animal will warn away the food upon which it depends for subsistence, or develop so intricate a mechanism for the altruistic protection of the innocent passerby. But it is equally clear that warning devices that tend to safeguard their owners are common in nature; and there is no more reason to question the purpose of the rattle than that of these other devices. It is only necessary to show that the rattle is used for this purpose, is often effective, and that its disadvantages are not so important as have sometimes been supposed.

Is the rattle used as a warning? On this point I think enough has been said: the rattle is so used—this is the reaction anyone familiar with these snakes invariably expects when he approaches one. Is it effective? The answer, of course, depends on the circumstances, as it would were we discussing the growl of a dog, the hiss of a gander, or the earth-pawing of a bull. These are all warning reactions, and they may result in success or failure, depending on the character and purpose of the trespasser. They may be followed by more direct action or by the retreat of the warner.

One of the mistakes made by early writers, and repeated in later natural histories, is in the general characterization given the rattler's warning posture. Usually the snake is portrayed in its resting coil with rattle sounding. To the possible inadequacy of such a warning mechanism one may agree, but this is not the entire story by any means. Just as the cat arches its back, fluffs its tail, opens its mouth, and spits and squalls, so the rattler has more than a single warning reaction. It is true that a snake found in its resting coil may first sound the rattle without changing position. But if this preliminary warning fails to halt the trespasser, it quickly adopts more spectacular methods, which concurrently place it in a better position either for defense or escape. For now, still sounding the rattle furiously, it raises the anterior part of the body above the ground in an S-shaped spiral, with the head and neck held like a poised lance ready for a forward lunge; the posterior part of the body is flattened to stabilize the anchorage or to facilitate mobility; the tongue, with tips widely spread, is protruded to the utmost and is alternately pointed upward and downward with intervening pauses; and the snake inhales and exhales with a violent hiss. Now the rattler is ready for whatever may come; it can strike, if the enemy comes within range, or it can retreat (still facing the intruder) toward the nearest rock crevice or bush that might serve as a refuge. In this composite picture, the rattle serves as the alarm bell.

When a rattle is heard in such circumstances, no prior experience or mental process is necessary for a realization of threatened danger; both the stridency of the sound, and the other actions of the snake to which the sound draws attention, will have an immediate effect on any creature capable of the most elementary reactions of self-protection, impelling caution or retreat. One early writer while climbing a rocky peak had his attention suddenly drawn by the noise of a rattle to a creature whose like he had never seen before. He didn't know it was a rattlesnake, but it "looked so venomous" that he backed away and fell down the cliff.

In judging the effectiveness of this warning as a means of protection, one must have in mind the kinds of carnivores and birds of prey that seek the rattler as food, and the ungulates that might tread on it fortuitously, or with destructive purpose. Against many of these, such as wolves, coyotes, bobcats, and the like, the rattle would be a valuable protective adjunct — not unfailing, of course, but still of major importance in frightening these creatures into looking elsewhere for a meal. One has only to watch the reaction of a dog or cat with a rattler to see how effective this is, despite the qualification that some of them become experienced rattler killers.

An experiment has shown that a weasel would not attack a sidewinder while the latter possessed and sounded its rattle, but did attack when the snake's rattle had been removed. And the observation has been made that most animals scamper off when they hear the sound of the rattle. Hawks, owls, and ravens might be similarly frightened.

As for the larger herbivorous mammals, they are not usually of such a disposition as to go out of their way in search of trouble, although deer and antelope do show an inclination to attack rattlers by jumping on them. From observations of the reactions of horses and cattle it is quite evident that the rattle has a definite protective value.

Although some writers speak of animals being shocked into immobility, while others say they are driven to flight, it is probable that most of them simply adopt a policy of avoidance. Horses sometimes shy at rattlers, but the reaction is by no means so universal as some of the earlier natural histories indicated. However, they do tend to avoid a snake that suddenly rears and rattles in their path.

Summarizing this phase of the theory, it may be concluded that the rattle is of definite adaptive value when sounded as a warning to animals which, with intention or unconsciously, might injure the snake.

We come now to this query: is the rattle ever a detriment by advertising the presence of the rattler to potential enemies, from which it might otherwise escape?

It has been suggested that the use of the rattle may serve only to invite destruction by certain animals, such as deer and hogs, which do kill rattlers upon occasion. Now certainly this would be true if the rattle served to invite destruction by advertising the snake's presence, when inconspicuous inactivity, or a silent withdrawal would be a safer policy. This, however, misjudges the rattler's ordinary sequential response to the presence of an intruder. No field collector knows how many rattlers he may pass, that escape by merely lying quiet. But I have spied enough of them, fully aware of my presence, yet making no sound or movement, except possibly a telltale flicking of the tongue, to realize that they will depend on this method of escape whenever possible. The rattle is the reaction of a suddenly startled snake, or of one aroused to action by the persistent encroachment of an enemy.

This criticism also misjudges the reaction of animals to the sound of the rattle. They are pictured as saying to themselves when they hear it, as a man would, "Here is a dangerous creature that I had best destroy, for the safety of others who may pass this way." But no wild animal reasons thus; there is always a more specific purpose in attack, such as a desire for food or the protection of young. Against such creatures, keeping in mind the fact that the rattle is not sounded until procrypsis has failed, its possession is definitely protective, for it often causes them to turn aside and avoid the snake.

With man, an enemy who can destroy at a distance without endangering himself, the rattle is a disadvantage, for man is poorly equipped with senses, and the rattler may sometimes advertise itself too quickly when it might otherwise lie undiscovered. But this is no valid argument against the adaptive value of the rattle as a warning mechanism, since rattlers long antedated man in the New World, and the rattle was developed without regard to the novel conditions imposed by this addition to the local fauna.

Conditions of Use

The conditions under which a rattlesnake will sound—or fail to sound—its rattle as a warning are discussed elsewhere. It is pointed out that it is by no means invariably sounded prior to a strike, as is often thought to be the case—a widespread belief that should be strongly discouraged in the interest of safety. For whether a rattlesnake will rattle when disturbed, before undertaking a more retaliatory defense, depends on a number of conditions, such as the species and the individual temperament of the snake; the suddenness with which the intruder comes upon it, and the closeness of his approach; whether the snake was startled out of sleep; whether an injury to the snake was involved; the temperature; the availability of hiding places; and other, similar variables. No one can guess a snake's course of action in advance. It will usually rattle before striking, if the danger is not too imminent to permit delay; but surely no one is justified in depending on the rattle as a trustworthy or invariable advance note of caution.

Although the S-shaped defensive coil is the usual position of a rattling snake—this being a part of the general defensive posture of which rattling is another element—it can rattle when outstretched, either at rest or crawling. It often rattles while trying to escape to some nearby refuge. Disturbed snakes may sometimes be heard rattling in the restricted confines of a hole or rock crevice.

Annoyed snakes have been reported to rattle for several hours without a moment's interruption. I have never made any accurate records of long runs, but in the laboratory, when endeavoring to concentrate on scale counting, I have frequently found it necessary to remove live specimens, so annoying is the continuous rattling of especially nervous snakes not yet accustomed to captivity.

Many conditions affect the distance at which a rattler may be heard—the size of the snake more than any other—which explains some of the conflicting accounts that have been published. Another important qualification is the length of the rattle string. As I indicate elsewhere a string of about six to eight segments is probably most efficient as a sound producer; too short a string produces too few contacts for full stridency, while the outer segments of extra-long strings interfere with, or damp out, the full vibrations of the segments next to the tail.

The distances at which the sound of the rattle may be heard have been reported as from three to 160 yards.

Several writers comment on the greater danger from the pigmy rattler *(S. miliarius)* because its tiny rattle can hardly be heard at all. However, it should also be mentioned that the minute size of the snake, with its meager supply of venom and short fangs, greatly minimizes any danger from it, regardless of the slight audibility of the rattle.

I experimented in a closed room with adults of three species of rattlers having notably small rattles. A long-tailed rattler *(C. stejnegeri)* was audible at from four to ten feet. One central-plateau dusky rattler *(C. t. triseriatus)* could be heard at ten feet, but only if attention were focused on it. Certainly it would not have attracted attention at that distance. Another individual could not be heard beyond four feet. The rattle of an Omilteman small-headed rattler *(C. i. omiltemanus)* was audible as a faint buzz at seven feet. It is evident that such rattles as these tiny affairs have retained little value as warning devices, at least as far as man is concerned. (In the case of mammals with keener hearing, such as dogs, even the pigmy rattling could be very effective.)

Much has been written of the variations in the sound of the rattle that a rattlesnake may produce at will. While the speed of vibration is largely dependent on temperature, it is probable that the snake has some control over both the frequency and the amplitude of vibration, and therefore of the tone and intensity of the sound. That there is some degree of control over this intensity can be proved if we further aggravate a snake already rattling, thus causing it to increase its muscular activity. However, most of the varying sounds attributed to the rattles are exaggerated or fictitious, being cited to prove a duality of rattle use that does not exist.

One writer believed a rattlesnake able to sound three cadences with the rattle: first, an alarm; second, a milder sound when crooning to its young; and, finally, a conversational tone when mating. As mother rattlesnakes do not remain with their young, and mating rattlers do not rattle, there is no verification of these variations in sound for different purposes.

It is commonly observed of captive snakes that many of them — perhaps a majority — become so accustomed to the presence of people that they must be deliberately annoyed to be made to rattle. In fact some wild rattlers will often rattle only after being prodded with a stick.

Over the years several writers have commented on the increased danger resulting from wet rattles, wetness supposedly rendering them inaudible.

As a matter of fact, there is a considerable variation in the degree to which the rattles are muffled by dampness. When the snakes have been swimming, with the rattles under water — those that I have watched made no attempt to keep the rattles above water, as often reported — the rattles usually become so water-soaked that they make little noise until they have had an opportunity to dry out or until the water has been thoroughly shaken out of them.

It has been stated that shedding also interferes with the snake's ability to rattle, but it seems to have little effect. Actually, a snake whose sight is hampered by the so-called "blue-eyed" stage of shedding may be the more likely to rattle upon sensing an intruder.

There is a belief in Georgia that rattlers lay their heads on their rattles when sleeping to keep the dew from dampening them. Another myth is that rattlers will not rattle after dark, a yarn in which no credence will be placed by anyone who has hunted them at night.

Rattle Nomenclature

Some of the misunderstandings and doubts regarding rattle development and structure have been caused by a lack of consistency in terminology. To facilitate the discussion that follows I shall list some of the usages currently applied.

The complete caudal appendage of the rattlesnake is called the *rattle*, although it may alternatively be referred to as the *rattle string* to distinguish the entire set from the several parts, each of which may be referred to as a *rattle*. To avoid this possible source of misunderstanding it is probably best to use the term *segment* for the individual parts of a rattle. The segment, in turn, comprises from one to three (occasionally four) *lobes*. The segment joining the string to the tail of a rattlesnake may be referred to as the *proximal, anterior,* or *attached* segment, that at the opposite end being the *distal, posterior,* or *terminal* segment. The posterior is

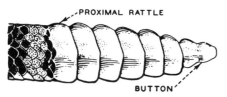

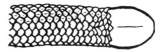

Fig. 13 A prebutton

Fig. 11 A complete string

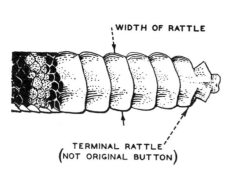

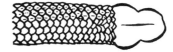

Fig. 14 A button

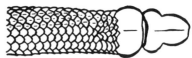

Fig. 12 A broken string after the loss of several segments

Fig. 15 A rattle-string comprising a button and a No. 2 rattle

the oldest remaining segment, and the anterior is the one most recently formed.

The *prebutton* is the segment that comprises the entire rattle that the snake has at birth. This is invariably shed with the snake's first exuviation a few days after birth; it is never retained as a part of the permanent string.

The *button* is the first retained rattle; it invariably comprises the posterior segment of every rattle string, provided the string is complete. No term has been more misunderstood. Occasionally the word *button* has been used to designate the proximal segment, because this "buttons" the string to the snake's tail. More often it is incorrectly used to designate the posterior lobe of the distal segment of a string regardless of whether or not the string is complete. By a *complete* string is meant any string from which no segment has been separated by breakage or loss, subsequent to the invariable loss of the prebutton.

The *rattle-number* is the serial number of any segment in a complete string, beginning with the button as rattle No. 1, the next as No. 2, etc. Rattle-numbers can be determined positively only if a string is complete.

The Character of the Sound

The term "rattle" is to some degree a misnomer, when applied to the noise made by a rattlesnake; for "rattle," by definition, implies discontinuous or discrete sounds, whereas the separate sounds emanating from a snake's rattle are much too closely spaced to be perceived by the human ear. The result is a toneless buzz, or, in the case of the larger snakes, a strident hiss. Without mechanical means, such as an electric vibrator, it is quite impossible to imitate, by shaking a snake's rattle, the sound made by a live snake, for one cannot approach the snake's speed of vibration.

As previously noted, misunderstandings with regard to its sound have resulted from the names applied to the organ. For the rattle or rattle lobes, the earliest account (1554) used the Spanish *cascabel*, so it is not strange that an early English report should refer to the snake "which hath a bell in his tayle." Some eighteenth-century descriptions liken the sound to a child's rattle. If the children's rattles of those days were hollow, non-metallic devices, containing sand or seeds, the simile is apt.

Other resemblances have been mentioned, such as the ticking of many watches or a knife being ground. The sound has been referred to as whizzing, whistling, and sizzling. A resemblance to running water has been suggested. Several authors believe the sound to be similar to that of dried peas or beans shaken in the pod or melon seeds shaken in a box.

A number of writers liken the sound of the rattle to that of various insects referred to as cicadas, locusts, crickets, or grasshoppers. To anyone knowing the loose way in which such entomological names are employed, these beliefs are probably well justified. There are, for example, species of cicadas in southern California and Arizona whose strident buzz is so rattlesnake-like that they are quite deceiving to the rattlesnake expert, and presumably to the entomologist as well.

Several authors have likened the rattle to the buzzing of a large bee. The story has been told of a military camp in 1847 alerted at night because a buzzing bumble bee entangled in a buffalo robe was believed to have been a rattlesnake. I have never heard a bee that compared with a cicada as a rattle imitator.

Other sounds reported as similar to the rattle of a rattlesnake are blowing through loose lips, the wind in the trees, the hissing of a hard rain, or escaping steam. To my mind, I should choose, as those having the best likeness, either a small steam jet or some species of cicada. The best human imitation can be made by hissing through the teeth with the lips parted.

Throughout this discussion of rattle similitude, I have treated the subject as if all rattlesnakes of all species and ages made the same sound. This is, of course, far from the fact; my remarks are appropriate only respecting adult rattlers of the larger species, and even these only when equipped with rattle strings of some five to ten rattles.

Method of Vibration and the Production of Sound

When a coiled snake is rattling, the rattle string is pointed almost vertically, although tipped slightly backward. In this position, it delivers the maximum noise for a given expenditure of energy and with the least wear. The vibration is always transverse or lateral, at right angles to the center line of the tail and to the flat or side face of the rattle.

If the segments of a rattle are moved slowly, with a transverse motion similar to that seen under the stroboscopic light, the method whereby the sound is produced may be analyzed. Between each lobe of each segment, and the interlocking lobe of the segment within, a succession of contacts and separations is produced. No doubt each pair clashes at a number of different points, so that several distinct sounds are produced, as the wave passes each lobe. The total number of sounds per cycle equals twice the number of effective segments multiplied by the contacts per segment. Assuming only one contact per side of each segment, three lobes per segment, and a six-rattle string, at fifty cycles per second (an average number), we have a thousand contacts per second. It is obvious that a frequency such as this will result in a hiss rather than discrete sounds. A young rattler with only two rattles makes about a hundred contacts per second; the result is a much softer buzz than the hiss of an adult.* Since

*The Santa Catalina Island rattlesnake, restricted to a small, rocky island off the east coast of Baja, California, usually has a single rattle segment. There is no loose segment that can make a sound.

the rattle is hollow, dry, and the material somewhat pliable, like heavy parchment, the sound is unmetallic, lacking musical quality.

The snake, while rattling, often waves the tail slowly from side to side in addition to the ordinary high-speed vibratory motion.

One myth current in Kentucky is that a rattle, when cut off from a snake's tail, will continue to vibrate for about three hours, if held in the hand.

Speed of Vibration

It was written in 1688: "The Old [rattlesnakes] shake and shiver these Rattles with wonderful nimbleness when they are any ways disturbed." No one can watch a rattler's tail in action without being similarly impressed, for the tail end, and the rattles as well, are blurred by the rapidity of their motion.

The speed of vibration has been found to be characteristic of the individual snake — its condition and disposition — and highly dependent on its temperature.

The most complete and accurate study of the variations of rattle speeds at different temperatures was made in 1954. Rattle speeds were determined by a mercury-arc stroboscope, and temperatures by a mercury thermometer introduced via the cloacal opening. The approximate speeds ranged from 21.1 cycles per second at 50 degrees Fahrenheit to 98.6 cycles per second at 104 degrees Fahrenheit.

The degree of alarm or anger seems to be the most important factor affecting the vibratory rate under constant temperature conditions.

FORMATION AND LENGTH

Chronology of Rattle Formation

Much has been written about the chronology of rattle formation, and the relationship between the number of segments in the rattle and the snake's age. The rattle-per-year idea, first put in print by Hernández in 1615, is still widely believed. No snake myth was ever more indelibly imbedded in the public consciousness, as may be seen by frequent newspaper reports down to the present day. And it is small wonder, for the great majority of the early accounts of the rattlesnake reiterated this entertaining myth, and there is something peculiarly satisfying in having so concrete a record of a creature's age.

Denials of the age-from-the-rattle theory were not slow in coming, though they have never caught up with the myth and never will. The earliest disavowals emanated from two supposedly controverting facts, one of which depended in itself on another myth, the other on a discrepancy that is valid. To some it appeared impossible that a snake could be as old as indicated by reputed rattle strings of forty to eighty segments; to others it was evident that rattles were subject to breakage and therefore could not invariably give the correct age. Later, observations of captive snakes showed that snakes could, and often did, acquire more than a rattle a year.

As early as 1790 the conclusion was reached from the rattle structure and its method of formation, that the acquisition of each new rattle must be coincident with shedding (which occurs from one to four times a year, depending on the age and activity duration of the snake); and there has been a gradual acceptance of this belief, now fully confirmed by recent research.

Regarding the chronology of the rattle, there were, and still are, some special myths affecting only young rattlers, which in turn modify the age-from-the-rattle theory. One school of thought adhered to the belief that a rattler has no rattles until it is from one to three years old. Equally fantastic, and just as persistent, is the belief that rattlesnakes are occasionally born possessed of two or three rattles.

Another myth, resulting from misidentifications, reports that young rattlers often throw off their rattles when they first use them, which accounts for the snakes with sharp tails that are really rattlers but are thus disguised.

There is a yarn of a collector who gathered rattlesnake eggs, and claimed to hear the young rattling in their shells before they hatched, a story rendered somewhat doubtful by reason of the fact that rattlesnakes do not lay eggs but give birth to living young.

Upon the acceptance of the theory of the coincidence of shedding with the acquisition, or at least the disclosure, of a new rattle, one seemingly inexplicable and discordant feature appeared, namely, the fact that, after their first shedding, young rattlesnakes still possessed only one rattle, when they should have had two, one to match the original skin with which they were born, and the second, the newly acquired skin. This problem was solved by the present writer in 1940 upon observing that the rattle with which a snake is born (the "prebutton"), being too pliable and weak to remain, is invariably shed with the first skin.

Rattle-String Length

"James Morton yesterday killed a big rattlesnake on the bank of Trout Creek. It measured nearly five feet and had sixteen rattles and a button." *Blank County Weekly Herald.*

So reads a typical newspaper report, and from it the skeptic will conclude that, regardless of the snake's actual length—no doubt nearer three than five feet—the chances of its really having had sixteen rattles and a button were remote, indeed. For the statement implies the coincidence of two rare events: a rattle string on a wild* rattlesnake containing as many as sixteen rattles; and, superimposed on this, the still greater improbability that the string was unbroken, not a single segment having been lost since the first shedding, for this is what the retention of the button means. Citing the presence of a large number of rattles seems, to the average person, proof that the snake must have been truly a monster; but actually these long strings are so unusual that the herpetologist immediately doubts all the details of any story of which a long string is a part.

Besides plain lying, or, in milder terms, exaggeration, there are several ways in which these long strings become established beliefs. Sometimes there is a miscount. Then there are faked rattle strings, obviously not to be expected on wild snakes unless a rather intricate practical joke is involved, although often attached to skins that serve as ornaments or trophies. Again, many people count, as an additional rattle, each lobe of the terminal segment of a broken string, which adds at least two, or occasionally three, to the count. Sometimes the counter estimates the number of rattles that have been lost from an incomplete string and includes these departed wraiths, on the theory that the snake has played unfairly in losing them. Some people double the number by counting each side of the central crease. But mostly the long strings are just hearsay, the persistent echo of some campfire tale.

There were several long-string stories that became classics in their day. One was a report in 1723 of the killing of a rattlesnake having a string of between seventy and eighty rattles. Since this yarn appeared in the sedate *Philosophical Transactions* of the Royal Society (London), it was occasionally accepted; but it probably was not taken seriously by many, since who could conceive of a rattler seventy to eighty years old? Besides, the venerable creature not only had these proofs of a great age, but a sprinkling of gray hairs as well—a logical but last-straw addition, too much for even the most credulous.

Several extra-long strings have been pictured that can be seen to be fictitious, not through the evidence of unmatched sections, but by reason of their shapes. These show a series of rattles gradually increasing in size from the original button to the latest rattle joining the string to the tail, that is, with a uniform taper throughout. Actually, rattles do not follow any such growth sequence, for there is little or no growth in lobe size after the acquisition of about the tenth or twelfth rattle.

The faking of rattle strings is comparatively easy, and quite deceptive results may be attained if one has an adequate stock of separate short

---

*In discussing rattlers, the term "wild" is used, not in contrast with "tame," but to distinguish the snakes as found in the wild from captive specimens, especially those long in captivity.

Fig. 16 A famous long rattle string from *Columbian Magazine* or *Monthly Miscellany*, November 1786, often mentioned in subsequent publications. The irregularities of form indicate that this rattle was pieced together from the rattles of seven or eight different snakes.

strings from which to choose well-matched sections when piecing them together. Before being joined the sections are softened by moistening, after which they can be telescoped into each other without breakage. Some prefer milk for softening.

Of the many thousands of rattlers brought to the San Diego Zoo in the past twenty-five years, none has had more than sixteen segments in its rattle. This record string was on a Colorado Desert sidewinder *(C. c. laterorepens)*, a sand-inhabiting form characterized by long strings. None of my correspondents has reported any phenomenal string on a rattler caught by himself in the wild. Of two collectors of particularly wide experience, one reported a sixteen-rattle maximum, the other thirteen. Another collector had seen a string of thirty-two rattles reputed to have been found on a Mojave Desert snake. My informant did not say how closely he examined the rattles to detect a possible fabrication.

The longest unbroken strings I have seen on live wild rattlers were on a San Lucan speckled rattlesnake *(C. m. mitchelli)*, and on a red diamond *(C. r. ruber)* from Yaqui Pass, San Diego County. Each had thirteen rattles, including the original button. It is probably no coincidence that *C. m. mitchelli* has unusually large (and therefore strong) rattles in proportion to the length and bulk of its body, and that *C. r. ruber* is exceptionally peaceable in disposition.

Long strings are not scarce because rattlers do not grow so many segments during their lives (from this viewpoint twenty-five or thirty would not be exceptional) but because long strings are particularly subject to wear and breakage. And, in fact, breakage in rattle strings is both normal and beneficial. Very long strings, say beyond twelve rings, cannot be particularly serviceable from the standpoint of a sound-producing apparatus, and might be a positive detriment, for the very length and weight would damp out the vibrations. Furthermore, they would tend to arch over and drag on the ground, notwithstanding the fact that the normal crawling posture of rattlers causes the rattles to tilt upward to avoid contact with the ground.* Probably a string of from six to eight segments constitutes

*Strings reported of fantastic lengths, such as thirty to forty rattles, would certainly touch the ground as the snake crawled, and be quickly worn away.

the most efficient vibrator, and indeed breakage is rapid after there are eight in the string.

Rattles are normally lost either through the wear incident to use — that is, by being rattled so much that they wear out — or by being scraped or torn by interfering objects. For example, if a rattle be caught in the fissure of a rock or the fork of a branch, the snake will give a violent twitch of its tail and the offending rattle will be snapped off. At the San Diego Zoo, captive rattlers with long strings have been observed to break rattles off by bending a string at a sharp angle when the snake's own body was resting on it, or by pulling the rattles out from under its own coils.

The rapidity with which a loss of rattle segments can occur through excessive vibratory use has been observed in captive specimens. Those that are extremely nervous when brought in, and that fail to become inured to the presence of human beings, rattle for such long periods that the rattles quickly wear out, this being especially true of the terminal segments of long strings. But captivity does not invariably produce this effect; on the contrary, some strings longer than any carried by wild rattlers have been attained by captives, because those that lose their fear seldom rattle, and become so lethargic that they crawl about but little. As a result, the rattles are conserved, producing some phenomenally long strings not matched in nature, occasionally reaching twenty or even more before breaking. The record thus far at the San Diego Zoo is held by a timber rattlesnake (C. h. horridus) that had an incomplete string of twenty-nine rattles at the time of its death.

The records for complete strings are also likely to be made by captives. The longest complete string (with the original button) of which we have knowledge was on a Tortuga Island diamond rattler (C. tortugensis) that was received by the San Diego Zoo in 1937 with a complete string of five rattles. By August 1941, it had eighteen rattles, with the button still intact. Shortly afterward the rattle broke, eleven segments coming off in one piece. Complete strings of fifteen or sixteen rattles are not exceptional among caged rattlers.

Some early theories visualized an upper limit to the number of rattles, not owing to breakage, but to some natural limitation. One theory was that a snake acquired a rattle a year until a limit of about twenty was attained; another that there was some form of restriction affecting particular species — pigmy rattlers acquiring only three to seven rattles.

CHAPTER FOUR

Bodily Functions

Like all reptiles, rattlesnakes are under the dominant influence of external temperatures. The most familiar of
the larger animals about us—the mammals and birds—
are, to a considerable degree, independent of the temperatures in which they live, particularly of short-range
fluctuations, for they have internal heating and cooling
mechanisms that maintain an almost constant internal
temperature—a temperature optimum for their metabolic processes and muscular activity. But the reptiles—
ectothermic animals—are almost entirely lacking in this
control; with falling exterior temperatures, their own
bodily temperatures fall, with the result that the energy
derived from oxidation declines, as do muscular activity
and rate of digestion. (This is the reason why most reptiles, especially snakes, feel cold to the touch; they are
usually at a temperature well below that of the human
hand and are good conductors of heat.) At temperatures
that would not be uncomfortable to a mammal or bird, a

snake is reduced to enforced immobility. And, similarly, at above-optimum external temperatures, the lack of the cooling mechanisms of perspiration and a reduced effectiveness of evaporation from lung surfaces allow the body temperature also to rise, until at external temperatures that mammals withstand with no more than a moderate discomfort, the internal temperatures of reptiles rise to a dangerous degree; the muscle tissue hardens, circulation stops, and the animal dies. So important are these temperature effects on reptiles, so narrow the range of external temperatures in which their life activities may be carried on successfully, that their habits are largely circumscribed by their having to take the greatest advantage of favorable temperature conditions, and to avoid those detrimental or dangerous. When some objective — food, for example, or a protective refuge — can best be secured under unfavorable temperature conditions, a compromise results, with some sacrifice of the optimum condition for either category separately.

Another point to be borne in mind is the relative inactivity of snakes. Put briefly, it may be said that most of the time they do nothing; no enterprise or action is required of them. Their food and water needs are low compared with those of mammals and birds of similar size, for reasons discussed elsewhere. Ordinarily they mate but once a year. With these two primal drives at a minimum, it follows that they have both long, and relatively frequent, periods of inactivity, extending, in summation, over a considerable part of their lives. In the interests of comfort, safety, and the preservation of energy, these times of rest are taken in suitable refuges, in holes, or under rocks or bushes. Thus they are secretive; they are not so often seen as other creatures having the same population density.

In the portrayal of the habits and reactions of rattlesnakes that follows, these themes of temperature effects and secretiveness will occur repeatedly.

LIFE PERSISTENCE AND FRAILTY

It is popularly supposed that snakes — rattlers included — are highly tenacious of life. This belief is no doubt founded on the persistence of reflex body movements that continue long after a snake has been fatally injured, or even decapitated.

Probably the earliest statement as to the persistence of life in the rattlesnake was made in 1615 when it was reported that the head, if cut off, would live for ten days or more.

Although this is an exaggeration, it is true that there is a surprisingly long continuance of body movement in a decapitated snake, and that it is dangerous to handle the head of a venomous snake for some time after it has been separated from the body. Serious accidents have resulted from picking up severed snake heads or fatally injured snakes, and I wish to voice a warning against any carelessness in this respect. The case of a farmer has been reported, who, in mowing, cut off the head of a rattler. While brushing the grass aside to locate the head, he was bitten in the thumb with dangerous, but not fatal, results. Some two or three inches of the snake's neck had remained attached to the head. There was a fatal case of a child bitten by a decapitated snake in Ottawa County, Kansas. The snake's body had been severed three or four inches behind the head; it is not stated how the child happened to be bitten. A newspaper report of a fatal accident at The Geysers, Napa County, California stated that a man was bitten by a northern Pacific rattler that had been shot into three pieces. I was myself a member of a hunting party on which a man was bitten by a small southern Pacific rattler that had been shot in the neck with a .22 loaded with dust shot. Such a charge will break a snake's back and prevent it from crawling, but it will not be harmless until some time after death supervenes.

Based on experiments I have made with decapitated rattlesnakes, I would say that most severed heads are probably dangerous up to twenty minutes, and possibly dangerous for almost an hour. With only a short stub of a neck, the head is, of course, quite immobile; any danger, therefore, must result from deliberate handling. As to a fatally injured snake, the duration of the danger would depend on the nature of the injury; the only safe rule is not to touch the head until the snake has actually begun to stiffen in rigor mortis.

Despite their seemingly strong hold on life, as indicated by the persistence of movement in decapitation tests, rattlers are relatively frail creatures and are easily killed. And though they can be dangerous as long as capacity for movement continues, their backbones are both delicate and vulnerable, so that a smart blow of no great force will produce a fatal injury, although at first it will seem to interfere only with locomotion.

LONGEVITY

Such data as we have on the age that rattlesnakes attain must come from observations of captive snakes, for no program of field observations by capture and release has reached such a duration as to encompass the probable normal life of rattlesnakes. At the San Diego Zoo, twenty rattlers have lived in captivity for ten years or more. Two western diamonds (*C. atrox*) were still alive after eighteen years in captivity, the longest-lived captive rattlesnakes known up to this time. One of these snakes was an adult when received, so that it was still alive at an age of not less than twenty-one years.

From these examples I should presume that rattlesnakes in the wild occasionally attain an age in excess of twenty years, longer than either their prey or competitive mammal and bird predators.

Much that is mythical has been published on the growth and longevity of snakes, rattlers included. Some of these ideas have evolved from certain ancient myths as to the great age that snakes attain because they are thought to renew their youth when they shed their skins. A common figure in these myths is a thousand years.

SKIN SHEDDING

Rattlesnakes, like other snakes, shed their skins periodically. This shedding — or exuviation — is an important requisite to the health of the snake; presumably it is necessary to provide for both growth and wear. A rattlesnake adds a rattle to the string each time the skin is shed.

The Shedding Operation

A healthy snake sheds its skin by crawling through it in such a way that the skin peels backward over the snake's body, turning inside out as it proceeds. The old skin is shed in a single, unbroken piece, representing every exterior part of the snake, from the edge of the lips to the tip of the tail. Even the outer layer of the transparent scale over the eyeball is shed; and, in the rattlesnakes and other pit vipers, the surfaces of both the anterior and posterior chambers of the facial pit are evident as tiny bags attached to the skin.

The snake about to shed shows some nervous activity. Finally it rubs its snout on some rough object, thereby loosening the old skin along the edges of the upper and lower lips. Then the coverings of the top and lower surfaces of the jaws are loosened and turned back over and under the head. More scraping or rubbing against rocks or shrubbery, together with skin movements, and body kinking and writhing, continue the process, pushing the old skin back over the body. Sometimes there may be a temporary impediment and the old skin will gather into a bunch of accordion pleats. But soon this difficulty is surmounted, and, once the narrow neck-section has passed over the thickest part at mid-body, the snake quickly crawls through the remaining part of the skin and finally emerges with its bright, fresh surface fully disclosed. The old skin, sometimes bunched, but more often fully extended and inside out, is left to be destroyed by the elements or to be used as nest material by some fearless bird. Many kinds of lizards, and a few snakes, are known to eat their shed skins, but this is not the practice of rattlesnakes.

The normal shedding process requires from a few minutes to several hours, depending on the health and activity of the snake, the temperature and humidity, the availability of rough objects on which to scrape, and similar conditions.

Exuvial glands aid in the shedding, presumably by discharges that loosen the outermost layer from the layer below.

Effect of Shedding on Sight

The skin-loosening discharge is exemplified, at a certain stage of skin shedding, in the exudate that appears between the new and old covering of the eyeball of a snake, causing the milky or bluish appearance of the eye—the well-known "blue-eyed" stage that causes partial blindness for about a week, but which clears up some days before the actual shedding takes place. Many have thought that the blindness is caused by a thickening and resulting opacity of the old skin, but this is not true, as is evident from the fact that the eye clears again before the old skin is shed.

The belief has been expressed that rattlers with their vision impaired by incipient exuviation are seldom seen abroad, but stay down holes or in other refuges. There is little doubt that this is true, for a snake would be less likely to expose itself to its enemies while lacking the full use of one of its senses.

The scale that covers the eyeball is quite clear in a shed skin—clearer, in fact, than any other element of the skin. This is further proof that the opacity is not caused by a change in the skin itself.

Water Requirements during Shedding

It has been stated that rattlers are especially in need of water at shedding time. Snakes having difficulty in shedding have been observed taking to the water; and some believe they shed more frequently in rainy weather. Another view is that healthy snakes can shed equally well with or without water available; however, if one has difficulty in shedding, exuviation can be facilitated by putting it to soak in water for a night. I was once informed of a captive sidewinder whose condition became so bad that it was decided to destroy it. For this purpose an attempt was made to drown it in warm water. The following morning it was found to have shed several layers of skin (about five), and this was followed by its recovery, so that it lived for some time thereafter.

In 1932 it was found that the onset of the shedding process in snakes leads to an increased loss of water by evaporation through and from the skin. Researchers are now of the opinion that snakes seek water during ecdysis to avoid desiccation, and not to loosen the skin.

The Season of Shedding

There are various folklore beliefs regarding the mechanism of skin changing. Pliny stated that snakes eat, or rub themselves with, fennel preparatory to shedding. The Pennsylvania Germans believed that snakes shed by using brier thorns as hooks wherewith to pull the skin off. One writer thought the rattler's skin loosened at the tail first. But much of the recent folklore concerning shedding has to do with the seasonal regularity of shedding and its effects on the habits and dispositions of the snakes. There is an almost universal belief that rattlesnakes are blind — from the preparation for shedding — in August, and, at that time, are particularly dangerous.

As a matter of fact, there is little evidence that the snakes adhere to a rigid chronological schedule in changing their skins. Probably the weather, success in capturing prey, and other variable conditions tend to change, at least to some extent, the elapsed time between exuviations.

It is true that there is one shedding that runs quite true to schedule — that of the young rattlesnakes after birth. This usually occurs from seven to ten days after birth.

If there is any other regular shedding time, it probably follows immediately upon emergence from hibernation. Further studies on this matter of seasonal regularity are desirable.

One myth has it that rattlers shed annually for seven years and then shed no more.

Shedding Frequency

Coming now to the intervals between exuviations, I must admit that much of our available information is of somewhat doubtful validity, based as it is on observations of captive specimens. There is no denying that the conditions of captivity seriously modify the shedding schedules, particularly because of the relatively high temperatures in the reptile houses during winter. But even so, we can at least show that the growing young shed more frequently than adults. Further, because the rattles themselves are records, we are able to assemble the statistics of unbroken rattle strings and gain some idea of the natural sheddings in the wild during the first year or two of life.

As to the shedding frequency of captive rattlers, statistics compiled from studies of adults at the San Diego Zoo show that sheddings per year vary from a low of one, in two Colorado Desert sidewinders *(C. c. laterorepens)* and one southern Pacific rattler *(C. v. helleri)*, to a high of 3.9 in a Mexican west-coast rattler *(C. b. basiliscus)* and a timber rattler *(C. h. horridus)*. Skin shedding in captive young can take place as many as six or seven times per year.

From all available data I should judge that juvenile and adolescent rattlers in the wild probably shed from two to four times in their first full growing season, and from one to four times in their second; the number of sheddings depends on the duration of the season of activity. Adult rattlesnakes in the wild probably shed from about three times per annum where the climate permits activity almost all year, down to once per year where activity is limited to about six months of the year or slightly less.

LOCOMOTION

Since man saw his first snake he has pondered the puzzle of its motion. To see it flowing over the ground with effortless grace gives no indication that the propelling force is coming from within the creature itself.

"There are three things which are too wonderful for me, yea, four which I know not: the way of an eagle in the air; the way of a serpent upon a rock; the way of a ship in the midst of the sea; and the way of a man with a maid." Proverbs xxx: 18–19.

According to one nineteenth-century observer the snake has "neither fins, nor feet, nor wings; and yet he flits like a shadow, he vanishes as if by magic, he reappears and is gone again, like a light azure vapor, or the gleams of a sabre in the dark." The puzzle of the snake's motion lies in the lack of any apparent force exerted on the ground, such as is seen when an ordinary creature moves its alternating legs; there is no evidence of its pushing itself forward.

Although one method of crawling is both more common and less readily explicable than the rest, snakes actually employ four different kinds of propulsion to which the following names have been applied: horizontal undulatory, rectilinear, sidewinding, and concertina. Rattlesnakes use all of these at times, although only one species (*C. cerastes*) is adept at sidewinding—a distinction shared by only a few other desert snakes the world around; and the concertina method is of relatively restricted utility. I shall now describe each of these methods, the mechanical principles involved, and the circumstances under which they are likely to be used. Although the four methods are sharply distinguished when practiced in a typical manner, they can be employed in such a way as to change, by degrees, from one to the other; and, indeed, a snake can simultaneously use one method anteriorly, while another is employed toward the tail.

Horizontal Undulatory Progression

Most snakes normally crawl by means of the horizontal undulatory method—or, as it is sometimes called, the serpentine or sinusoidal. Yet, while the commonest of the four recognized methods, it is the most difficult to describe and to understand.

The snake, with its body in a series of side waves, glides smoothly along, each part of the body unerringly following the wavy path first

Fig. 17 Southern Pacific rattlesnake (*C. v. helleri*) in crawling position when using horizontal undulatory method of progression at slow pace

taken by the head and neck. The waves are as nearly horizontal as may be permitted by the unevenness of the ground over which the snake is crawling and upon which the waves rest; it never uses one or more vertical loops such as that affected by an inchworm, which was the snake motion most commonly depicted in the prints in the old natural histories.

The puzzle in the horizontal undulatory method of crawling is the conversion of the body waves—the lashing of the body from side to side—into longitudinal motion. One might expect that this lashing, if effective at all, would cause the entire body to slide across the ground, leaving a track as wide as the maximum width of the waves. Indeed, on a very smooth surface, this result is produced, for the lashing is highly ineffective. But give a snake a firm and rough substratum upon which to travel, and the lashing becomes converted into longitudinal motion, so that it ceases to be evident as lashing at all. The motion is not longitudinal in the direction of the objective, but along the body of the snake, as a file of ants might follow a winding path. Actually, the snake seeks and pushes backward against any irregularities in the ground that will prevent back-slippage, but the push is evident only if the objects pushed against are movable, as is the case when a snake traverses sand.

Some research has been done on the mechanical problem of how the lateral pressure against a pivot or anchorage point can be exerted by the snake's musculature in such a way as to convert the pressure into longitudinal forward motion along the body of the snake. And, although the fundamental kinematics of sinusoidal movement are now understood, much still remains to be done on this problem that puzzled the proverb-maker so long ago.

Rectilinear Progression

Rectilinear progression is a method especially useful to large thick-bodied snakes, such as pythons, boas, and the larger vipers. Rattlesnakes of all kinds, particularly adults, employ it extensively, chiefly when they are prowling and unhurried.

Most early descriptions of snake motion, whether undulatory or rectilinear, have pictured the snake as "walking on its ribs." However, even in this method of propulsion, the ribs are not employed in the sense of "rib-walking," for the tips of the ribs remain unmoved relative to the vertebral column. On the contrary, the crawling is effected by reciprocating movements between the snake's skin and its body.

In operation, a section of the skin of the belly is drawn forward so that the ventral scutes in that section appear to be bunched. This part of the body is then pressed on the ground so that the sharp rear edges of the ventrals may engage any available imperfections of the ground surface and thus be prevented from slipping back. Then the body slides forward within the skin, pulled by appropriate muscles, until it once more is in normal alignment with the skin. Then once again the skin slides forward and the process is repeated. During the forward motion, the skin is raised so as to clear the ground, or at least enough to reduce the friction of its advance. When it comes to rest, it is pressed into the ground to secure an anchorage against slipping back, while the body is being pulled forward.

The straight trail of a snake using rectilinear progression has sometimes led to the assumption that the snake was moving by means of vertical undulations. It is true that vertical undulations are involved, but only to the very slight extent that the forward moving ventral scutes are made to clear the ground, as they are pushed forward. They make no high vertical loops such as were pictured in so many of the early illustrations of moving snakes.

Sidewinding

Sidewinding is the natural way of crawling of several kinds of desert snakes, a method followed whether they are hurried or merely prowling. It is also used by other snakes when under the necessity of negotiating smooth surfaces on which their usual means of progression are found to be ineffectual.

Our own sidewinder, the small desert rattlesnake *(Crotalus cerastes)* with the alternative name of horned rattlesnake, attracted attention from the early days of the travelers who met it while crossing the southwestern deserts. They were quick to note its peculiar motion and to invent the appropriate term "sidewinder," which became current as early as 1875. The term is often misused in applying it to rattlesnakes of various species that sidewind only rarely and imperfectly, or to snakes that have the reputation of striking sidewise.

The sidewinding progression is difficult to describe, but motion pic-

Fig. 18 Track of an adult sidewinder traveling from left to right. The lower marks (rather less J-shaped than usual) were made by the head, the upper marks by the tail. The small tracks were made by beetles. (Photograph by Dr. Raymond B. Cowles)

tures of a sidewinder at a natural gait, when slowed up on the screen, are quite effective in demonstrating the sequence and coordination of the motions. Essentially, sidewinding involves a side-flowing or looping motion whereby only vertical forces (rather than transverse) are applied to the supporting surface. The track comprises a series of short, separated straight lines, set at an angle of about thirty degrees with the direction of progression, each line approximating the length of the snake. If the track be in fine sand, and undisturbed by wind, the impressions of the ventral scutes can be clearly seen, for there is no transverse or sliding motion to obliterate them. Each section of the track has a fairly evident J-shaped mark (made by the head and neck) at one end, and a T-shaped terminus (made by the tail) at the other. The direction of progression, if one desires to track the snake, is that toward which the hook of the J is pointing; however, the T-shaped mark left by the tail is nearer to the destination than the J made by the head.

Sidewinding represents the most efficient use of a loose supporting medium, such as sand, which can offer little resistance, and therefore little reaction, to transverse forces directed across the surface (as do the sticks, stones, and irregularities of ordinary firm ground by which the typical snake aids its sinuous progression), but can exert a considerable resistance to forces applied vertically.

A person watching a sidewinder will see no resemblance between the

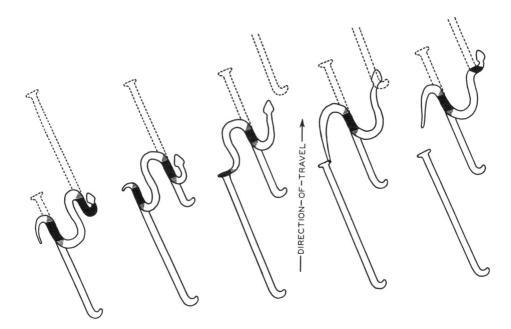

Fig. 19 How a sidewinder makes its tracks. Consecutive positions of a sidewinder's body in relation to the tracks. The solid track-outlines have already been made; the dotted outlines are yet to be made. Only the solid-black sections of the snake's body are in contact with the ground; the rest of the body is raised sufficiently to clear the ground.

motion as I have described it — throwing out the head, laying down the body, throwing out the head again, etc. — and what the snake seems to be doing; for the snake telescopes these operations by executing several simultaneously. Long before the tail has been placed, at the end of the laying down sequence, the head has already reached out for a new anchorage, so that the moving snake is never, even for an instant, fully outstretched along any one of its tracks. As a matter of fact, it is always touching at least two tracks at once, with that portion of its body between tracks arched slightly to clear the ground surface; and at the time the head is first touching its next anchorage, the tail-tip is just leaving the track two steps behind, so that for a moment in transition the snake may actually be in contact with three tracks at once.

Why *C. c. laterorepens* should have become so adept at sidewinding must remain a matter of speculation, but whatever conditions of sand surface, body contact surface, and bulk caused these particular snakes — and not other desert dwellers — to adopt sidewinding, it is so deeply ingrained as not to be abandoned when the snake lives on surfaces where sidewinding is not superior to the more customary means of progression. A young sidewinder that has never experienced sand sidewinds at birth.

Concertina Progression

The concertina type of progression has been called the "earthworm" type, but I prefer the designation of concertina movement; for the earthworm progresses through alternately lengthening and compressing portions of its body, a method impossible for snakes, although occasionally attributed to them.

In the concertina movement, the central part of the body is alternately gathered into two or more sinuous curves, and then restraightened. While the longitudinal distance occupied by the central part of the body is being shortened, by change from a straight to a wavy form, the snake anchors its head and neck, so that the shortening results in drawing the tail forward. Then the central curves are again straightened so that the longitudinal space occupied is again lengthened; but, as this is being done, the tail is anchored so that the head and neck are thrust forward. Hence, with each complete cycle of movements the entire snake advances by the difference between the lengths of the snake when in its straightened and waved positions.

Actually, the concertina method is not the normal means of progression of any snake; rather, it is used under particular circumstances as, for example, when a slow, hesitating advance is desired. It may be employed by a snake investigating an object whose nature is obscure, in which case there may be considerable pauses between the alternations. I have seen it used by a king snake stalking a young rattler; and I believe it a particularly satisfactory method for stalking because it has two advantages over the more rapid undulatory method: first, the head is not subject to lateral movement and therefore can concentrate better on the object approached; and, second, a part of the body is always stationary. This latter is important with snakes patterned in brightly contrasting colors, since the moving part of the body has a confusing effect of disappearance, while the stationary part fixes the attention on that very stationary quality, giving an illusion of safety to the prey.

Experiments with surfaces indicate that rattlers probably use concertina progression in their explorations of mammal holes, throwing out side loops to engage the walls whenever the ground is too smooth or soft to facilitate rectilinear progression.

Rattlesnake Speed

The speed with which rattlesnakes can travel might be thought to be of importance in connection with the snake-bite problem, as a criterion of how successful one might be in escaping from an attacking rattler. Actually, they so rarely attack, and they move so slowly—this refers to their crawling, not striking—that no one with his wits about him would have the slightest difficulty in evading a rattler moving at its top speed.

The persistence of an erroneous idea in natural history is exemplified by the way in which the notion that rattlesnakes are swift-moving was carried down through book after book. The first extended account of the rattlesnake ever printed (1615) says that the rattler moves over rocks and precipices at great speed, and that the Mexicans called it *ocozoatl* after a wind. Audubon claimed he saw a rattler chase a gray squirrel and gain on it. In 1791 the opinion was expressed, based on field observations, that a rattlesnake could move no faster than a man or child could walk, which is substantially correct. Since then, accounts of rattlesnake speeds have been more realistic, although several of the ancient fantasies regarding rattlesnake speeds, including the tale of the rattler that outdistanced a man on horseback, have recently been revived. It is now known that rattlers, at their highest speeds, would have difficulty in attaining three miles per hour, even for short distances.

With regard to snake speeds, it should be observed that the creatures suffer from a number of physiological handicaps and deficiencies, quite apart from their lack of legs. Even at the most favorable body temperatures, snakes are capable of high degrees of activity for brief periods only, because of the physiological effects of their heart and lung structures and their limited energy storage. Furthermore, since muscular energy operates at an efficiency of about twenty percent, about four times as much energy must be dissipated in the form of heat in a moving creature as is useful in producing movement. This would be a serious handicap to an animal having no sweat glands or other perfected means of cooling by evaporation, to say nothing of having almost no surplus water. And it should not be forgotten that the sinuous course of a snake's path involves a further disadvantage, since the snake must travel from twenty-five to fifty percent farther than the straight-line distance in order to reach its destination. The body of a snake reputed to be traveling at thirty miles an hour really would be moving at forty or more.

Snake-speed tests are difficult to carry out in a satisfactory manner, which is why so few valid data are available. It is almost impossible to cause a snake to proceed directly from one point to another at its best rate of speed. If chased, or kept continually frightened in order to keep it at top speed, it will dodge and feint, endeavoring to escape by craft rather than by speed. And a venomous snake, such as a rattler, if it is being overtaken, will pause, throw itself into a defensive coil, and endeavor to save itself by intimidation, backed by a readiness to strike. This causes the failure of such demonstrations as the well-advertised "rattlesnake derbies." In one such test in 1939 the winner was reported to have run the thirty-foot course in twelve minutes and twelve seconds, or at a rate of a mile in about a day and a half.

Misunderstood Features of Crawling

In addition to the exaggerations of speed, and the misinterpretations concerning the several kinds of locomotion, there have been other errors or exaggerations with respect to rattlesnake crawling.

One writer stated that a rattler could stand on its tail and move along with the body held perpendicularly. Actually this is only an exaggeration, for it is true that a rattler can move with the anterior part of the body raised well off the ground. Just as a rattlesnake can merge one type of crawling into another, so also it can quickly change from a crawling position to one of defense. Characteristically, one can crawl while assuming a defensive attitude, raising the anterior part of the body and facing an intruder, while the posterior part propels the snake backward or sideways toward some refuge or concealment.

A prowling rattler, if it has detected prey near at hand, will sometimes raise the head and neck well off the ground, while the body progresses by the horizontal undulatory or the concertina method.

One writer believed that rattlesnakes could ascend smooth surfaces by erecting their abdominal scutes as hooks.

Two people who had been bitten noted that the rattlers scampered away with every manifestation of delight at what they had done, and progressed in vertical waves. One may hazard the guess that these victims had been fortunate in imbibing the standard Western snake-bite remedy in advance of the accident.

Snake Tracks

Much can be learned by a study of snake tracks, particularly in sandy, desert areas where they may be followed for long distances without a break. Indeed, professional collectors in such areas, setting out early in the morning before the wind has obliterated the tracks, use them as the principal means of locating their quarry.

From the track, an experienced collector can determine, with moderate accuracy, the kind and size of the snake that made it. He can follow its night wanderings and learn much respecting its habits. The direction an undulatory or sidewinding snake has taken is easy to discover; and even a snake proceeding by rectilinear progression will revert to the other methods often enough to reveal its direction by telltale piles of sand or at least will drag indicatory accumulations of sand into any natural depressions in its path.

SENSES

Sight and the Eye

The eye of the rattlesnake has always had a fascination for people, particularly because it was, indeed, thought to be the source of the snake's ability to fascinate animals by those who have conceded it that mystical — and mythical — power. To many people the eye has a sinister appearance, a result, no doubt, of its fixed and unwinking stare, the narrow, catlike pupil, and the effect of the overhanging scale — the supraocular — that gives it a scowling and beetle-browed aspect. Many early writers found every evidence of malignity in the rattler's glance, making statements such as these: the eyes shine like fire as the snake levels its dreadful glare upon its prey; when the snake is threatening an enemy, the eyes become as red as flaming coals. One author noted that the eyes of a rattler blazed with fury when a stone was thrown at it. Another warned hunters to beware of the rattler's hypnotic eye. There are references to the magnetic influence of the rattlesnake's eye.

Several writers remark on the dilation of the eyes when the creature is angered: one claims that, in exerting its mesmeric power, it dilates its eyes to the size of a ten-cent piece. The fearful quality of the rattlesnake eye is summarized by the statement that it is so dreadful no one can look at it for more than a moment. (The mythical use of the eye in fascinating prey is mentioned elsewhere.)

But all of these writers are, in fact, concentrating into an imagined appearance of the rattlesnake's eye, the sinister threat of the snake's defensive posture. For the eye is singularly inexpressive of the snake's feelings and deportment. Far from flashing fire, it remains virtually unchanged, whether the snake is asleep or threatening some trespasser upon its seclusion. The snake has no movable eyelids; the scale-sheathed skin that surrounds the eye has neither the power of dilation nor of other voluntary movement; the pupil itself has little range of lateral motion; and, finally, unless the snake be asleep, the pupillary opening is responsive only to the incident light. So the expressiveness of the eye that we are likely to attribute to other creatures, because we observe it in ourselves, is certainly not evident in rattlesnakes.

As to the acuity of a rattlesnake's vision, there have been many guesses by observers, but there have been the usual difficulties in segregating the effects of the several senses to learn which one has produced some particular reaction that may have been noted. This is especially the case with pit vipers, such as rattlesnakes, since they have two organs — eyes and facial pits — responsive to radiation. Some of my correspondents, as a result of their field observations, believe rattlers to have good vision, but only at close range, and this seems to be a sound conclusion. Other correspondents bring out the fact that rattlers are especially responsive to moving objects.

My own tests, under moderate illumination, indicate that a rattlesnake can detect movements by sight alone at distances of fifteen feet at least. These experiments were made on a nervous snake housed in a glass jar. Such a snake is particularly useful in tests of this kind since it will react to external stimuli by rattling, whereas after it has become accustomed to captivity, it will make no such response. By the use of a piece of white paper on the end of a thin rod that could be made visible to the snake through a door, it was easy to ascertain under what circumstances the snake became aware of the paper. No other sense but sight — not hearing, smell, the pit, or air motion — could possibly have been effective in causing this glass-encased snake to rattle. Yet it did rattle whenever the white paper was brought within its range of vision, although not always before the paper was waved about.

Experiments have shown that rattlers deprived of both pits and tongue are able to catch mice by sight alone. With cold-blooded prey the pits would be of little use in any case. But even with all their senses intact, it is surprising how frequently captive rattlesnakes are seen to miss their prey when they strike at it, even at close range. So although we have little doubt as to the importance of sight in rattlesnake activities, tests show the eyes to be far from perfect instruments.

Certain generalities with regard to the eye structure in snakes are of interest. No snake has a fovea, or retinal center of high sensitivity, hence there is no fixation or precise aiming of the eye at objects. In most snakes the eye sockets are so far back on the sides of the head that the range of possible stereoscopic vision is limited to a very narrow field directly in front of the snake. For both of these reasons eyeball movement is a relatively unimportant feature of eye usage.

The most obvious external characteristic of the rattlesnake's eye is the vertical pupil. This shape of pupil, although often thought characteristic of poisonous snakes, is, in fact, quite independent of a snake's venomous quality; it is more an indication of nocturnality than anything else, although it is not a universally accurate criterion even of this feature of snake adaptation, for despite the fact that the elliptical pupil and other structural characteristics of the rattlesnake eye show it to be particularly adapted to nocturnal use, rattlesnakes are far from being exclusively nocturnal. It is not even to be supposed that they see better at night than in the daytime, for they do not; they are merely equipped to respond better to the limited radiation available at night than a diurnal animal would be.

As we consider the actual serviceability of the sense of sight to rattlesnakes, we observe that it might be used to guide them toward three primary objectives: mates, food, and protection. But for the first of these the rattler is probably dependent on other senses, smell particularly. As to the second, the rattlesnake is a slow-moving creature; and although he occasionally trails his prey, he is much more accustomed to lie in wait for it, or to seek it down holes. The ambushing rattler, then, needs sight to aid him only in detecting and striking prey but a few feet away, and in this he has the assistance of the facial pits, if the prey be warm-blooded.

As to protection from enemies, whether birds or mammals, the rattler is too slow to outrun or dodge even the slowest of them, hence a superior or long-range sense of sight would not be important in the detection of danger, except that it might warn him to avoid discovery by freezing. For a rattlesnake's escape must first depend on concealing coloration; and, if this fails, on a noisy and threatening posture while in retreat, backed up by the venomous strike. Here again he has need only for short-range vision, particularly responsive to movement. So it would appear that the purposes for which a rattlesnake might use his eyes bear out the supposition that the range of vision is relatively short.

When a rattlesnake is asleep, it apparently closes the iris to the minimum opening, and only a thin black line of pupil remains visible. That the snake is truly asleep, and not merely reducing the aperture to a point adjusted to the impinging light, can be proved by disturbing the snake without changing the illumination, whereupon the pupil will dilate. In fact, I have, on one or two occasions, awakened a rattlesnake by shining a flashlight beam into one eye, whereupon the aperture widened, notwithstanding the increased light intensity. A snake accustomed to captivity, if disturbed only slightly, will go to sleep again almost immediately.

The pupil of the rattler's eye appears black. The iris is flecked or punctated with colors, rather metallic in quality, generally in gray, brown, or yellow. There may, in fact, be an approach toward a pattern, a horizontal dark line across the iris being evident in the sidewinder. A study of the irises of some fifteen species shows that there is a definite tendency of the iris color and pattern to match that of the adjacent head so that the eye is made inconspicuous.

The eyeball has some range of horizontal rotation. Such motions are not easy to induce, as the snake does not ordinarily follow an object with the eye. The most effective way of producing motion I have discovered is to place a snake in a glass jar and rotate it slowly. The snake's eye will first attempt to compensate for this motion, and will then twitch back to its normal orientation. Thus, as the jar turns, one eye turns forward, the other back. The readjusting twitch is made in both eyes simultaneously. I should judge the rotational range of each eyeball (in a horizontal plane) to be only about twenty degrees. The two eyeballs can rotate independently, as can be demonstrated by watching one eye directly, while that on the far side of the head is observed in a mirror.

If the head of a rattler be tilted up (or down) in front, the eye will rotate so as to keep the major axis of the pupillary ellipse vertical. If the tilting be in the nature of a side rotation, the eye on each side will attempt to compensate for the rotation, by maintaining the optical axis of each eye as nearly horizontal as possible. Both of these reactions to tilting the head indicate that the rattler has a good sense of balance.

Eyeball motions cannot be brought about by bringing some object up toward the eye, either from the front or side, or even by touching it. The snake seems to place full confidence (if the anthropomorphism may be excused) in the ability of the spectacle to protect the eye, and does not flinch in any way as motions are made toward the eyeball.

Observers have noticed the method by which some reptiles wave the head from side to side in order to get impressions of an object directly ahead in each eye alternately. I have seen this done by a sidewinder *(C. c. laterorepens)*.

We may safely conclude that the victim of a campfire tall tale reported that rattlers could discern a vulture or falcon passing overhead, and take refuge; or that rattlers were seen prowling about with their heads cocked sideways on the lookout for birds' nests in the trees.

It is doubtful whether a rattlesnake is in any way influenced by color in the acceptance of prey. Captive rattlers show no preference between gray and white mice; in fact, mice artificially colored blue, green, yellow, and red were readily accepted at the San Diego Zoo.

Summarizing our knowledge of the rattlesnake's eyesight, we may conclude that it is fairly good at close range and is particularly responsive to moving objects. The structure of the eye shows it to be relatively more perceptive in faint than strong illumination. It is probable that the eye affords a general impression of the surroundings rather than a sharp image of any particular object.

Hearing

Rattlesnakes, in common with all other snakes, lack external ears. There would seem to be a considerable difference of opinion among herpetologists as to whether or not snakes are deaf. Yet this apparent difference of opinion may be more a matter of definition than otherwise, for all agree that under certain circumstances snakes are made aware of external vibrations first carried by the air as sound waves, and then transmitted to the substratum upon which the snake rests, for the snake is particularly sensitive to the faintest tremors of the substratum. Certainly, as far as rattlesnakes are concerned, they do not appear to be conscious in any way of the direct impingement on any part of the head or body of the types of air vibrations that we know as sound waves, although they may recognize air movements via the pit. But if the type of the substratum upon which the snake rests is in any way responsive to sound waves, as is the case, for example, with an ordinary wooden or fiberboard box, then the vibration is, in turn, transmitted to the snake, which is thereby alerted. This is not to suggest that snakes are only responsive to these foundational vibrations if the air serves as an intermediate avenue of their transmission; on the contrary, ground tremors emanating directly from the disturbing source are much more often the cause of stimulation.

Rattlesnakes have neither an external ear opening nor an ear drum. The residual auditory apparatus that they still retain comprises a slender bone, which probably does not have the ability to transmit sound waves.

The most complete set of hearing experiments ever conducted on snakes—and these were principally on rattlesnakes—were made in 1923. They concluded that rattlesnakes are deaf, in the usual sense of the word, and that no rattler ever heard another's rattle.

I made a series of experiments on a large red diamond rattler *(C. r. ruber)* that would rattle upon the slightest disturbance—it had been slightly injured in the course of capture—and therefore was a good experimental subject. The snake taught me how necessary, yet difficult, it was to eliminate the possibility of sense impressions reaching it by unsuspected avenues, before one could really judge whether it was susceptible to the direct air waves that we call sound. When I clapped sticks together, with my hands beyond its range of vision, it reacted, but I found it was watching the swinging of my feet beneath the table, for I was sitting on a high stool without a footrest; and when a screen was interposed it caught my movements reflected in a nearby window. It was so extraordinarily sensitive to ground vibrations that footfalls fifteen feet away on a cement floor alarmed it, and this whether it rested on sand or on a blanket. Placed in a closed fiberboard box and suspended by a rubber band from the center of a stick that in turn rested on a pillow at each end—a precaution against vibration reaching it through the suspension—it reacted quite readily to sounds of clapped sticks or a radio; but clearly these were transmitted through the box acting as a sounding board, despite the deadening quality of fiberboard. Placed in a muslin sack with the same support it failed to react, but I found it was because this bundling of its body seemed to cow it. Finally, I placed it in a Chinese basket formed of woven bamboo withes, using the rubber-band and pillow method of suspension. Again it showed normal sensitivity by rattling at the slightest movement seen through the weave of the basket or to the dropping of a few grains of sand on it from above. But the basket evidently failed to respond to air-borne sounds, and the rattler in turn did not react to clapped sticks or a very loud radio but a few feet away, provided the radio was never turned completely off; if it was, the rattler sometimes sensed the heating of the tubes when it was turned on. I eventually tried alternating the radio between silence and full power at a distance of only six inches from the snake's basket, and with the noise at times so loud as to be quite distressing to a person ten feet away. Once or twice the snake reacted—to low notes, it seemed to me—but only hesitatingly and briefly; yet with this fearful clatter continuing, it would rattle violently if I came in sight, or touched the supporting crossbar, showing that it was still on the *qui vive.* I finally reached the conclusion expressed by others: that rattlesnakes are deaf to air-borne vibrations unless these serve to vibrate the solid material on which the snake rests or which a part of its body may be touching. To any vibrations so conducted the creature shows extraordinary sensitivity.

With snakes so sensitive to vibrations of the substratum that they react to many kinds of sounds that we hear with our ears, it may seem almost academic to attempt to determine the means by which they become aware of the disturbances. But this is hardly the case, since the threshold of response is clearly dependent on both the character of the disturbance and the agency of transmission available, and it may be of some practical value to understand these in order to know the best means of frightening snakes away. Hindus are said to wear sandals that make a creaking sound and also to beat the ground with sticks to warn snakes away. A custom of Mexican sheepherders is to shuffle their feet because they believe rattlers won't strike a moving object, which, of course, is not true. But foot shuffling might be of some utility in alerting a rattlesnake through ground vibrations, causing it to rattle.

The Tongue and Sense of Smell

As one approaches a snake in the wild, the first evidence that the presence of an intruder has been recognized is likely to be given by the tongue. The eyes are without the possibility of expression or winking; there are no ears to cock forward. Instinctively endeavoring to remain undiscovered through its blending coloration, the snake will seldom move until it seems to detect that it has been noticed by the trespasser. And so the flicking tongue is the only outward sign that the snake has noted something amiss. Clearly the tongue is being employed in some way to investigate the stranger; to give some indication of his character and intentions. Until quite recently, the nature of the impression conveyed by the tongue and the course of its transmission to the snake's consciousness were unknown. Now the tongue is believed to be an adjunct of the sense of smell, although not the sole vehicle of that sense.

From the earliest days, the snake's delicate, bifid tongue has been a fruitful source of superstition and folklore. Aristotle believed that the tips of serpents' tongues were double in order that the snake might get a double pleasure from what it tastes. It was a myth of the ancient Greeks that Cassandra and other prophets secured their powers of divination through having their ears cleaned by the tongues of serpents; and one Roman cure for ulcers was to have them licked by the tongues of sacred serpents. Noting that a snake used its tongue only when active, one writer thought it might be to facilitate cooling by evaporation, in the manner of a dog; this would be virtually ineffective because of the minute surface of a snake's tongue.

Two mythical purposes of the tongue have been especially persistent, notwithstanding the fact that it is obviously unfitted for either, and that both can be disproved by a most superficial investigation. The first is that the tongue is used to lick the prey before swallowing. No shape of tongue could be more ill suited to such a purpose than the delicate, almost thread-like tongue of the smaller snakes; yet the myth that snakes cover their prey with saliva to facilitate swallowing has persisted from ancient times down to the present day.

The other myth—that the tongue of a snake is a stinger, the source of the danger in venomous snakes, is even more fantastic, yet just as indelible. It is a very ancient idea—as evident from Biblical and Shakespearean quotations. But the most surprising thing is that this is one of the most frequently heard beliefs today, as can be verified if one listens to comments by the visitors to the reptile house of any zoo. Let any snake protrude its tongue and someone is sure to exclaim, "Look at it stick out its fangs!", or "Did you see its stinger?"

Slightly more plausible purposes of the tongue have been a source of argument for many years. Some are: the tongue serves the purpose of keeping the animal's nose clean; it is used to catch small creatures or insects. (Two of my correspondents claimed to have seen rattlesnakes do this, a quite impossible trick.) Allied to this is the idea that the tongue may be employed as a decoy, to attract insects, or even birds, to their destruction. The statement has been made that the tongue might be used to lap up fluids. I have never seen rattlesnakes use the tongue when drinking, even when sucking drops of water from some solid object—a stone, for example—as they sometimes do. One writer thought the tongue to be concerned with the function of voice—with hissing. Such is not the case, although rattlers, like other snakes, can and do hiss, with or without the simultaneous protrusion of the tongue. Another writer was so disturbed by the conflicting and fantastic theories that he concluded the busy tongue had no purpose whatever, which may well be considered the most fantastic conclusion of all. A dual theory as to the purpose of the tongue has also been expressed: upon the part of a resting snake it is a warning device, whereas, in the case of a prowling snake, the prey would have its attention so concentrated on the movements of the tongue—a sort of mesmerism—that it would fail to notice the gradual approach of the snake itself, until too late to escape.

In contrast with these supposed physical uses, it was natural that some type of sense reaction should be attributed to the tongue, since anyone watching an exploring snake would be convinced that the flicking tongue must surely be conveying to the creature some impression of its surroundings. Some early writers thought it an organ of taste, and many others have made the same natural assumption. But it was later shown that there are no taste buds in a snake's tongue. Many herpetologists, some of the greatest prominence, believed it was used as a feeler—that is, an organ of touch. Occasionally it has been suggested that the tongue is used for sensing air vibrations, an equivalence to an organ of hearing.

All of these theories of the purpose of the tongue have lately been displaced, at least in large part, by the proof, through extensive experiments, that the tongue is an accessory to the sense of smell, in that it serves as a conveyor of external material particles to two pits known as Jacobson's organs in the roof of the mouth. These, in turn, convey to the brain the sensations resulting from the character of the particles. As the sensation-affecting particles are more often secured directly from the air than from contact of the tongue with material objects, it is probable that the tongue should be considered an accessory to the sense of smell rather than taste. However, Jacobson's organs are not the sole source of the snake's olfactory knowledge, for it can likewise detect odors through the nose and the more normal organs of scent. The new theory of the tongue's use served to solve another problem as well, for Jacobson's organs had also been a puzzle.

Observations of rattlers in the field, as well as theoretical considerations, indicate that rattlesnakes have a keen sense of smell. As I discuss elsewhere, they have been observed trailing their prey, and they seem able to determine by scent whether a mammal burrow is occupied. As is the case with other venomous snakes, it is their custom to strike their prey, withdraw to avoid any chance of retaliation, and then methodically to seek out the animal, which has usually run a short distance before dying, by following its scent.

This interesting comment on the use of the tongue and sense of smell by captive rattlesnakes was recently made by a zookeeper:

When a cage door is opened, a rattler will assume that it is time to eat, and, if it happens to be the time to clean cages instead, it might strike the shovel used for cleaning and hurt its mouth. If the shovel is moved very slowly close to the head of the snake, to give it a chance to find out what is going to happen, the rattler will test the shovel with its tongue, and then coil up in a corner and pay no further attention to the cleaning.

Superficially, the rattler's tongue does not seem to differ from that of other snakes. The base comprises two parallel, conjoined cylinders, distally branching into separate pointed tips. The tips are so thin and delicate that they can hardly be felt when flicked against the hand. Rattlesnake tongues are black at the outer end and flesh-colored inwardly. It is quite easy to stimulate a rattler to extrude the tongue by the slightest disturbance, or by introducing some odorous substance into its cage. The mouth is opened very slightly to permit the egress of the tongue.

A man's reaction, or that of other mammals, indicates that he usually becomes conscious of a new odor in the course of normal breathing, following which he sniffs rapidly to gain a better sense of the stimulus. But snakes breathe much less frequently than mammals, the rapidity depending, not only on bodily activity, but on temperature as well. Their reaction to a new odor never results in sniffing, but is followed by a rapid flicking of the tongue.

We may conclude that the rattlesnake's olfactory sense, being innervated by way of two separate and somewhat independent paths, is quite acute, superior, in fact, in importance to every other sense, except possibly sight, with respect to the primary objectives of food, mates, and protection from enemies.

An argument can be advanced, especially with regard to rattlesnakes, for one other purpose of the tongue besides its connection with Jacobson's organs and olfaction—that of an action calculated to alarm an enemy.

A rattlesnake alerted by some movement, odor, or tremor of the ground, responds by flicking its tongue in and out repeatedly and rapidly. While the tongue is out it will be seen to pass quickly through a considerable vertical arc. If the snake be further alarmed, it then throws itself into its striking coil; its mien changes completely, for now it threatens by rattling and hissing, while the head is drawn back like a poised javelin. And simultaneously the use of the tongue changes. There are now much longer intervals during which the tongue is extruded to its limit and much shorter intervals of withdrawal. Not only is the outward extension greater than when bent solely on investigation, but the quivering tips are more widely separated and the tongue is first pointed vertically downward and then erected vertically with considerable pauses in each position. Sometimes it will change from one to the other of these opposite positions several times before withdrawal. Often the extrusion is so great that the tongue cannot be held exactly erect but falls off slightly to one side. It does not seem possible that this handling of the tongue can apprise a snake of the nature of an intruder any better, if indeed as well, as the more rapid flicking. But that it adds to the snake's pugnacious appearance there can be no doubt, and I am therefore of the opinion that it is a part of the threatening posture designed to frighten an enemy. It certainly does add to the spine-tingling picture.

The Facial Pits

The pit vipers comprise a family of venomous snakes, the Crotalidae, characterized by, and called pit vipers because of a deep pit, or facial opening, on each side of the head.* In the rattlesnakes, the pit lies somewhat below a line from the nostril to the eye and slightly nearer the former. It is larger and more conspicuous than the nostril.

Since the earliest days there have been many theories as to the purpose of the pits. Their location on the sides of the head naturally led to the supposition that they comprise some kind of a sense organ. They have been called ears or extra nostrils. (In Latin America the name *cuatro narices,* four nostrils, is still widely used for pit vipers.) They have been thought to be tactile or olfactory in purpose; to bring air into contact with the venom, thus affecting its chemical properties; to be tear sacs; or to detect pressure variations.

Finally, in 1937, the theory was advanced that the pit is a temperature-differential receptor, by which the snake is enabled to determine the direction of objects having higher temperatures than their surroundings, a warm-blooded creature such as a bird or mammal being a good example. Thus the pit aids both in locating prey and striking it. The theory was proven by experiments with snakes having all other senses destroyed or blocked off. For targets incandescent lamps covered with black paper were used, whereby the heat could be readily controlled. It was found that snakes, with only the pits available as sense organs, were able to, and would, strike quite accurately at the covered lamps while heated, but lost their ability to detect them when they cooled. The snakes were able to locate a moving, heated lamp at a distance such that it produced a temperature less than half a degree Fahrenheit above the surrounding air at the pit. The maximum recognition distance was found to be about fourteen inches.

From these experiments, it appears that the thermoreceptive function of the pit is essentially of short-range value. It should have a somewhat greater range at lower air temperatures than higher, since the detectable temperature differential produced by the warm-blooded prey would be experienced at a greater distance.

Touch

Since snakes have no tactile appendages such as legs, the sense of touch is not so important to them as it is to many other kinds of animals. Yet, despite their being sheathed with scales that might be expected to dull their sensitivity, they are apparently responsive to the slightest external contact. I have already mentioned their susceptibility to ground tremors, allied as this is to hearing.

Dropping particles of sand on rattlers or brushing them with feathers

*Only one author known to me misunderstood the meaning of the term; he stated (in 1873) that pit vipers received the name through their frequenting pits or caves.

show them to be sensitive even to these light contacts; if not sufficiently aroused to coil they will at least indicate attention by flicking out the tongue to discover the source of the annoyance.

It has been reported that rattlers exhibit a pleasurable response to petting. The story is told of a tame rattlesnake that so enjoyed being stroked with a brush that it would roll on its back like a cat—a story needless to say, that has been relegated to the realm of rattlesnake folklore.

Distant Detection

It is a common experience in the field to have a rattler suddenly sound off when the observer is still at some distance. The sense by which the snake has become aware of the intruder is generally difficult to determine. One would first suspect sight, yet it often happens in these cases that there are intervening objects. Since rattlesnakes are deaf to air-borne vibrations, as every evidence indicates they are, then either ground tremors, or the sense of smell is the avenue of perception that serves to warn the snake. I deem the pit too short a range to be useful in such a situation.

Through the years, I have, myself, had similar experiences with a variety of species of rattlesnakes—with southern Pacifics, red diamonds, speckled, and tiger rattlesnakes. In one instance a snake rattled violently in a thick sumac bush a hundred or more feet from where we were hunting. In another, a friend and I were walking down a creek where erosion had made deep cuts so that the banks were high above us on either side. A southern Pacific rattler on a grassy shelf on the bank above us rattled as we passed below. It was found to have bluish eye coverings, preparatory to skin changing. Since we were hidden from it, it must have sensed our presence by vibration or odor. Similar accounts have been published.

One writer reported that a rattlesnake in a barrel could detect a man's approach, although he walked in stocking feet.

From many experiences, we are certain that, whatever the limitations and imperfections of a rattlesnake's senses, their cumulative value is such as to serve well in advising them of the approach of danger, at least in the form of a creature as large as man. It will be noted, in the cases of the field notes sent me by correspondents, that men on horseback were usually detected at greater distances than those afoot. This, it seems to me, seems to support a ground-tremor theory of detection.

Aggregation and Sex Recognition

Without doubt it is the rattlesnake's acute sense of smell that enables it to follow its fellows—the leaders must have some homing instinct—to a general gathering place for winter hibernation, and to find a mate in the spring. But, with respect to the source of the odor that is trailed, there remains some uncertainty. There appear to be family differences, and the rattlesnakes have not been adequately investigated to determine the facts respecting this group.

In some kinds of snakes sex recognition is through some odorous substance contained in the dorsal skin or through discharges from the anal scent glands. Whether either is true of the rattlesnake is not known.

Intelligence

Although no snakes have the degree of intelligence with which the myths of bygone days would credit them — the wily serpent aura — they do, of course, have enough intelligence — if that is the proper word for a creature of such reduced mental capacity — to satisfy the needs of food, self-protection, and reproduction, and thus to survive. The degree of intelligence that a snake exercises can be judged only when it is alert in fulfilling one of these primary motivations, for at other times it seems so dull as to be almost comatose. Because they are more active, the slimmer kinds of snakes have an appearance of being more alert than slower, heavier kinds like rattlers; but whether they are more intelligent we cannot be sure.

A rattler accustomed to feed in captivity shows, upon the approach of its keeper, an alert attentiveness that gives one a sense of intelligence. The lively way in which it rises up and faces the door of its cage when it is opened, the recognition of its regular attendant or his routine, the ability to distinguish a food forceps from a cleaning shovel, all testify to an appreciation of procedures.

Similarly, a rattlesnake discovered in the wild seems to exercise some acumen in determining when to abandon passive concealment in favor of flight, and when to turn and endeavor to intimidate an enemy. Rattlers are often found basking in front of holes or crevices that will furnish a refuge in emergency, another indication of a primitive astuteness. They know well how to threaten, and at the same time to retreat toward some place of safety.

That rattlesnakes come eventually to know their keepers, there can be no question. They often show no fear of handling by their owners but adopt a defensive attitude when strangers come near.

This naturally brings up the question of whether rattlesnakes can be tamed. The answer is that they can, as has been demonstrated by many handlers. This is not at all surprising, for rattlers are not inherently vindictive or vicious. Like all wild animals they are fearful of unaccustomed situations and wary of strange creatures, toward which they adopt defensive tactics that appear to us to be violently aggressive. To whatever extent they can be cured of fear, they may be considered tame — that is, they will generally not bite.

As early as 1615, the statement was made that the Indians tamed rattlesnakes by wrapping them with fine linen—just how this was done is not made clear. Writers of colonial days mention tamed rattlers, always gilding the stories with tales of their coming when called. The fact remains that many will offer no objection to gentle handling, and thus may be called tame. Of course, there are species differences: some species are more nervous than others. And there are individual differences within each species. Some become quickly accustomed to people, while others will rattle at the approach, even of a keeper, after years in captivity. Tamable or untamable, I hope nothing I have said will suggest to anyone that tame rattlesnakes are really safe to handle. One can visualize a dozen kinds of slips or accidents that might occur, while holding a tame snake, that would frighten it into biting. Only scientific and controlled investigations for a definite purpose can ever justify what would otherwise be a foolhardy and unnecessary risk. The experiences of countless snake handlers have shown that no snake is ever tame in the sense that it will not bite if accidentally frightened or injured.

Returning to the topic of rattlesnake intelligence, I would like to mention a few of the embroidered and fantastic stories surrounding this aspect of rattlesnake life. We are told, for instance, of the man, who, while hiding from the Indians, found himself lying beside a big rattler. He played dead, and the snake, after satisfying itself of his demise, crawled away. Behind this there is the sensible advice to remain immovable if one finds himself within striking distance of a rattlesnake in its striking coil, at least until the situation has been surveyed to see which way to jump. Anyone close to an outstretched rattler or one in its resting—pancake—coil can easily step out of the danger zone to safety without being particularly concerned.

Then there was the rattler that had the run of its master's cabin. Upon the advent of a stranger it was told to "git in yore bed" which it promptly proceeded to do.

A number of interesting but impossible yarns concern the intelligence of the diamondback. One, observing from the edge of a large field that a rough spot in the center, where it was accustomed to hunt, had been cleared, turned back into the woods. Another used a drain pipe regularly to cross a sandy road, in order to avoid making a track where it could be seen. One wise diamondback had no difficulty in selecting, for attack, from a ring of men surrounding it, the particular individual who had traitorously divulged its hiding place to the others.

CHAPTER FIVE

Behavior

Since rattlesnakes, like other reptiles, are fundamentally dependent, for muscular activity and the processes of digestion and gestation, on external heating rather than on chemical heating derived from their own metabolism, it follows that they can live only in places where the daily and seasonal persistence of adequate temperatures is of sufficient duration to permit the successful completion of these processes. Thus reptiles are considerably limited to tropical and temperate zones, and even there, at the colder limits, they have been forced to a two-year reproductive cycle, for the young cannot be brought to term in one.

With creatures so dependent on external conditions, it is natural that the activities of rattlesnakes should be greatly influenced by the temperatures of their surroundings, including the radiant heat from the sun, and so we find that their seasonal and daily habits are strongly affected by thermal conditions. Since suitable temperature ranges are met at different times of day and in different seasons, in the various latitudes and longitudes inhabited by rattlers, we naturally find notable habit differences within a single species. Of course, any rattlesnake activity is the result of some phase of the primary rattlesnake objectives, which are food, reproduction, and self-preservation, and their habits must be so arranged as to fit the pursuit of these objectives into zones of permissible temperatures. No generalities with regard to the seasonal or daily activities of rattlesnakes are valid without our taking temperature into consideration.

It must be noted that air temperatures at which certain responses have been observed are not so important as might be presumed, since they seldom represent the actual body temperatures being experienced by the rattlesnakes; for body temperatures are the result of all thermal factors effective at the moment, including metabolism, conduction from the ground, convection from the air, and radiation from surrounding objects and the sun.

If confusion is to be avoided in a consideration of the effect of body temperatures on rattlesnakes, it is essential to distinguish between two viewpoints. The first involves purely physiological considerations: what are the temperatures that a snake can withstand without death or permanent injury; to what temperatures will it voluntarily submit without seeking amelioration; and, finally, what is the temperature zone best suited to its life processes and well-being? The second phase has to do with the habit adjustments of the rattler, whereby it fits the attainment of its necessities of food, reproduction, and protection as best it can into the available temperature conditions. For, in almost none of its habitats are the temperatures at an optimum level continuously.

PHYSIOLOGICAL TEMPERATURE LIMITATIONS
Maximum Temperature Limits

Although snakes, as well as other reptiles, are popularly supposed to prefer and seek the hottest weather and exposures, lying out in the sun whenever possible, the fact is that such basking is limited to times of moderate temperatures, for they would be quickly killed by environmental temperatures not particularly uncomfortable to man.

Although experiences show some differences in the time required to kill a rattlesnake in the hot sun, no doubt correlated with differences in ground and air temperatures, the intensity of the sun's radiation, and the size of the snake (and therefore the rapidity of heat absorption), there is a general agreement that the time of survival under extreme conditions is about ten to twelve minutes. Even the sidewinder, which inhabits the Mojave and the Colorado or Sonoran deserts—including the floor of Death

Valley, where the highest air temperatures ever recorded in the Western Hemisphere have been attained—seems able to withstand no higher temperatures than other rattlesnakes, and for no longer periods. Probably the body temperature at which irreversible muscular heat-rigor ensues is virtually the same in all rattlesnakes, at about 113° F., although damage from which the snakes may succumb later may be entailed at lower body temperatures, even down to 110° F., or below, with some variation depending on the duration of the unfavorable temperature.

Minimum Temperature Limits

At the other extreme, the lethal minimum, the transition from at least a permissible temperature to a fatal one is more gradual than at the upper limit; for long before the danger point is reached a zone of lethargy is entered that drives the snake, by instinctive response, to some refuge. This, although it may not improve the temperature experienced, will at least safeguard the animal while helpless and immobile through the stiffening of its muscles. However, it is true that the advent of falling temperatures is likely to be more sudden and unexpected than rising. Rattlers subjected to a falling temperature experience a benumbing lethargy, whereas those in a rising temperature approaching the danger point are active and alert, and should be able to reach a safe refuge. So it is probable that many more rattlers are killed by freezing than by heat, having been rendered incapable of motion when only a few feet or rods from safety. Rattlers enroute to their dens are particularly subject to this danger.

Just what minimum temperatures rattlesnakes can withstand and still survive is not known; probably much depends on the rapidity of freezing and thawing, for it is generally believed that, if the thawing be gradual, they will survive.

A correspondent in La Jolla, California, told me of throwing some frozen young prairie rattlesnakes into a garbage can and later finding that the sun had revived them.

An important temperature stratum in the life history of the rattlesnake is that which is called the critical minimum—the temperature causing a cold narcosis that prevents locomotion, and thus renders the snakes unable to defend themselves against enemies, or to relieve a thermal impasse, if they have not already reached a sanctuary.

I experimented with several rattlers of various species, placing them in the vegetable room at the San Diego Zoo where the temperature was maintained at about 40° F. and found them quite capable of movement. This, however, is by no means proof that they would escape if caught on their way to the dens at this temperature, for it generally required persistent annoying to stimulate them into movement. It would seem that, although fully capable of motion, they would probably lack the necessary urge or will to reach a refuge more than a few yards away. I should judge that at temperatures of about 46° F., or below, there would be little voluntary motion even if their safety depended on it, although they would certainly escape if the temperature subsequently should rise again.

From all available data, I conclude that adult rattlesnakes, if thawed slowly, may recover from a few hours of subjection to a temperature of 4° F., and can endure 37° F. for some time, measured in days or more.

The minimum voluntary tolerance—resulting in retreat underground—I judge to be about 61° F. After all, the word "voluntary" depends on circumstances and the snake's necessities. I expect that a rattler that had not yet obtained food following emergence in the spring would be voluntarily active at a lower temperature than under other circumstances.

Optimum Temperature Range

Finally we come to the normal activity range—the range of greatest comfort and bodily well-being. I should place this—at least for our Nearctic rattlesnakes—at between 80° and 90° F. At this temperature the snakes are alert, their muscular activity well toned, and bodily processes—digestion and gestation—at an optimum. Again, these temperature levels refer to the internal temperatures of the snakes—not to some single exterior criterion, such as the air temperature in the shade.

TEMPORAL VARIATIONS IN HABITS

Having discussed the physiological temperature limitations of rattlesnakes and their temperature preferences, I shall proceed with the second phase of the discussion, namely, the effect of temperature on their habits; how their seasonal and diurnal activities are influenced by temperature; their various methods of taking advantage of differences in microclimate; and the extent to which they endure uncomfortable and even dangerous temperatures to attain their life necessities of food, mates, and security from enemies.

To this end, they have evolved a variety of expedient instincts and may react in various ways to take advantage of favorable weather and to protect themseves against its deficiencies. This requires both seasonal and daily adaptations or adjustments to thermal conditions; it involves going underground when it is too hot or too cold* on the surface; and involves taking advantage of whatever time of day or night produces the most favorable temperatures for surface activities, always provided, however, that the main objectives are not thereby neglected. Where food or mating

*It may also be too cold for activity below ground, but at least the snake is protected from enemies while comatose from cold, and, if it goes deep enough, from death by freezing.

urges require activity under unsatisfactory temperature conditions, a compromise is made and the snake endures discomfort and even, to some extent, danger, to secure the primary objective.

It should be pointed out that, regardless of any willingness to pursue its activities under unfavorable temperature conditions, a minimum duration of adequate temperatures must be available if the species is to survive; for such temperatures—and they must emanate from external conditions—are necessary for the life processes of digestion and gestation. To secure these periods of adequate temperatures the snake indulges in basking, although with some sacrifice of safety from enemies.

Seasonal Variations in Activity

Although rattlesnake activities are strongly affected by temperatures—using the term in the broad sense of the thermal impact or flux from the surroundings that produces a specific body temperature—this does not mean that rattlers are most active when the weather is most propitious. On the contrary, spring is the time of greatest activity, for they are hungry then after their long winter fast, and, in most areas, this is the mating season as well. So, although spring is a period of below-optimum temperatures in many areas, the rattlers' activities are then at a maximum, and in some areas they not infrequently imperil their own safety by becoming active too early.

The popular supposition that rattlesnakes are lovers of hot weather, whereas spring is really their season of maximum activity, leads to reports, almost annually in the press, that "rattlesnakes are out unusually early this year." This is particularly true in southern California, where the rattlers do not congregate in dens, but are first seen singly, or in pairs or trios, lying about the cactus patches and granite outcrops, sunning themselves quite carelessly and conspicuously. As this is coincidentally the wild-flower season, it is natural that they should be noticed by motorists and hikers.

Effects of Reproductive Cycle on Seasonal Activity

The reproductive cycles of rattlesnakes affect their seasonal activities in three principal ways: the increased activity of the mating season; the decreased roaming of the females when heavy with young but under the necessity of maintaining a temperature adequate to gestation; and the roving of the young themselves after their appearance.

Along the southern border of the United States, and in northern Mexico, rattlesnakes normally mate in the spring, soon after emerging from hibernation. In some instances they have been observed mating at the dens; but it is probable that mating more often occurs after they reach their summer ranges. At any rate, mating coincides with their heaviest feeding period, so that these two conditions combine to produce the annual season of maximum rattlesnake activity.

In the northern latitudes where biennial broods are the rule, the mating season may be largely confined to autumn, and seasonal activities are correspondingly affected.

When the females are heavy with young in the summer, they become increasingly secretive so that many more males are encountered than females. At such times about twice as many males as females are seen abroad, whereas at the dens the males outnumber the females by only about ten percent.

The young rattlesnakes in most areas are born between August 1 and October 15, centering in mid-September. The young are quite active immediately following their appearance, for the securing of food is almost essential to their surviving the following winter. They are both hungry and inexperienced, and consequently careless of concealment, so that they frequently are found roaming in the daytime.

Denning Dates and the Risk of Immobility

It hardly needs to be stated that, with the close dependence of rattlesnakes on exterior temperatures, the dates on which they enter and leave hibernation—the length of their active season—depend on latitude and altitude and other features that affect local climatic conditions.

Under the mild conditions that exist in coastal southern California, rattlers are out of sight most of the time from December 1 to March 1; but whenever there is a warm spell, raising the temperatures into the 80s for several days, the snakes may be found sunning themselves on the rocks. Similar conditions occur elsewhere along the southern border of the United States. In cooler climates they may enter their dens as early as September or October and emerge as late as April or May.

Short temperature drops resulting in immobility are of little importance to a snake unless dangerously low levels are reached or the snake is attacked by some enemy while helpless. It is by no means unusual to find snakes chilled into complete lethargy in the early morning during their active seasons. This is even the case in mountain areas in the tropics and is of frequent occurrence in desert mountains. (Rattlers are even occasionally seen on, or close to, snow banks.) But a sudden prolonged cold snap can prevent them from attaining entrance to even a nearby den, with fatal results. (Enforced immobility through chilling in a refrigerator is sometimes used in handling snakes, as, for example, when a minor operation is required, or to quiet a rattler for photographing.)

In summary, the active season of rattlesnakes in the United States varies from five months, in the north and at higher elevations, to nine or ten months, or even longer, along the southern border. Where the seasons are less severe, the time limits of activity are less sharply drawn. There is a sharp peak of activity of rattlesnakes in the spring; this peak is not the result of favorable temperatures, and, in fact, in many areas, occurs despite unfavorable temperatures, but it is the time of mating and of the heaviest food requirements.

Rattlesnakes as Weather Indicators

Though rarely, rattlesnakes have been considered weather indicators — reptilian ground hogs of a sort. Although their dates of entering and leaving hibernation are undoubtedly affected by seasonal variations, there is no real evidence that rattlers can anticipate adverse conditions in the manner so often attributed to animals. In fact, the frequency with which they are caught and immobilized on the way to their dens by a sudden cold snap should be convincing evidence that they lack this reputed power.

Summer Migrations and Estivation

Migrations of rattlesnakes toward their dens in the fall and away from them in the spring are discussed elsewhere. It is probable that they must sometimes travel as much as two miles or more to reach their dens, but ordinarily the distance is probably less than a mile. Since suitable rocky outcrops are usually on hillsides, the spring migration is likely to be downhill to the valleys below.

That rattlesnakes estivate — using the term to indicate a complete suspension of activity — is to be doubted, although a reduction of activity during the summer is quite normal and has been noted by many observers. Hibernation is forced, in most areas of the United States, by exterior conditions that would be fatal at any time of the day or night. Such a condition rarely exists in summer for more than a few days; for while the diurnal temperatures, for long periods, reach levels beyond the ability of any snake to withstand, there is nightly a sufficient fall of temperature, both of ground and air, so that snakes can and do emerge from their subterranean refuges if their bodily necessities require. They are not prevented from doing so owing to any lack of adaptation for nocturnal forays, for they are, in fact, well equipped for night activity, and adopt such a regime in any season when daytime temperatures are less favorable than night. Another reason for the long periods of subterranean seclusion is the increased humidity down holes, which reduces moisture loss.

It seems, therefore, that to whatever extent rattlesnake activities reach a low point in summer, the reason is to be sought in a lack of need for activity rather than uninterrupted lethal temperatures. By summer the mating season is over and the spring hunger to replace tissue and fat

lost during hibernation has been satiated. Feeding intervals have now become matters of weeks if not months, and food may, indeed, often be
secured in the mammal holes or rock crevices in which both the snake
and its prey have sought refuge.

Effects of Humidity

It is probable that, next to temperature, humidity may be most important in controlling rattlesnake activity.

In coastal California, winter rains are the rule, and are coincident
with cool weather, hence tending to inhibit snake activities. But in sections of Arizona, Sonora, and adjacent areas of the Southwest where summer is the season of rainfall, our experience seems to bear out the popular
belief that snakes—rattlers included—are more active after the rains begin than before. It is probable that the rain has a sufficiently cooling effect
in this hot season to make it feasible for the snakes to forego their nocturnal habits for a time, and in any case rain and clouds bring the temperature down nearer to the snake optimum. Sometimes snakes may be
flooded out of their holes. One writer noted that a rattler he observed did
not seem to mind a hard, beating rain, and made no effort to seek shelter,
although bushes and holes were available nearby.

As is to be expected, rattlesnakes sometimes cool themselves by resting in the infrequent pools and seeps of the arid West. A Nevada correspondent reports:

On July 20, 1934, I found a medium-sized Great Basin rattler (*C. v. lutosus*)
floating placidly in a tiny pool of a slow-seepage spring under an overhanging
ledge of one of the drier and more barren draws of the eastern Peavine Mountain,
northwest of Reno. On this particular occasion I slipped in under the ledge and
drank my fill of water, before noticing the snake whose head was not more than
eight inches from my face during the entire proceeding. As I drew back (it is
necessary to crawl prone the last couple of feet to reach the spring) I noticed the
snake. Only his head and a short portion of the neck were resting at the edge of
the two-foot pool, the rest lying three-fourths submerged. I lay there and watched
it for about fifteen minutes, then went on my way, leaving the animal in my
drinking water. All the while the snake made no movement, and I did not attempt
to startle it. My impression was that the snake was enjoying the cool water, for it
was a record hot day outside, and everything else was under cover.

Diurnal Variations in Activity

I have discussed the seasonal effects on rattlesnake habits and shall
now touch upon the diurnal effect. Necessarily the two are closely interrelated, yet an attempt to separate them will clarify the discussion. The
snakes are somewhat more independent in choosing daily than seasonal
periods of optimum temperatures, for they can take refuge underground

against an unfavorable day, but not always against an unfavorable season, if they are to survive.

Their most obvious expedient, of course, is to become active at the time of day or night when temperatures are nearest the optimum, and this is done to a certain extent. But it is evident that, other things being equal, most Nearctic rattlers prefer nocturnal to daylight activity, even though the temperature may be less favorable then. No doubt this is because the small mammals that constitute their principal food supply are also primarily nocturnal, although their choice may be partly induced as a protective measure against enemies. It is to be remembered that the rattlesnake's eyes are well adapted to night vision, and that the pit, as a high-temperature receptor, is probably more effective at night because of the greater temperature differential between the warm-blooded prey and the surrounding objects.

The following generalities are valid: rattlers are more nocturnal in summer than in spring or fall; desert rattlesnakes are more nocturnal than those in more humid areas or where the brush cover is heavier; lowland rattlers are more nocturnal than montane individuals or species; and adult rattlesnakes are more nocturnal than juveniles. The greatest surprise that awaits the person who believes the rattlesnake to be essentially a warm-weather or diurnal creature is experienced when he comes upon side-winders and Mojave rattlers active at night on the desert with a bitterly cold wind blowing — a wind so strong one must lean against it in walking, while it cuts through a heavy overcoat to the accompaniment of whirling sand and debris.

From my discussions of rattlesnake danger with such people as hunters, fishermen, campers, mountain climbers, and others making excursions into the wild, I am surprised at the extent of the belief that rattlesnakes are universally diurnal, or, under extreme circumstances active only at twilight, so that with darkness any danger vanishes. This has led to accidents in walking about a camp at night without the precautionary measure of a flashlight or even shoes. For this reason, I shall repeat: rattlers do not turn in for the night when darkness falls.

Whether the moon has any particular effect on the night activities of rattlesnakes is not known. As to desert reptiles in general, such evidence as has been accumulated indicates that they probably prefer dark nights and are less active under conditions of strong moonlight.

It is not my intention in citing these instances of the night activities of rattlesnakes to give the idea that rattlers are exclusively or even essentially nocturnal. They can be found abroad at all times of the day, from early morning to late evening, for much depends on the season and temperature. My reason for stressing the night phase is because it seems to be little known to the public generally.

The Basking Range

In the process of determining the temperature-activity interrelationship of rattlesnakes, some distinction should be made between mere basking in the sun and true activity—that is, the pursuit of some definite objective other than warmth. When ground and air temperatures are such as to result in a suboptimum body temperature, snakes often increase their comfort and well-being by basking in direct sunlight, thus securing the increased warmth of direct radiation. Basking is likely to be indulged in whenever the air temperature is below about 75° F., the ground is cool, and the sun is shining. Rattlesnakes probably seldom bask when air temperatures are below 55° F., unless the surface of the ground has already been warmed well above this level by the sun's rays. Wind conditions— or the availability of a sunlit nook protected from the wind—are important factors, since a cool breeze will cause a greater loss of heat through convection than that accumulated from impinging radiation or from conduction.

Taking advantage of basking is important in increasing a species' geographic range, for it lengthens the season during which body temperatures adequate to such processes as digestion and gestation are available.

Basking is not restricted to times of severe weather conditions, such as early spring or late fall, but may be utilized for brief periods at some time of day at any season when temperature conditions are suboptimum and the sun is shining. Thus in the late spring and at higher elevations in the summer, rattlesnakes will be found basking in the sun in the early morning, since the later midday air and ground temperatures will be above the optimum. This is highly characteristic of sidewinders in the desert.

Other Temperature-Control Expedients

In addition to attempting the adjustment of their activities to periods of favorable seasonal and diurnal temperatures, rattlers adopt various expedients to secure the advantages of favorable differences in microclimate.

One of the simplest of these methods is to take advantage of temperature differences within a narrow space. Rattlesnakes make a practice of lying in the edge of a patch of shade, where, by varying the relative amounts of shade and sun upon the body, the snake can attain any temperature between the cool of the shade and the maximum produced by the full solar radiation. A movement from full sun in the morning to full shade at noon, with a return to basking in the afternoon is a common maneuver.

Rattlers have another way of averaging temperatures when one position is too high and the other too low for comfort. In the morning in the desert, the ground is often quite cool only an inch below the surface; for sand, with its air-filled interstices, is a relatively poor conductor of heat. Sidewinders have a way of bedding themselves down by manipulating their coiled bodies until their dorsal surfaces are level with the sand. When the snake is in this position the dorsum is heated by the sun's rays while the ventrum is cooled by the substratum. One expedient that the lizards can employ to control the absorption of radiation is to vary the body hue and therefore reflectivity. These color changes have long been known in the lizards, but it is not so well known (and, in fact, has been denied) that rattlesnakes have some power of color change. I have observed it in the southern Pacific rattlesnake *(C. v. helleri)*, the Arizona black rattler *(C. v. cerberus)*, and the sidewinder *(C. v. laterorepens)*.

Of course, the principal means of temperature regulation available to rattlesnakes, when the external temperature is above the optimum, is to seek refuge in a mammal hole or a rock crevice.

Since the advent of black-top pavements in our southwestern deserts, rattlers and other snakes have been observed warming themselves at night on these roads, after the desert air and ground temperatures have fallen below the preferred level. The storage of heat in the pavement is considerable. This is one of the reasons why night-driving on desert roads is so successful as a means of collecting snakes.

It is probable that under natural conditions rattlesnakes are seldom killed by heat. Although the warning zone of danger is relatively narrow, and the critical zone of muscular damage quite quickly attained by an exposed snake, only in the rare contingency of a snake being caught in a bare expanse of desert would the situation be likely to prove fatal. At times when diurnal temperatures reach these dangerous extremities, the rattlesnakes have largely become nocturnal; and while the early morning often finds them in the sun, it will usually be close to the shade of some bush. When surface temperatures attain dangerous extremes even in the shade, mammal burrows and rock crevices are sought. In the morning, rattlers are usually found near such refuges.

DEFENSIVE AND WARNING BEHAVIOR

Rattlesnakes have many kinds of enemies. Escape from these is one of the three most important requirements necessitating definite action upon the part of the snakes if they are to survive, the others being the pursuit of food and mates. Naturally, the reaction of a rattlesnake to an enemy depends largely on the character of the enemy and the accessibility of a refuge, as well as on the disposition of the rattlesnake. But, in general, there are three successive phases of defense: an endeavor to avoid detection through quiescence and protective coloration—the method of procrypsis; an endeavor to escape by flight; and finally, if these fail, an active defense by coiling, threatening, and even striking. This last resort may be quite spectacular, and is the particular rattlesnake attitude upon which its sinister reputation is largely based.

Disposition and Temperament

Since a rattler's response to an exterior threat is by no means stereotyped, but depends on both the species of snake and its individual temperament, it will be desirable, first, to touch on some of the factors that affect the nature of the snake's reaction to intrusion.

It hardly needs to be said that the actions of rattlesnakes in defense can be judged objectively only by those who have had experience with them and are not unduly frightened by their threatening attitudes. The uninitiated are likely to be so startled that they see a violent attack even in a snake retreating toward some hiding place.

There are undoubtedly species differences in rattlesnake temperaments, for some are more nervous and excitable than others. They show this by throwing themselves into a striking coil more readily and quickly—they are "on the prod," to use an expressive Western phrase. If kept in captivity, such species retain a menacing demeanor longer than others, although almost all rattlers eventually become inured to the presence of human beings.

At the top of the list of rattlers quick to anger, I should put the western diamondback (C. atrox), followed closely by the eastern diamondback (C. adamanteus), and then by several of the subspecies of the speckled rattlesnake (C. mitchelli). At the other end of the scale would be a notably peaceful rattler, the red diamond (C. r. ruber), a surprisingly mild-mannered snake in view of its close relationship to the western diamond.

Considerable individual differences have been found also, some within a particular species only trying to escape, while others defending themselves violently. As one writer puts it, rattlers are not to be trusted, for some violate all rules.

Various reasons have been suggested as the causes of these differences—sex, age, weather, environment, etc.—and some of these may be valid. Several authors consider the males more belligerent than the females, and maintain that they are particularly pugnacious during the mating season.

Although the large males are probably the most ready to stand their ground, particularly during the mating season, young rattlers, like the young of almost all snakes, are in some ways more belligerent than adults. Baby rattlesnakes only a few minutes after birth—indeed, as soon as they are free of the encumbering fetal cases—will coil and strike, in a pose quite similar to that of the most experienced adult. Even the end of the tail, although only equipped with the soundless prebutton, will be seen to vibrate as if in anger. The young of all snakes are subject to more hazards from enemies than the adults, and the threatening poses they adopt may, in some cases, scare off the animals that would prey on them.

The environments in which rattlers are discovered will affect their reactions toward trespassers, for it is natural that they should be more prone to adopt a fighting pose if refuges are distant and difficult of access.

Several of my correspondents make the point that a startled rattler, such as one awakened from sleep, is likely to be rather aggressive; several have found rattlers more ready to defend themselves if disturbed while feeding or drinking.

The idea that the belligerency of rattlesnakes varies with the season is an old one, and may have some factual foundation, in that rattlers are certainly not active when cold. However, it is doubtful that the variation in temperament is as noticeable or uniform as commonly supposed. Since, in most areas inhabited by rattlers, they are largely nocturnal in summer, it is possible that their reputation for being quarrelsome at that season may result from their resentment at being disturbed, when found resting or sleeping in some cool spot in the daytime.

Finally, there is a theory that has virtually become folklore, to the effect that rattlesnakes are particularly vicious when blinded by the imminence of skin shedding; and that this occurs each year at the same time — usually stated to be August — for all snakes, which is consequently a season of particular danger.

As I have shown in discussing the shedding operation, snakes usually shed more frequently than once a year — this is particularly true of the adolescent period of rapid growth — and the shedding season is therefore not restricted to a particular time of year. Furthermore, the period of partial blindness is a relatively short one; it occurs a few days before, rather than immediately before, the skin is shed. For their own protection, snakes probably seek seclusion during this period of defective eyesight and thus should not be especially dangerous. Captive snakes show no particular irritability during the "blue-eyed" stage when the spectacles over the eyes are dulled by the accumulation of liquid beneath. Doubtless a partially blind snake would be more likely to strike, owing to a feeling of helplessness, if closely approached in its seclusion during the short period of dull vision.

Protective Coloration and Concealment

Having discussed the temperament of rattlesnakes and some of the causes of its variability, I shall now describe the successive actions that most individuals follow when confronted by a dangerous enemy such as man.

The first reaction is almost always to lie quiet in the apparent hope of escaping discovery — to "lie doggo," as the British slang phrase has it. This dependence on procrypsis, or protective coloring, is no mere policy of doing nothing, nor does it result from the snake's not having discovered its enemy. On the contrary, it is deliberate — the result of instinct, intelligence, or whatever term may be applied to the snake's attempt to attain a particular goal. That the rattler is alert to the trespasser is sometimes shown by a flicking out of the questioning tongue, or a premonitary click as the tail is adjusted to sound the rattle should this be found necessary. That this instinctive ruse is often successful there can be no question. On several occasions I have tried the experiment of informing a companion not experienced in hunting rattlers, that one was in plain sight in a patch of rocks or brush within a circle, which I roughly indicated, and then have asked him to point it out, without, of course, approaching the spot. It was each time surprising to see how long it took him to locate the snake. This affords a convincing evidence of the effectiveness of protective coloration.

To some early writers, the idea that rattlers depend upon procrypsis for protection would have seemed impossible for two reasons: there was, first, the widespread idea — it has now become folklore — that rattlers give forth so strong an odor that their presence is easily detected, even when they are resting undisturbed. The second is that the slightest movement of the snake causes the rattle to sound involuntarily, also an ancient misunderstanding upon the part of those who had not seen a live rattler in action.

Flight

If the enemy shows, by coming nearer the snake or by adopting a belligerent attitude, that lying motionless has failed, then the rattler may attempt retreat, or it may throw itself into a striking coil and threaten the intruder. In fact, it frequently does both at once; for the striking coil permits a snake, while facing the interloper and ready to lunge if he comes within range, to retire at a good rate toward any available sanctuary; in military parlance, it executes a rear-guard action.

Rattlesnakes, when lying about in the open, generally remain near places of refuge; of this there can be no question, and their prompt and direct manner of availing themselves of concealment in the refuge leaves little doubt that this is an intentional expedient or an instinctive habit. It is a common observation when the snakes are scattered about outside their hibernating dens, taking advantage of the last sun of the autumn or the first of spring. Many collectors have reported how necessary it is to get between the rattlers and their holes, toward which they streak as soon as they become aware of an intruder.

Sometimes a refuge-seeking rattler adopts a ruse that may well be dangerous to an inexperienced pursuer, as is evident from the following observation from a correspondent:

One rattler crawled in a hole, leaving his tail protruding several inches. I nearly fell for this trick, but noticed his head just inside the hole, apparently ready to strike when I reached for his tail.

It has been stated that an escaping rattler on a hillside will almost always head downhill, thus making better speed. This is certainly not invariable; it is probable that the nearest bush or rock crevice that might serve as a refuge more often guides the direction of escape.

Defensive Attacks

Though I have suggested that rattlers usually go through two phases of passive defense — procrypsis and flight — before adopting a fighting pose, this is by no means universal, and no one should be so foolish as to depend on their adherence to this sequence. My correspondents cite numerous instances of various deviations from this program — of snakes that adopted the final fighting pose without preliminaries, and others that even pursued an active offensive. Here are some reports of rattlesnakes attacking men:

□ Upon one occasion I was traveling on foot early in the season and encountered a small rattler, twenty-one inches long, which struck at me when I was more than three feet away. As I backed up and started to pull my gun, the snake came at me and struck again. I backed up again, and this was repeated until I had backed up eighteen feet, when I shot the snake.

□ Usually rattlers try to crawl away, but occasionally they will coil and hold their ground. We have even noted a few cases in which they would start toward a person, striking as they came, making it necessary to retreat or kill the snake.

□ In Sheep Canyon, off of Lytle Creek, I had a rattlesnake deliberately pursue me. I changed the direction of my retreat several times to check on this, and each time the snake kept right after me.

Obviously some rattlesnake attacks result from an initial hostile motion of the person making the report, but at other times it is impossible to determine what first aroused the snake. If people pass in single file along a path, for example, the head of a party may disturb a rattler sufficiently to cause it to strike one of those who follow. Sometimes a snake is disturbed by passing animals just as a person comes along. Of course, it is to be expected that an injured snake will attack.

With regard to these instances of attack, no one should presume that any rattler could move fast enough to overtake a person who really wished to get out of its way. The only danger would be that, while backing up in order to watch the snake's movements, one might trip over something and fall in its path. And even after a rattler has taken an offensive attitude it is quite possible to cow it so that it will no longer attempt to strike, but will only hide its head in its coils.

It may be said, in passing, that instances of apparent aggression are not restricted to rattlers or even to venomous snakes.

It would be regrettable if, owing to their sensational nature, these reports of occasional rattler aggressiveness should lead some of my readers to impute a great deal more vindictiveness to rattlesnakes than they deserve. I could cite plenty of evidence to the contrary, although naturally much less exciting.

Warning Mechanisms

The Rattle. One of the principal elements of the rattlesnake's standard warning behavior is the use of the rattle, for the primary function of the rattle is to warn away or frighten animals that might be injurious to the snake.

In the early days, those who wrote on natural history were so impressed with the unique character of the rattle that they stressed its use as a God-given benefit to mankind, believing no rattler ever struck without first giving fair warning by rattling. Later observations failed to confirm this, and from time to time controversies on this point have raged in the questions-and-answers columns of the sporting magazines in which both sides have bolstered their theories with field observations. Actually, the evidence points to the general conclusion that rattlers usually warn before striking, and that a clarification of definitions would compose some of the apparent differences between those who have maintained opposing views. For the question naturally arises as to what one means by the term "strike." If this is thought to be synonymous with "bite," then certainly rattlers do not invariably warn an enemy (they never warn prey); for if one steps on a rattler it will quite likely bite without even striking; and if a rattler is suddenly and violently alarmed it may strike instantly, and may or may not rattle simultaneously. The procedure is largely a matter of timing; if a rattler is annoyed by an approaching enemy, and it is given time to shift from procrypsis and flight to active defense, then in nearly every instance it will throw itself into its menacing, S-shaped coil, and at the same time it will sound the rattle. But if an attack comes suddenly, the rattler may retaliate with an instant strike, and there will be no interval of rattling. Certainly no one should expect a rattler to advertise its presence by rattling when there is a chance that it may escape detection by depending on concealing coloration and quiescence, since this would obviously defeat its purpose. Yet many take this as a failure to deliver the much-heralded warning.

The rattle, in fact, is the poorest kind of a defense against man, for in his case it invites, rather than discourages, attack. We may be sure that had there been missile-throwing and weapon-carrying creatures on earth during the aeons when the rattle was being developed, the ancestors of the rattlesnakes would never have survived to bequeath this queer appendage to their descendants of today.

There is a strange theory that rattlesnakes always rattle three times before striking. This has now become a part of the folklore of Maryland and of the Pennsylvania Germans.

The Hiss. A second element of the rattler warning posture is the hiss. It is surprising how many writers have stated that rattlers cannot hiss. The assumption has evidently been that they do not need a warning hiss, as do other snakes, its place being taken by the rattle. As a matter of fact, rattlers can and do hiss quite loudly, although the sound is largely overshadowed by the more strident rattle. But it is only necessary to still the rattle by tying a bit of cloth around it, and the unquestioned hiss will be heard from any rattler in its striking pose. The hiss compares favorably in volume with that made by most other snakes, although not so loud as some, such as the bull or gopher snakes, which have a special formation of the epiglottis to increase the sound.

It is to be remembered that snakes are deaf to air-borne sounds and that the hiss is a threat—a part of the defensive pattern—and not a means of communication between individuals. All sounds made by snakes, as far as known, including the rattle of the rattlesnake, are warnings; for necessarily they must be made for their effects on creatures that can hear, which eliminates other snakes. Some Brazilian natives believe venomous snakes decoy small birds and mammals by imitating their calls. One writer tells of some harmless snakes that lived under his house and, while undisturbed, kept up a kind of hissing conversation.

Similar imaginary sounds have been imputed to rattlers, particularly the eastern diamondback (*C. adamanteus*). One was reported bedeviled until its hiss turned into growls. Another was seen coming down the trail with its mouth open, making a noise like the complaining of a distant bull. Still another was seen on the trail of a rabbit, weaving from side to side, and emitting a strange, humming moan. One author makes much of a mysterious mating call he attributes to the diamondback. The snake is smart enough to conceal this from all but a few human beings; when it knows men are about, the wailing call is changed to something like a bird's chirp. Stories of this kind usually emanate from sounds heard at night, when the source of the noise cannot be ascertained or identified. It may be pointed out that not only is there no evidence that rattlers make such sounds, but anatomical examination fails to show a possible source.

Although a rattler usually hisses only while in the striking coil and when simultaneously rattling, this is not always the case. Sometimes it will stop rattling yet continue to hiss.

Use of the Tongue. Another feature of the rattler's warning posture is the tongue, for a rattlesnake on the defensive uses it in a manner quite beyond the necessities of its function as a sense organ. As previously mentioned, as an adjunct to the sense of smell, the tongue is repeatedly flicked out horizontally, wavered slightly, and then retracted to make its depository report. But a rattler on the defensive holds its tongue, extruded to the utmost, alternatively vertically erect, and then downward, and this for considerable periods, ten to fifteen seconds or more, before retraction. The black tips are widely separated, and these, with the pink central column, make an imposing show, for they are the most mobile element in the entire posture. This use of the tongue seems to be primarily for its visual effect; for, although it is frequently withdrawn, the long pauses between retractions must actually interrupt, to some extent, the sensation of smell.

The Scent Glands. Some items in the rattler's defense arsenal are used rarely and are not a part of the customary defensive posture. Most snakes are equipped with a pair of musk or scent glands in the tail. The purpose of the glands is not known with certainty, and the use may differ in different kinds of snakes, although from the exceedingly offensive odor of the discharge in many genera and the fact that they are excreted when the snakes are handled, it may be assumed that they comprise a defense mechanism.

In rattlesnakes they are occasionally used as a part of the defense mechanism, especially if a snake is being handled or mistreated. Rattlesnake scent is to me pungent but not particularly offensive (and therefore not particularly effective), much less so than the disgusting odors emitted by garter and king snakes, for example.

Body Flattening. One element of the defensive posture—the extreme flattening of the posterior part of the body—may be largely utilitarian, for it results in a more adequate base from which a strike may be launched. Also, the flattening is so marked that it seems to increase the apparent size of the snake and may thus accentuate the threatening posture. It is more characteristic of a snake standing its ground than one retreating, while still facing the enemy in the striking coil.

That the flattening is, indeed, partly an intimidating pose may be judged from the fact that rattlesnakes striking prey do not flatten to anywhere near the same extent. Guided by the difference between the attitude toward an enemy and toward prey, we may distinguish between those features of the pose that are largely utilitarian, such as the S-shaped wave in the neck, and those that are part of the pattern of intimidation, such as the rattle, the hiss, and the vertical tongue. For the latter are not a part of the posture preparatory to a strike for food.

The question may be asked whether rattlesnakes distinguish between different classes of animals and adopt different defense mechanisms to repel them. There can be no question that they do recognize king snakes and other snake enemies. No doubt rattlers can distinguish mammals from reptiles by use of their facial pits, but whether this induces a variation in the defense reaction is not known.

Mythical Defensive Actions

With creatures so provocative of sensationalism, it is not surprising that some rather fantastic stories have been printed concerning the defense reactions of rattlesnakes. In 1778 it was reported that when a snake is animated by resentment, every tint rushes from its subcutaneous recess, giving the surface of the reptile a deeper color. Some reports alleged that the scales on the snake's back were erected when the reptile was enraged. One author records that a disturbed snake bares its fangs, and the tongue becomes the color of the hottest flame. Another tells of a rattler rearing its snout no less than four feet above its coils, with the head swelled to a violent degree, and its throat shining with beautiful and vivid colors. One traveler was told by his guide of a rattler that pursued a young man in a series of leaps.

The picture of a rattlesnake threatening an enemy with gaping mouth and protruding fangs is an ever-recurring one in the literature, but it is not true of an uninjured snake. Such stories gain credence from seeing a snake that has injured its mouth in striking something, or whose head has already been damaged by a blow.

The story is told of rattlers going a mile from their dens to make a combined attack on their enemies. In Arizona they raided a ranch house one night and drove out several sheepherders. Then there are the fantastic yarns of the attacking methods used by diamondbacks: the man who was chased by a rattler and was searched for here and there as he hid behind a tree; the theory that a rattler, sneaking away on the defensive, can shrink to half its normal length like a worm; and finally the story of a man on horseback who was outrun by a diamondback.

Distinguishing the Resting from the Striking Coil

One would expect no confusion between the resting and striking coils of a rattlesnake, for they are markedly different. But the fact remains that when some hunter, who has reported almost stepping on a rattler

Fig. 20 Southern Pacific rattlesnake (*C. v. helleri*) in resting coil

"coiled ready to strike," is asked to describe the creature's position, in nearly every instance he reveals that the rattler was in a flat, pancake-like coil — in other words, the snake was certainly resting and was probably asleep. For this motionless coil, this flat spiral of the body, with the head lying on the outer edge, is in notable contrast with the true striking coil. In the latter, the body is in a vertical spiral, with a widely opened loop as a base on the ground; above this, the anterior part of the body is erected, culminating with the head and neck in a loose S-shaped wave, like a poised and threatening lance. The striking coil is in no way static — it is alive with motion, incipient or actual, a portent of disaster. For there is not only the wave in the neck that requires only straightening to produce a forward lunge — there are the other evidences of belligerence and men-

Fig. 21 Western diamond rattlesnake (*C. atrox*) in striking coil

ace: the sounding rattle; the hisses emanating from the swelling body; the flattened posterior affording a base from which the strike may be launched; and the protruding tongue, alternately pendent and vertically erect, with the forked tips widespread. It is alert, hostile, and menacing. And back of the noise of rattle and hiss, the waving tongue and poised head, is the threat of the really dangerous strike. At such a time a rattler seems almost to deserve its reputation for malevolence.

To return for a moment to the resting coil: except for the fact that the body rests on the ground in several turns of a flat spiral, there is no particular uniformity of the arrangement. The head and neck, more often taking the form of an S than the arc of a circle, generally rest on the other coils. The rattle is sometimes at the outer edge of the coil, but more often sticks up near the center.

A rattler in its resting position will often be found settled down in grass or ferns as if they had been pushed aside to form a depression or bed. If rocks are nearby, a snake will usually coil with one edge of the body against a rock. It is by no means unusual to find sidewinders (*C. cerastes*) bedded down in sand and partly covered over.

A rattler can strike only a short distance from its resting coil—only the distance involved in straightening the S-shaped section of the neck. But in any case it will almost never do this, preferring, if disturbed, to throw itself first into its striking coil, which it can do very quickly by raising the anterior part of the body above the ground and at the same time spreading the after part into a wider loop to serve as an anchorage. Experiments with a piece of hose coiled to simulate a rattler in its resting

Fig. 22 Sidewinder (*C. c. leterorepens*) bedded down in sand. The scale marks in the sand show that the snake used its neck to pull sand over its outer coils. (Photograph by Dr. Raymond B. Cowles)

coil will show that a forward lunge for any distance will twist the head laterally, thus illustrating the unsuitability of this type of coil for a strike. Anyone wishing to prove that a rattler in its resting coil is not "coiled ready to strike" has only to disturb one found in this position to note the quick change to the true striking posture.

Some have pointed out that, as a rattler cannot strike unless suitably coiled, it is virtually harmless when outstretched. This is hardly true, since a rattler can bite in any position, and besides can assume a striking coil with great rapidity.

From the standpoint of defense, the rattler's striking coil has several advantages over any other posture: the S-wave in the neck (a longer wave than the short one of the resting coil) permits a forward lunge of the head without any lateral twisting that would disturb the sight or aim; the elevated position of the head permits a good view of the enemy, and the vertical section of the body allows a quick change in orientation; and, finally, the wide circle of the posterior part of the body not only provides a base from which a lancelike drive of the head may be launched, but likewise permits a retreat toward any available refuge while the snake still faces and threatens the enemy. A retreating snake faces backward, toward the enemy, across its own body, so that the length of the strike required to reach the foe is usually greater than in a snake standing its ground.

It should be understood that the striking coil of the rattlesnake is not unique; many other snakes, both venomous and harmless, adopt defensive postures that differ from that of the rattlesnakes in only unimportant details.

There is a widespread myth to the effect that rattlesnakes invariably coil clockwise in the Northern Hemisphere and counterclockwise in the Southern. There are also various beliefs as to the consistency with which the rattler, in its resting coil, has the rattle at the center of the coil or the periphery. But there is no uniformity in the position taken.

The Strike

One of the best of the early descriptions of the strike of a rattlesnake was made in 1822; it observed that the strike was only a forward lunge of the elevated head — a straightening out of the neck, the posterior part of the body remaining stationary.

The forward lunge of a rattlesnake from its striking coil, if it be thoroughly aroused, is delivered suddenly and with considerable speed — so fast, indeed, that the motion of the head cannot be followed with the eye. The withdrawal is slower and the motion can be observed. The snake does not threaten or start a strike with open mouth, but a glimpse of the open mouth can usually be faintly seen at the end of the stroke as the head reverses direction. Since the invention of high-speed motion pictures, it has been possible to photograph the rattler's strike so that the details of the mouth opening and the advancement of the fangs may be seen. Further, by means of successive stroboscopic exposures taken of a striking rattlesnake, with a timing device in the backgound, the speed of a rattlesnake's strike at the midpoint of the strike was determined. It proved to be not nearly so fast as popularly supposed — far from the "fastest thing in nature" as has been stated. An adult prairie rattlesnake (*C. v. viridis*), in twenty separate tests, struck with an average speed of 8.12 feet per second. The variation was from 5.2 to 11.6 feet per second. There was some correlation of speed with temperature, although it was not high. These speeds are slower than those of a man striking with his fist.

One author tells of a prairie rattler that struck a thrown rock in midair. However, the stories of how rattlers unerringly strike bullets shot at them, so that it is impossible for a marksman to miss, are, of course, mythical.

The rapidity with which a snake will repeat a strike depends on conditions, and particularly the continuance of the threat imposed. If the enemy does not retreat, a rattler may strike again and again at short intervals. However, if the threat continues or the rattler misses several times, it may become cowed.

A large rattler can strike with considerable force. One man who was struck in a rubber boot between knee and hip said the blow felt like a thrown rock. In experimenting with fang penetration, I have caused large rattlers to strike objects such as shoes attached to a stick. The blow is a sharp one, but hardly sufficient to throw a person off balance as the Osages stated to Audubon in 1827.

The height at which a rattler's head is poised for a strike, depends, of course, on the size of the snake. A four and one half-foot western diamond will strike from a height of about eight to ten inches and will lunge slightly downward. Puttees or other adequate protection below the knee will greatly reduce the danger in hikes through snake country.

Much has been written about the distance to which a rattlesnake can strike, which is also obviously dependent on the size of the snake. This is a matter of some practical importance since it defines the zone of danger. As a matter of fact, much depends on circumstances, such as the position of the snake, the species, and the degree of its excitement.

From my own experience I should say that a rattler will rarely strike more than half its length, measured from the front of the anchor coil, and almost never beyond three-quarters of that length. However, I should not like anyone to take this as a guarantee of safety, if only for the reason that one cannot accurately judge the length of a coiled snake.

Although rattlers in their striking coils usually strike on the horizontal or slightly downward, if the target be a large one, so that they need not aim at a particular spot, they can strike directly upward, or at any angle between. A vertical strike rarely exceeds a foot or foot and a half, even though the snake be a large one.

A frequent source of controversy in the sporting and nature magazines is whether a rattlesnake can strike without coiling. As is so often the case, the argument really stems from a misunderstanding or confusion of terms. Some persons seem to consider the terms "bite" and "strike" synonymous, which they are not. A snake, to bite, needs only to open its mouth and imbed its fangs in any object within reach. The muscular power of the jaws is quite sufficient to drive the fangs into tissue without the momentum of a strike, as many a man has learned to his cost when holding a rattlesnake behind the head so carelessly as to permit the snake to free an inch or so of neck and thus turn its head and reach its captor's

Fig. 23 Central American rattlesnake (*C. d. durissus*) in threatening pose. (Photograph by Dr. C. Picado T.; used through the courtesy of the R. L. Ditmars estate)

hand. No strike is here involved, if, by strike, we mean a forward lunge of the head permitting a snake to reach an object otherwise beyond its range.

It is true that the momentum of a strike does aid in imbedding the fangs, which point almost directly forward as the head reaches the end of the stroke, but the strike is not an essential part of a bite.

Another source of dispute lies in a misunderstanding of the nature of the snake's coil. To many the word implies a flat, tight, pancake-like spiral—that is, the resting coil. Seeing a snake strike from other than this type of coil—and it would be impossible to strike more than a few inches from one—the observer says that the snake struck without coiling. But in almost every instance the snake has indeed struck from its typical striking coil—the open anchor loop, elevated forebody, and S-shaped crook in the neck—although this may seem to the observer to be too loose or open a formation to deserve the term "coil" at all.

But, aside from these arguments that flow from a lack of agreement upon definitions, it is true that rattlesnakes sometimes strike without the formality of coiling, or possibly it should be said that they pass in a single violent movement from a stretched-out position, through the striking posture, into the strike itself.

Now we pass to accounts quite beyond the realm of possibility. One writer says a rattler jumps as a coiled spring would when compressed and released. Another tells of a guide who aimed a sword cut at a six-foot rattler. The sword missed and the rattler jumped over the man's arm and lit eight feet beyond. Another claims a diamondback is so active that it could strike between the eyes of a man standing erect. One has seen rattlers launch themselves through the air and strike a foot beyond their lengths. All of these are capped by the author who tells of the diamondback that struck double its length by diving across its own tail; the rattler that struck past a man ten feet away, just missing him; another that struck and reached the knee of a man on horseback; another that just missed a horse's nose, to do which he leaped at least eight feet clear of the ground; and finally the account of the snake that jumped three feet straight up out of a hole in the ground.

CLIMBING PROCLIVITIES

Whether rattlesnakes climb trees is a question often appearing in the queries columns of the sports and nature magazines. From the observations of many field men it is now known that they do so occasionally, but they are not as adept climbers as many other normally ground-inhabiting forms such as rat snakes, king snakes, bull snakes, racers and whip snakes, all of which frequently climb trees in search of birds and their nests. The rattler is handicapped in climbing by its relatively stout body and short tail; it cannot climb with the facility of the more slender-bodied snakes.

In the zoo, rattlers occasionally climb the small shrubs in their cages, but do not remain there at rest for long periods as do many of the other snakes. Possibly they may climb to find some avenue of escape.

It is thought that some species of rattlers are more likely to climb than others, and this may be true, the timber and black-tailed rattlesnakes being among the more persistent climbers. Yet observations show how widely prevalent the custom is, for nearly every species common enough to be under frequent observation has been observed upon some occasion up in a tree or bush.

Various reasons have been advanced to account for tree-climbing by rattlers; most observers believe it is to secure food, such as small mammals, birds, or eggs. One writer mentions an instance in which a rattler was found coiled around a nest fifteen feet above ground.

Some observers report that rattlers occasionally climb into bushes to avoid water or to dry themselves. When the Everglades are flooded, for example, the pigmy rattlers *(S. m. barbouri)* are often seen in small trees, or lying coiled on cabbage-palm leaves eight or ten feet high.

On the distinctly mythical side we have the statement that rattlers in northern California catch insectivorous birds by imitating the buzzing of a bee with their rattles. Its author saw this done by rattlers concealed in the dense foliage of trees twenty feet above ground.

Some think their climbing is not so much to search for birds' nests as for the purpose of finding an advantageous perch for spying prey or enemies.

When the belief in the power of fascination was at its height, it was thought that snakes had only to fix their baleful eyes upon some unfortunate bird or squirrel, whereupon the helpless prey fluttered or fell into the waiting mouth below. Hence it was unnecessary for the snakes to exert themselves to climb trees.

SWIMMING ABILITY

Although rattlesnakes are not so aquatic as the sea snakes found in the tropical oceans, or even as accustomed to water as are the water snakes, garter snakes, or water moccasin of our own country, nevertheless they are good swimmers. Some species frequent marshy areas, as does the eastern massasauga *(Sistrurus catenatus catenatus)*; and the eastern diamondback *(C. adamanteus)* is sometimes found at sea swimming to and from the keys along the Florida coast. Some species have been seen crossing swift rivers.

It is clear, both from published accounts and from the statements of my own correspondents, that all rattlers will take to the water and swim readily enough when their travels in search of food, refuge, or mates require them to cross intervening water. Snakes—rattlesnakes included—are quite buoyant, since the pause in respiration is long and comes after inhalation, instead of expiration as in mammals. Furthermore, the rattlesnake lung includes a bladder-like prolongation that increases its air capacity. Thus they float readily, and a sinuous motion of the body and tail propels them at a fair rate of speed.

How the Rattles Are Held

A number of authors and correspondents have stated that swimming rattlers always swim with their tails elevated to keep their rattles dry. In order to test this theory, I experimented with twenty-one species and subspecies of rattlesnakes by placing them in a fresh-water pond. The rattlers seemed unafraid of the water, and were so buoyant that their heads and necks could be raised well above the surface—twelve inches in large specimens—so that they could look about and get their bearings. They could easily climb a wall six or eight inches high to get out of the pond.

As to trying to keep their rattles dry, the actions of all of these different kinds of rattlesnakes were remarkably uniform; they made no apparent effort to hold their tails above water, even when floating motionless; and as soon as they wished to move, they regularly used their tails for propulsion, with their rattles below the surface. Occasionally when resting or floating, the rattles were held above the surface, but this seemed to be more a repetition of their usual land-crawling posture with the rattles pointed upward than an attempt to keep the rattles dry. Even when resting on lily pads, they seldom avoided trailing the rattles in the water, although they might easily have kept them out.

Actions in the Water

Rattlers shed water readily—that is, the skin seems protected from wetting by an oil film, but this does not apply to the rattles. If annoyed, the snakes of my experiments rattled, or tried to, while the tails were immersed, and the rattles became thoroughly wet and at least partly filled with water. When a snake was removed from the water, the rattle was first heard faintly, but the sound increased as the water was shaken from the rattles. Several of the rattlers with which I experimented drank water while floating.

The question of whether rattlesnakes can bite when in or under water is occasionally asked. The answer is that they can bite readily enough, although, lacking the anchorage of a solid base, they cannot strike efficiently.

Rattlesnakes do not drown easily, as evidenced by the report of a correspondent:

I have thrown prairie rattlesnakes in the water and they can swim quite well. In fact, we tried to drown one under water for nearly half an hour but it wouldn't drown.

This endurance would be expected in a cold-blooded animal with extra air storage capacity.

Several correspondents comment on the fact that rattlers frequent the shores of streams or ponds to drink, particularly in summer, and that occasionally they take to the water to keep cool.

Undoubtedly the champion salt-water swimmer among rattlesnakes, at least of those found in the United States, is the eastern diamondback (*C. adamanteus*) of Florida and the adjacent states. Here are some observations on their seagoing propensities:

☐ Two friends of mine killed a diamondback rattler in the Gulf of Mexico, two miles from land. This snake evidently was going from an island to the mainland.

☐ The Florida diamondbacks are often found swimming wide waterways, and, on one occasion, one was found floating in the Gulf of Mexico, twenty-two miles from land, apparently having floated out with the tide at night.

Tales of Swimming Rattlesnakes

Among the decidedly doubtful stories of the prevalence of rattlers in water are the following: rattlers were seen in Pennsylvania to cross rivers and lakes by the hundreds and thousands on the way to their dens; a man rowing out in Lake George, New York, killed twenty-nine rattlers; rattlers take to water only when chasing prey at night; rattlers will not cross running water unless a bridge is provided.

A story requiring confirmation is that of a writer who reports that a timber rattler, swimming in a river, paused and coiled when he whistled at it at a distance of thirty feet. Swimming moccasins will stop at a whistle nine times out of ten, according to this author.

A backwoodsman in upper New York pleasantly combined two myths when he told of a mother rattler which protectively swallowed her young and then transported them across Lake George with her tail in the air, rattling like an outboard motor.

Danger from Swimming Rattlesnakes

Although there can be no doubt as to the swimming ability of rattlesnakes, they can be easily avoided by anyone in a boat having a weapon or means of propulsion. An author was greatly exaggerating when he stated that they swim with the speed of an arrow, and endanger people in small boats. Conceivably a rattler, tired from swimming, might try to climb into a boat, but it would be easy to thrust it off with an oar or stick.

Fishermen on the freshwater lakes of San Diego County are sometimes badly frightened by southern Pacific and red diamond rattlers that swim near the boats. Although the most sinister motives are attributed to these naval attacks, we may be sure that a rattler, if it approaches a boat, is only seeking a temporary resting place.

Although a rattlesnake could bite under water, a person must be careless indeed, who would get near enough to a swimming rattler to be within its striking range and then proceed to annoy it into biting.

CHAPTER SIX

Populations and Ecology

Who, in wandering about the hills in search of rattlesnakes, has not wished for a special Geiger counter responsive to some unknown *Crotalus* emanation, whereby he might locate every rattler in the area, hidden though it might be, in the weeds, under a stone, or down some mammal hole? For the secretiveness of these snakes keeps them in hiding much of the time, and even when in the open their procryptic color patterns make them difficult to see. Altogether, the most experienced hunter, except at denning time, or in a sandy desert where they may be tracked, will never find more than a small fraction of the rattlers in his area.

Necessarily our knowledge of rattlesnake population densities is quite inadequate, for the technique productive of the most accurate data, that of tagging, releasing, and recapturing as many individuals as may be found, requires a long-range program, and has only begun to be applied. Also, with venomous snakes such as rattlers, this experiment should never be countenanced except in remote areas where there will be no danger to persons or livestock from the snakes released. But at least, from the data at hand, we may learn something as to the proportion of snake populations that rattlesnakes comprise, and whether two or more species may commonly occupy the same ecological niche. For much is being learned of the character of the surroundings frequented by the different kinds of rattlers.

The Importance of Ecological Factors

In a study of rattlesnake populations, it will be desirable, first, to discuss their ecology, their habitats, and the conditions that govern their lives and limit their spread. A preliminary survey of rattlesnake ranges makes it evident immediately that the various species and subspecies are subject to quite different ecological requirements and limitations.

Rattlesnakes are restricted to the New World; they have a vast range in the Americas from Lat. 51°N. to Lat. 35°S. In this area they occupy many different types of terrain—deserts and jungles, valleys and mountains, sand dunes and forests, mainland and islands. But no single species includes all of these habitats in its range; this wide variation appertains only to the rattlesnakes as a group. Each species has requirements of its own, sometimes quite restrictive, in others more inclusive.

Although there are specific exceptions, most species and probably more individuals, are to be found in arid places. Deserts, grassy plains, and brushy or rocky hills appear to be more suitable for rattlesnakes than areas clothed with trees. It is for this reason, as well as because of varied topography, that Arizona contains more species and subspecies than any other state, and that many species are restricted to the southwestern United States and northwestern Mexico.

The primary factors affecting distribution are food supply, climate, and refuges. Rattlesnakes live to a large extent on small mammals, with lizards as the second most important component of diet—indeed, the most important in the case of some species and the juveniles of others. Ranges and population densities are first of all dependent on food availability. Having no metabolic temperature-control, rattlers must confine their activities to times and places where air and ground temperatures, and radiation, are such as to produce body temperatures within a relatively narrow range—65° to 95°F. They cannot live in places where the season of adequate temperatures is too short for fat accumulation, growth, and reproduction.

Changing climates of the distant past have produced the discontinuities in the ranges of the rattlesnakes as we see them today, such as that of the Arizona black rattlesnake *(C. v. cerberus)* in Arizona, where it is now restricted to the timbered areas of the central section and a few mountain ranges of the south, this population having become separated by intervening deserts from its close relative, the southern Pacific rattler *(C. v. helleri)* along the southern California coast.

Ecological changes produced by man, particularly in agriculture, generally result in conditions less favorable to rattlers than the original primitive conditions, even though there may be an increase in the food supply. And there are not only the changes in surroundings and the destruction of refuges to be counterbalanced, but the ever-present enmity of man and his domestic animals. For example, the massasauga, once common on the prairies of the upper Mississippi Valley states, has now retreated into the marshes or wastelands; the sidewinder has disappeared from the irrigated sections of the Imperial and Coachella valleys in southern California, although still common in the adjacent desert; and the prairie rattlesnake has been driven out of the farm lands of Kansas and Nebraska.

Although human occupation, with agricultural and industrial developments, has now eradicated rattlers from many extensive areas they once frequented, it is of interest to note that in primitive times few sections of the United States were not tenanted by at least one species of rattlesnake. Aside from the mountain peaks—above eleven thousand feet in California and nine thousand feet in most other areas—the only unoccupied places were northern and eastern Maine, northeastern Wisconsin, upper Michigan, northern Minnesota, northeastern North Dakota, northwestern coastal Oregon, and western Washington. It is probable that in pre-Columbian days at least some part of every state was inhabited by rattlesnakes.

Rattlesnakes have been quite successful in maintaining themselves on some islands, particularly off the coasts of California and Baja California. Some of these island populations have gradually diverged in character from their mainland congeners until several have justified specific or subspecific recognition.

RANGES

Range Limits

One should never presume, from a range map, that the animal whose range it purports to show is to be found everywhere within the cross-hatched area, for there is great ecological variation and consequent habitability or inhabitability in almost every area. Look, for example, at the map of such an expanse as the Mojave Desert. Here we have a vast plain, up through which protrude numberless scattered mountain peaks and chains. Any range map would show the sidewinder, the Mojave rattler, and the southwestern speckled rattler as occupying this territory, but it would fail to disclose the fact that the first two live almost entirely on the flats, and the last near or in the mountains.

Red diamond rattlesnake,
Crotalus ruber ruber

Western diamondback rattlesnake,
Crotalus atrox

Eastern diamondback rattlesnake,
Crotalus adamateus

Mojave rattlesnake,
Crotalus scutulatus scutulatus

Maps redrawn from *Snake Venom Poisoning*, by Findlay E. Russell (J.B. Lippincott, Co., 1980).

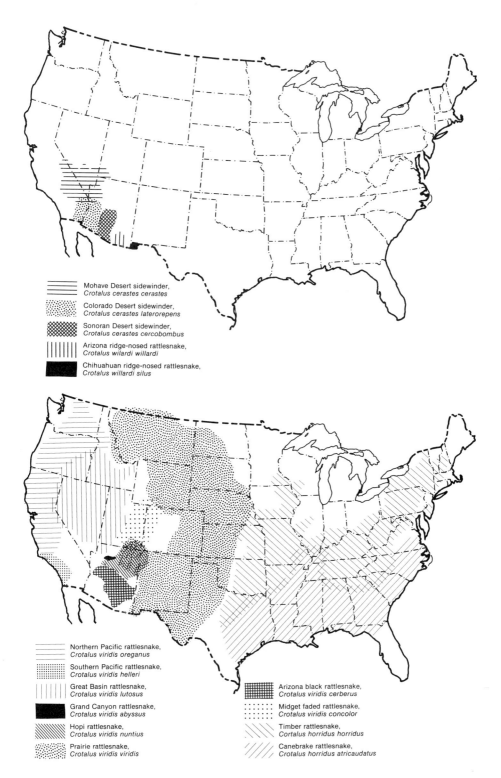

Mohave Desert sidewinder,
Crotalus cerastes cerastes

Colorado Desert sidewinder,
Crotalus cerastes laterorepens

Sonoran Desert sidewinder,
Crotalus cerastes cercobombus

Arizona ridge-nosed rattlesnake,
Crotalus wilardi willardi

Chihuahuan ridge-nosed rattlesnake,
Crotalus willardi silus

Northern Pacific rattlesnake,
Crotalus viridis oreganus

Southern Pacific rattlesnake,
Crotalus viridis helleri

Great Basin rattlesnake,
Crotalus viridis lutosus

Grand Canyon rattlesnake,
Crotalus viridis abyssus

Hopi rattlesnake,
Crotalus viridis nuntius

Prairie rattlesnake,
Crotalus viridis viridis

Arizona black rattlesnake,
Crotalus viridis cerberus

Midget faded rattlesnake,
Crotalus viridis concolor

Timber rattlesnake,
Cortalus horridus horridus

Canebrake rattlesnake,
Crotalus horridus atricaudatus

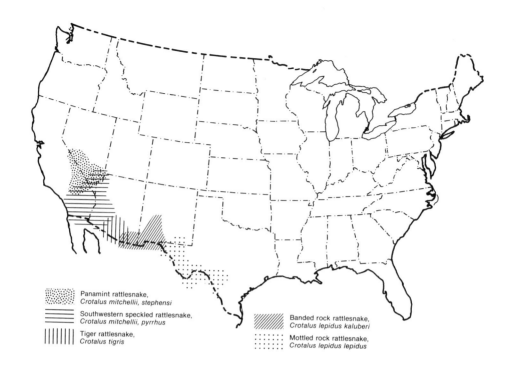

Panamint rattlesnake,
Crotalus mitchellii, stephensi

Southwestern speckled rattlesnake,
Crotalus mitchellii, pyrrhus

Tiger rattlesnake,
Crotalus tigris

Banded rock rattlesnake,
Crotalus lepidus kaluberi

Mottled rock rattlesnake,
Crotalus lepidus lepidus

Northern black-tailed rattlesnake,
Crotalus molossus molossus

Twin spotted rattlesnake,
Crotalus pricei pricei

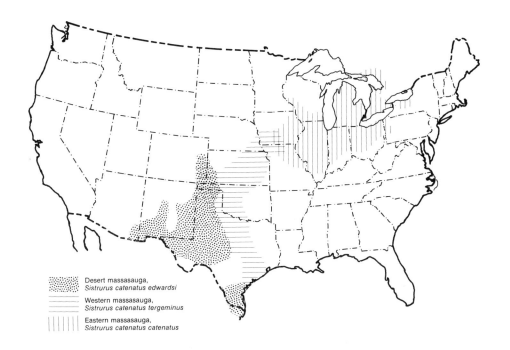

Desert massasauga,
Sistrurus catenatus edwardsi

Western massasauga,
Sistrurus catenatus tergeminus

Eastern massasauga,
Sistrurus catenatus catenatus

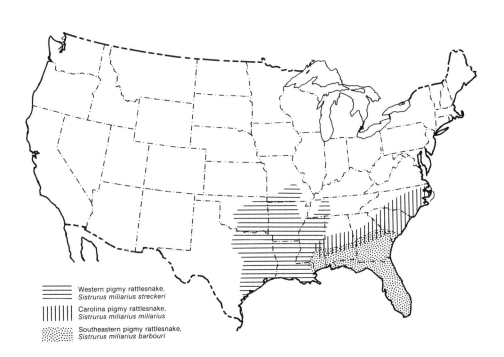

Western pigmy rattlesnake,
Sistrurus miliarius streckeri

Carolina pigmy rattlesnake,
Sistrurus miliarius miliarius

Southeastern pigmy rattlesnake,
Sistrurus miliarius barbouri

In addition to these broader habitat selections, there are others more local in nature, for some species of rattlers prefer brush, whereas others in the same district may be largely restricted to rocks or to sandy washes. These preferences are often difficult to determine with accuracy except by continuous local observation, and even then an occasional stray will complicate the record. When these local preferences are taken into consideration, we find that there may be less competition or overlapping between rattlers than might be presumed from the maps. Despite the fact that some sixteen species and subspecies of rattlesnakes are found in Arizona, I know of no place where more than six occupy the same ecological niche in any area.

Where range limits involve rather steep ecological gradients, it is, of course, fairly easy to fix them, though, when one is actually on the ground, it is often difficult to see why the next mile would not have been equally habitable. But sometimes the limit seems to be quite inexplicable; here the species is to be found—over there in what seems to be the same kind of surroundings it is absent. We can only guess that there is effective some deterring feature of climate, food, or shelter, whose presence or effect is not apparent.

Altitudinal Range

The altitude attained by rattlesnakes is of considerable interest to sportsmen and campers, for they feel reassured when they are at an elevation at which they no longer need be on the watch for rattlers. While not common at high altitudes, rattlesnakes do reach heights above those where they are popularly presumed to exist. The only safe general rule is to be on the lookout for rattlers in the warm season on any mountain in the United States up to 9,000 feet, except in California where the range is extended to 11,000 feet. In Mexico, one small subspecies attains an elevation of at least 14,500 feet.

ISLAND HABITATS

A considerable number of islands are inhabited by rattlesnakes. Some of the island populations, by their differences from those of the nearby mainland, suggest that they have persisted there since the original separation of the islands from the adjacent mainland; others clearly represent more recent colonizations, having reached the islands by swimming, transportation on floating debris, by being carried by birds of prey, or by some similar means of dispersal. In some instances involving narrow straits, swimming produces a continual interchange of individuals between island and mainland. Occasionally islands have been found to be advantageous habitats, offering a better food supply or greater freedom from enemies. This, in some cases, has resulted in a heavier population on the island than on the adjacent mainland, or persistence after the mainland population has been exterminated.

Some of the inhabited islands are but little more than dry areas in swamps, or mounds in a general lowland; others are isolated by the permanent water separations of rivers and lakes; while some lie off the Atlantic, Pacific, or Gulf coasts.

On the Pacific Coast, the offshore waters are both deeper and colder than on the Atlantic or Gulf of Mexico coasts, and the islands are ecologically more differentiated from the mainland than are the Atlantic and Gulf islands. Therefore, rattlers are found swimming much less frequently in the Pacific, and it is extremely doubtful that there is any interchange—or has been for a long time—between the insular and mainland populations.

POPULATION FACTORS

Population Densities

Trustworthy figures on actual population densities of rattlers are scarce indeed, owing to the obvious difficulty of making accurate censuses. The most complete studies thus far have been those on the northern Pacific rattlesnake on an experimental range in Madera County, California. In this study the conclusion was reached that there was somewhat in excess of one rattlesnake per acre on this rolling cattle ranch. The studies were made by capturing, marking, and releasing rattlers and subsequently recapturing them. It was then possible, by calculations involving the ratio of the recaptured snakes to those newly taken for the first time, together with data on the distances traveled by the recaptures in the interim, to estimate the total population with some accuracy. Such a program, to yield dependable results, requires several years for completion.

Reports giving some indication of rattlesnake population are these: a rancher in Texas in clearing brush killed sixty rattlers per square mile; another killed seventy-seven. Over a period of twenty-eight years one collector received from one thousand to five thousand eastern diamondbacks per annum at his snake exhibit at Silver Springs, Florida, with a grand total of about fifty thousand. These figures give some idea of the commonness of this large rattlesnake in parts of Florida.

Some of my correspondents, on the other hand, have commented on how few rattlesnakes they have seen, although their work frequently took them into rattlesnake-infested country.

Although many islands are reputed to be "alive with rattlesnakes," such reports are generally exaggerated, and it is usually found that the population density is not conspicuously different from that of similar areas of the mainland. However, there are three islands in the Gulf of

California that have really heavy concentrations, as has been demonstrated by several collectors, for considerable numbers have been taken on each island on relatively short visits. On a visit to Tortuga Island in December, 1934, the indigenous diamondbacks *(C. tortugensis)* were seen every 100 to 150 feet. This concentration was so great that a bird-nesting expedition actually left the island for fear someone might be bitten.

Effect of Dens on Population Estimates

In those areas where rattlesnakes congregate from considerable distances for purposes of hibernation, particularly if the den locations are accessible to people, it is very difficult to get truly objective estimates of population densities on the summer range. It is rarely understood that the conspicuous concentrations observed at the dens are drawn from wide areas, aggregations to be noted only for a few days in the fall or spring, and that these are by no means representative of the population density anywhere during the seasons of normal activity. It is but natural that a person happening upon a den at the appropriate time would conclude that the entire country must literally swarm with rattlers, whereas he might travel for days, after they had scattered for the summer, and rarely see a single one. There is little doubt that the published reports of the great numbers of rattlers in some parts of the West in the early days were based on such observations at dens. Even today the stories of increasing populations of rattlers seem usually to be premised on the greater publicity now being given to the discovery of dens.

Rarity of Particular Species

It goes without saying that some rattlesnakes are less common than others, less common even in their areas of greatest abundance. But it is doubtful that any is rare in the sense of being really few in numbers throughout its range.

There are subspecies of rattlesnakes—*miquihuanus, vegrandis,* and *oaxacus,* for example—of which but two or three specimens have ever been secured for study by scientific institutions. But, so far as we know, these subspecies are rare only because the places where they live are difficult of access; there is no reason to believe that specimens could not be collected in considerable numbers if one could readily reach the localities where they occur.

Population Trends

A downward trend in rattlesnake population has been noted in published statements as well as in observations from my correspondents.

The persistent enmity of man and his domestic animals usually results in a sharp decrease in rattlesnake populations about human habitations. This has been a common observation since colonial times. It is only where there is some inaccessible refuge that can serve as a reservoir— some rocky cliff, impenetrable marsh, or dense expanse of chaparral— that rattlers are able to maintain themselves near populous places. When they fail, a decline is to be expected, even though the agricultural activities of man may serve to sustain an increased food supply in the form of a larger rodent population.

The more important population-reducing agencies incidental to agriculture and other human activities are as follows: elimination of refuges and cover, such as rock piles, fallen trees, brush, marshes, and rodent holes; increase in domestic animal enemies, particularly dogs and hogs; reduction in food supply by rodent control; direct destruction by agricultural machinery, especially the disc plow and mowing machine; and the installation of smooth-surfaced, high-speed roads, fatal traps for almost every snake crossing when the traffic is heavy.

Proportion of Rattlesnakes in Snake Populations

In many areas, rattlesnakes are the commonest snakes; indeed in some they may exceed all other kinds combined. Unfortunately, accurate population statistics covering snakes of all kinds are lacking except from a few areas, and even from these, the results can only be approximations, so different are species of snakes with respect to size and conspicuousness. One study, conducted in San Diego County, California, showed that although the relative proportion of the several species varied from zone to zone, the total proportion of rattlers remained about twenty percent of the entire snake population. Censuses in other areas have produced percentages ranging from eight to eighty.

Composite Populations

There are extensive areas of the United States and Mexico in which two or more species of rattlesnakes occur together. Sometimes they occupy the same ecological niche; in other cases a careful survey will reveal slight separations. It is particularly profitable to study the conditions along the edges of the ranges of two overlapping species, for here the slight territorial differences that permit one species to extend its range beyond that of another may be clearly manifest.

Certain composite populations of rattlers are only slightly competitive, because of differences in size, prey, or ecological preferences, while others are fully competitive. Yet I know of no evidence of their obstructing each other in any way or attempting to protect their feeding territories. With respect to denning proclivities, it is to be noted that most composite populations are found in territories where the season of hibernation is so short that individual refuges, rather than extensive dens, are the rule; or they occur in mountain areas where suitable refuges are so close together that the snakes need not travel far to find one. Such meager evidence as is at hand seems to indicate that, in areas inhabited by two or more species, they seek separate winter refuges. This, if it be verified by future observations, would indicate a greater antipathy between rattler species than there is between rattlers and harmless snakes, for that the latter often hibernate with the rattlesnakes there is no question.

However, the enmity between rattlesnake species, to whatever extent it exists, is not an active one. I was told that a southern Pacific and a red diamond rattler were found coiled together near Laguna Beach, California; and at the San Diego Zoo a male southern Pacific mated with a female red diamond and produced hybrid offspring. Recently a *ruber* × *helleri* hybrid was found in the wild. Captive rattlesnakes of different species, whether of the same or opposite sexes, when caged together at the Zoo have never shown the slightest animosity toward each other. They dwell together in peace, although they may accidentally strike each other in the excitement of being offered food.

Areas having several different kinds of rattlers are usually characterized by a rough terrain and a complex vegetation. In such a place appropriate microniches may be found suitable to species normally preferring slightly different habitats; different microclimates may be at hand, with adequate sites for basking; and finally the diverse ecological conditions may produce a diverse food supply.

HIBERNATING DENS AND THEIR USES

One of the most interesting features of rattlesnake life is their practice of assembling at particular points—dens, they are called—for their winter hibernation. Rattlesnakes are by no means the only snakes that gather into restricted refuges from widespread summer ranges for the purpose of hibernation; yet the great numbers involved and their prominence, as they lie about the den entrances in the last sunny days of autumn or first warm days of spring—which I have termed the "lying-out" periods—have resulted in a marked focus of attention on the practice.

These denning proclivities are more conspicuous in northern areas, or at higher altitudes, for here fully protective refuges are more necessary and population concentrations are greater. In the south, where a shorter season of hibernation prevails, or where it may even be interrupted by occasional warm spells, the snakes take advantage of any convenient hole or rock crevice. In these more makeshift situations only a few rattlers may gather together for the winter; or they may even seek separate shelters.

The denning urge in rattlesnakes seems to be entirely predicated on temperature, a fall of temperature in the autumn starting them toward the dens.

Physical Character of Dens

In a discussion first of the physical character of the dens, it is to be noted that, where rocky formations are available, the snakes seek deep caverns or crevices; but in the plains areas they are forced to use the holes of mammals, particularly those of the gregarious prairie dogs.

The degree of cold reached in any area not only affects the duration of hiberation but also the nature — the depth, particularly — of the refuge. The following observation of the prairie rattlesnake from a correspondent will give an idea of the habits of denning snakes and the characteristics that a den must have to afford them adequate protection:

Fully ninety percent of the maturer rattlers congregate annually in groups of from fifty to several hundred to hibernate in the same den. I think about 250 is the average number hibernating together, though I know of several times this number using some dens. The first frost starts them toward their winter refuge. Slowly but surely they find their way back. The old snakes seem to lead the way and the young ones, that is, the one- and two-year-olds, follow their trails. I have watched them coming to their dens for hours. It is my opinion that the majority of the young-of-the-year do not go to the dens, except where the older snakes hibernate in prairie-dog towns.* I think these little ones seek ground holes and similar places of refuge. In fact, I occasionally find that even adult stragglers survive the winter in small burrows, such as are dug by the striped gopher.

There are two requirements essential for a den, depth and dryness. Snakes hibernate only in places that are not subject to floods, and in dens penetrating to a depth below the frost line. They must find such holes or cavities as nature provides. Prairie dogs and badgers are the only burrowing animals that dig deep enough holes; consequently on the open prairie many groups hibernate in dog towns. In more irregular or rocky areas, the best accommodations are offered by slides on steep bluffs, cracked-open banks along deep gullies, sinkholes or washouts that have ceased to conduct water, crevices in rock ledges, caves, or porous scoria deposits.

It is conservative to say that 90 percent of the dens are protected from the north wind, that is to say, they are on the southern slopes of the bluffs. Even in prairie dog towns, the holes are selected that are especially favored by the warm sun. Most of the large colonies occupy dens in the rough areas and near the tops of high ridges or buttes.

Where the soil is light and subject to drift, dens are likely to be more or less temporary, for rattlers do not keep the galleries cleared as would the original prairie-dog occupants. Prairie-dog holes extend for fifty feet or more, and go down to a depth of ten to twelve feet. The holes at a den are rounded out and worn slick by frequent entry, and any one with experience will readily recognize a den.

When prairie dogs abandon a town to rattlers, the burrows gradually fill up. Eventually they must be abandoned for new localities. Rattlers normally select the burrows that are at slight elevations, on southern slopes and well protected

*It may be suggested that the greater dispersion of hiding places in a dog town may protect the young from being crushed, as they would be in the crowded spaces of a rock-crevice den.

from flood waters. They usually take over a small section of a dog town and occupy the same holes as long as they are tenable. Then they take possession of other burrows and force the dogs to withdraw and dig new holes, and so on.

Den Populations

The number of rattlesnakes inhabiting a den depends on topography and climate. Where suitable sites are widely separated, populations at specific dens tend to run larger; but where adequate rock slides and crevices are numerous the snakes do not congregate from such distant points. Yet even where good refuges are close together, the gregarious nature of the snakes will lead to the selection of particular sites so that the concentrations are fairly large. It is only along the southern border of the United States, where the season of hibernation is quite short, that the selection of a refuge becomes more a matter of individual than group choice.

The several subspecies of the western rattlesnake *(C. viridis)*, especially in the northern parts of their ranges, undoubtedly attain the greatest den concentrations. Data on den populations supplied by my correspondents indicate that from fifty to a thousand rattlesnakes have been found congregated in a single den.

Persistence of Dens

That the same sites are used as dens over periods of many years, if the population is not exterminated by man, is evident from the observations of my correspondents.

These seemingly permanent dens are in rock formations. As has been pointed out, prairie-dog holes gradually become filled up by drifting sand or earth, for the new reptilian tenants cannot keep the galleries cleared as did the original excavators. Thus, from time to time, the rattlers must occupy new towns and dispossess the dogs. But in rocky sections the situation is different, and it is evident that the same site, once having been found satisfactory, may be used every winter for many years.

Several correspondents have observed dens that have been used for up to twenty-five years. We have no knowledge of what stimuli lead rattlesnakes to the same refuge year after year.

Spacing of Dens

The distance between dens, affecting the distance that the average rattlesnake must traverse to reach its winter retreat, is a matter of interest, but dependable data are not at hand. This will no doubt eventually be studied by the marking of snakes, thereby ascertaining their wanderings. It was once stated that rattlers travel from twenty to thirty miles to reach their dens, but that they travel such a great distance is to be seriously doubted.

I judge from the comments of several of my correspondents that in rocky areas where suitable crevices are likely to be closely spaced, the dens are so selected that the tenants seldom need travel over a mile to reach one. In severe climates, the distance traveled to a den varies with the intervals between suitable accommodations. The distance traversed is probably rarely over a mile or so at most, for otherwise the fatalities resulting from sudden cold snaps would be excessive.

The story is told of how a well-known trapper of pioneer days once took refuge from a violent thunderstorm under a shelf of cliff. Rattlesnakes started to go by, first in small groups, and finally in hundreds and thousands, on their way to a den. The ground was alive with them, and he and his horse were cornered until the swarm had passed.

One writer believes that rattlesnakes have courses to their dens that they persist in following, despite changes in surroundings. One such trail was through a pine forest that the canebrake rattlers used even after the forest was cut down; another was through a suburban section of Augusta, Georgia.

Life at the Dens

Since a rattlesnake den is a place of refuge and concealment, little is known of the life of the snakes during actual hibernation; but road and mine excavations that have opened dens in winter have supplied some observations. Safe in their retreat below the frost line, the snakes are found to lie torpid and virtually motionless in groups of masses—"balls" as they are often termed—until aroused by the spring warmth. These balls of rattlesnakes have also been observed on the surface during the lying-out periods. One writer saw masses of rattlers as large as wash tubs. Another saw a ball of prairie rattlers as large as a watermelon. These balls are said to contain from two or three to hundreds of snakes.

It has been suggested that these masses of rattlesnakes are composed of snakes indiscriminately mating or endeavoring to mate, but there is nothing to substantiate such a presumption. Balls of rattlesnakes within a den, by reducing the surface-mass ratio, would minimize heat and moisture losses and would save space as well. There is thus no reason to question such observations as balls one foot in diameter or which contain a hundred snakes.

However, there is less likelihood that the balls of snakes seen outside the dens during the lying-out period represent accurate observations. Presumably there is here some exaggeration, such as might be induced by the exciting circumstance of coming upon a pile rather than a ball of rattlers at the entrance to a den. For, under such conditions, balling up would defeat in some degree the primary reason for lying out, namely to be warmed by the sun. But that the snakes sometimes pile up in veritable masses at the entrance to a den there can be little doubt.

Upon their habits on the surface during the lying-out period, more complete field notes are available. For, at midday, when the weather is favorable, they gather about the dens, separately or intertwined in groups, to take advantage of the last warmth of the autumn, or the first reviving sun of spring. Usually they lie close to the holes so that they may quickly find sanctuary if danger threatens. Coming unexpectedly upon one of these concentrations of rattlesnakes is sufficient to startle even the hardiest hunter or fisherman. They are particularly striking sights at prairie-dog towns, where the topography is such that more can be seen from a single vantage point than is possible amid tumbled rocks and boulders.

Several of my correspondents have commented on the wariness of the rattlers during the lying-out period. At the first disturbance or threat of danger they take refuge in the holes or crevices. The slightest movement of an intruder causes one or more to rattle and the alarm spreads contagiously. Since rattlesnakes are deaf, it is probable that their sensitivity to earth-borne tremors suffices to transmit the alarm to those that have not actually seen the intruder.

Animals That Hibernate with Rattlesnakes

Prairie Dogs and Owls. That other animals hibernate with rattlesnakes is well known, several kinds of harmless snakes being the most frequent cohabitants. Before discussing these, it may be well to dispose of the ancient myth of that happy family of congenial lodgers, the prairie dogs, burrowing owls, and prairie rattlesnakes. The story started, of course, from the sight of prairie dogs, owls, and snakes scattered at the mouths of the prairie-dog holes, from which observation it was supposed that they were peaceful tenants of the same holes. But its truth was denied at a relatively early date by some of the western travelers, one of whom gave this particularly effective summary of the relationship in 1893:

> Rattlesnakes [are] dangerous, venomous creatures; they have no business in the burrows, and are after no good when they do enter. They wriggle into the holes, partly because there is no other place for them to crawl into on the bare, flat plain, and partly in search of owls' eggs, owlets, and puppies to eat.

One writer reported that a prairie dog would not enter a hole containing a rattler. He put carbon disulphide on a dog, upon which it started down one hole and then returned. It then went down another hole and stayed. He put the volatile, ill-smelling fluid on a corncob and shoved it into the first hole, whereupon a rattler came out. At least twenty-five ground owls' nests were dug out by another writer, who found rattlers in none of them.

Upon this subject my own correspondents have made the following observations:

☐ I have seen rattlesnakes take possession of a prairie-dog town in the fall; all the dogs, owls, and rabbits moved out when the snakes moved in.

☐ Recently I parked my car centrally in a small prairie-dog town where the snakes were dispossessing them. The dogs wouldn't enter their holes, that is, most of them wouldn't. I could see several snakes at a time, apparently exploring the various holes to determine which suited them best. After watching them for more than an hour, I slowly and quietly walked around and found that the dogs were so confused that I could walk right up to them.

Other Snakes. So inherently vicious are rattlesnakes popularly — but erroneously — supposed to be, that visitors at the Zoo are frequently heard to express surprise that more than one can be kept in a cage without a fight ensuing. But not only are they peacefully gregarious, other genera of snakes may join them in their winter seclusion. At the San Diego Zoo we keep a mixed cage of rattlers and bull snakes to illustrate this amicable association.

Some of the snakes reported denning with rattlers are bull snakes, gopher snakes, milk snakes, racers, garter snakes, and copperheads.

The benefits of heat and moisture conservation that rattlesnakes secure by aggregation would likewise accrue to any other snakes that den with them, which sufficiently explains their presence.

Other Animals. It is probable that the associations with other animals that have occasionally been reported are more fortuitous. It is to be doubted whether these go beyond the other animals' accidentally selecting an adjacent refuge. But here are some examples:

In one rattlesnake den we found thirteen rattlesnakes, four turtles, two skunks, and a swarm of bees with a lot of honey. We had trouble getting the snakes out because of the bees and had to abandon this den. We have also found mice, mouse snakes, and coachwhips in the dens with rattlesnakes.

AGGREGATION AND DISPERSAL

Advantages of Aggregation

The question naturally arises as to what advantages rattlesnakes gain by this gregariousness at the time of hibernation. Several possible benefits may be mentioned. One advantage is group rather than individual experience in the choice of suitable refuges; and the ability to locate the same den year after year, transmitted from adults to the young that trail them. There may be some protective advantage against enemies when the snakes are congregated about the dens, since they seem able, possibly by movement, to transmit a sense of alarm to each other. For those that mate in the spring, the winter concentration is of definite value in providing mates. And finally, as previously mentioned, the balling up within the den, since it decreases the ratio of area to mass, has the advantage of conserving both moisture and heat.

With large population increases, den concentrations might become detrimental. Some inmates might be crowded out into only partly protected fissures; others might have their bodies fatally entangled. Some individuals may be weak and emaciated when they issue from the dens, and some of the females carrying eggs into their second active season may have suffered injury. It is possible that overcrowded dens may serve to limit population growth. Certainly hibernation is a season of hazard.

Summer Ranges and Territoriality

Closely related to the spacing of hibernating dens, and their repeated use by the same individuals year after year, are questions of the homing instinct, territoriality, and similar factors that have been shown to be of such importance in the lives of many birds and mammals. It is known that rattlesnakes disperse from their winter quarters and spread out through their summer ranges. Even if the average distance traveled be less than a mile, it will be obvious how scattered the summer population must be compared with the lying-out concentration at a den. At a den, the population may reach one thousand rattlers, crowded within a space of less than an acre. If the same den represents the assemblage of the rattlers from a mile around, the population density in summer would be one rattlesnake on each two acres, which shows why summer snake collecting is so meager in yield compared with den raiding.

One problem that arises with respect to the summer activities of rattlesnakes is the extent to which they adhere to some summer operating headquarters—if, indeed, they have such a base of operations—and the space that they cover in their summer rambles in search of food. There is also the question of whether or not they exercise what is known as territoriality, that is, defend an area against other rattlesnake trespassers.

So far, little work in these areas has been done on rattlesnakes. One writer expressed doubt that the sidewinder had any regular range or even a permanent refuge, but thought rather that it traveled over the desert at random, moving as far as two thousand feet in a single night. Yet a professional snake collector operating in some Southern California sand dunes thought that each western diamond *(C. atrox)* had a more or less permanent refuge in a particular mesquite thicket. A snake would travel to and from this each night over a definite route from bush to bush, following almost the same course on its return, thus establishing a regular trail used repeatedly by the same snake. In one case a pair of rattlers traveled together over parallel tracks. Although marking and recapture studies do not demonstrate the existence of a homing instinct, that is, a tendency to return when removed from an accustomed range, they do indicate a definite territorial conservatism, that is a restriction of wandering or prowling to a relatively small area.

From the somewhat inadequate data available, it appears that rattlesnakes probably have home ranges, or at least favorite refuges to which they habitually or occasionally repair, but that they do not have a defended territory, from which other rattlers are driven away.

Accidental Dispersal. Migration

In addition to the seasonal population movements just discussed, there may be forced or natural migrations over greater distances. Rattlers are known to swim readily. Since they are quite buoyant and difficult to drown, it is to be supposed that many would survive a turbulent river-flood, even if there were no brush or debris upon which to ride. Thus there is little doubt that floods involve an important means of dispersal. It is, however, to be noted that the involuntary colonists often find the new habitat conditions unsuited to them and therefore a permanent range-extension is not established.

Aside from the dispersal by floods, presumably the migration of snakes into new territories largely results from their seeking an increased food supply, a much slower type of dispersal than that produced by a flood. Sometimes they follow the rodents when these are increased by more suitable conditions resulting from irrigation.

Range records are confused when tourists have hauled animals about, and then, having tired of them, have released them far from their natural haunts. A frozen diamondback was once found in Maine in December, evidently thrown from a passing automobile. This species does not occur naturally north of the Carolinas.

Others escape from zoos and circuses. A colony of western diamond rattlesnakes *(C. atrox)* was said to have become established far from the normal range of this species. It appeared that the snakes originally escaped from a circus.

There is little doubt that the presence of one of the two subspecies of the western rattlesnake *(C. v. viridis* and *C. v. nuntius)* in the vicinity of the Hopi villages in northeastern Arizona, has resulted from the gathering of snakes for the snake dances over a rather wide area—they may be brought in by visiting Indians from a considerable distance—followed by their release after the ceremony. Only *nuntius* is native there.

Public exhibitions of rattlesnakes were at one time forbidden in France because of the fear that a pair might escape and start a colony. In 1940 the city of Berkeley, California, adopted an ordinance making it unlawful to keep or transport venomous reptiles in the city without a permit. This was enacted for the purpose of preventing the escape and colonization of dangerous snakes not already indigenous to the area.

CHAPTER SEVEN

Food

KINDS OF FOOD

Mammals

It is probable that every species of mammal living in the same territory with some kind of rattlesnake, provided it is small enough to be swallowed, occasionally falls prey to these snakes. Even if the adults are too large, the young may form a regular part of the diet. For a hungry rattler, with his death-dealing fangs, can quickly subdue even so fierce and bloodthirsty a fighter as a weasel, although such carnivores are not regularly eaten.

Because of their convenient size and population density, the following genera of rodents probably comprise the bulk of the food supply of rattlesnakes: white-footed mice, pocket mice, grasshopper mice, meadow mice, harvest mice, kangaroo rats, wood rats, and chipmunks. To these there may be added, as important elements in the food supply of the larger rattlesnake species, prairie dogs, ground squirrels, squirrels, and cottontail rabbits, the young of these larger mammals being particularly subject to predation.

It is to be expected that most pit vipers, including the rattlesnakes, should feed on warm-blooded prey, since they have the facial pit to facilitate its detection.

The method by which rattlesnakes generally seek their food—that is, by lying in wait for and lunging at passing creatures—coupled with a not especially acute sense of vision, occasionally results in their eating creatures quite different from their customary prey. A few unusual mammals have been reported, such as minks, skunks, opossums, and bats.

As to the maximum size of prey, one writer states that a large western diamond *(C. atrox)* can eat a full-grown jackrabbit, and one of my correspondents has voiced the same opinion; but I think this needs confirmation. Certainly adult diamondbacks can swallow a full-grown cottontail.

Birds and Eggs

Birds comprise an appreciable part of the diet of many rattlesnakes, particularly those that live in areas where ground-nesting birds are plentiful.

Rattlers can become a nuisance around breeding farms where game birds are raised, although, since they do not excel as fence climbers, they are not nearly so troublesome as some of the harmless snakes, such as the gopher snakes, king snakes, and racers. Studies so far have shown it doubtful that rattlesnakes can be considered one of the important predators of wild game birds in any area.

Rattlesnakes occasionally eat eggs, but to a lesser extent than some of the other large snakes, such as those already mentioned.

One author told how a "yellow Pacific rattlesnake"—probably a Mojave rattler *(C. s. scutulatus)*—climbed five feet into a mesquite tree and robbed a bird's nest of its eggs; however, the rest of the paper is so inaccurate as to invite skepticism regarding this episode. Another found an eastern diamond rattlesnake *(C. adamanteus)* in a woodpecker's hole. Upon examination, the snake was found to contain the mother woodpecker and five eggs.

A number of my own correspondents have reported the eating of eggs by rattlers. A member of the U.S. Forest Service in Wyoming discovered a prairie rattlesnake *(C. v. viridis)* in his hen house; the stomach, opened because of a suspicious bulge, was found to contain an unbroken hen's egg.

Reptiles

To a considerable degree, lizards comprise one of the normal food items of rattlesnakes. This is especially true in the Southwest, where lizards are plentiful, and it is also characteristic of particular rattlesnake species. The smaller rattlesnakes are habitual lizard eaters, since the attenuated form of the lizard allows it to be more easily swallowed than a mammal, which is thicker-set for the same bulk. This is one of the reasons why the smaller rattlesnake species are confined to areas with adequate lizard populations.

The adults of some of the small species continue to prefer lizards even after they have attained such a size that they could readily eat adult mice. However, studies have shown that, in general, there is a trend from lizards toward mammals as a basic diet, as the snakes age and thus grow to a size such that mammals within their swallowing capacity are more plentiful.

Other reptiles are rarely eaten. One writer threw a small land tortoise into a rattler's cage. It disappeared and he assumed it had been eaten, but this is to be doubted.

Rarely will a rattlesnake eat another snake, though in captivity, rattlesnakes are occasionally cannibalistic, a result that may happen when two rattlesnakes start to eat the same mammal. When this occurs, one is quite likely to swallow the other, the swallower being the larger, or, if about the same size, the one which, in advancing its jaws along the prey, happens to take the first bite that puts its upper jaw over the edge of the cage-mate's. This danger is well known in zoos, and therefore care is taken, in feeding, not to permit two snakes to seize the same animal.

Sometimes cannibalism among captive snakes is not the result of two snakes seeking the same prey. At the San Diego Zoo, one prairie rattler *(C. v. viridis)* ate another that had been a cage-mate for several years. Both were adults and of similar size. The meal proved too large and was regurgitated. Sometimes one rattler will eat another when they are crowded together in a shipping container.

Cannibalism among rattlesnakes in the wild is probably quite unusual; and, in the rare instances when it has been reported to have occurred, the snake eaten may have been dead before it was swallowed.

Amphibians and Fishes

It is probable that all species of rattlesnakes occasionally eat frogs or toads. To several species inhabiting humid areas, they constitute a more important source of food than lizards.

Two of my correspondents have mentioned fishes as rattlesnake food. One claimed to have seen a prairie rattlesnake *(C. v. viridis)* in the act of eating a trout about eight inches long; the other observed a western pigmy rattlesnake *(S. m. streckeri)* catching small minnows in a shallow slough. Northern Pacific rattlers *(C. v. oreganus)* have been reported to catch salmon and trout. This report requires confirmation.

Arthropods

Insects and other arthropods are rather frequently mentioned as being among the creatures eaten by rattlesnakes. Of a few of the records there can be no question; but, as to most of the others, it is probable that the insects were ingested as the stomach contents of other prey such as lizards and toads, and then outlasted their original captors during the digestive process in the rattlesnake. It is significant that insect remains are most frequently found when there are lizard scales in the food residue.

Vegetable and Mineral Foods

No snake eats vegetable food, but occasionally small bits of vegetation, such as leaves, twigs, or sticks, are found in the intestinal tract, swallowed by accident or ingested with prey that had fed on vegetable matter.

When a rattlesnake is observed scraping prey on the ground to dislodge it, when it has become caught in the angle of the jaw while being swallowed, we can readily see how such foreign material may be ingested. It is not unusual to find coarse sand in the lower intestinal tract of rattlers.

METHODS OF SECURING PREY

If there is any predominant method by which the rattlesnake secures its prey, it is this: he lies in wait by a trail along which some small animal is likely to pass. Aided by a keen sense of smell, a heat receptor (the facial pit), and to a lesser extent, by sight, he becomes aware of a passing mammal. He makes a forward-lunging strike at the prey. At the end of the stroke his fangs seem barely to touch the victim. The head is drawn back and the snake waits. The bitten prey stumbles forward on its way for a few feet, or maybe for several yards, if it be large or the strike has not been fully effective. The victim seems to hesitate uncertainly, then loses control of its movements. A few convulsive kicks and it is dead, only a few seconds or, at most, some minutes after the fatal stroke; for the venom injection is large in proportion to the bulk of so small a creature. Meanwhile the rattler, after waiting for a short time, uses the sense of smell to follow along the course taken by the stricken animal. Reaching it, he carefully touches it here and there with his tongue, as if to find whether there remains any danger of a retaliatory bite or struggle. Finally, he seizes it by the nose and the actual swallowing proceeds.

A number of observers have recorded the actions of rattlers that were trailing stricken mammal prey. The snakes are described as being alert but unhurried, sensing their way along the trail of the prey with the flickering tongue that supplements the sense of smell.

Rattlesnakes not only lie in ambush along trails where prey may be expected to pass, but actively search for food down burrows or in rock crevices in which animals are accustomed to seek refuge. The snakes are evidently able to determine by scent whether a hole is likely to prove fruitful.

Sometimes hunting is thriftily combined with protection; a snake, while taking refuge in a mammal burrow from the sun or possible enemies, is advantageously located to make a meal of the owner.

While many animals are caught unawares by poised and waiting rattlers within whose striking range they pass, others are frightened into a fatal proximity by some other threat, a lurking coyote, for example, or a hovering hawk. If the prey fails to come within striking range of a hungry rattlesnake, the snake will sometimes trail it, trying to get close enough for a strike.

A number of creatures have been found dead in the field from rattlesnake bite, some because the rattler was unsuccessful in finding them after a strike, some because the snake was not hungry.

Some writers of bygone days have questioned whether poisonous snakes—rattlers among them—bite and inject venom into their prey, for it was thought that poisoning the prey would render it unfit for food. One claimed that he saw a rattlesnake swallow a bird up to the point where it had been bitten, but upon reaching this poisoned section the snake stopped and threw it up. But modern studies, both in the field and laboratory, have left no doubt concerning this purpose and use of the venom; and as to a possible detrimental effect on the swallower, it should be observed that venom is virtually harmless when taken internally—it must enter the blood stream to be injurious. The theory that the deteriorating effect of venom on the tissues of the prey actually aids in digestion, a theory advanced in 1787, is now generally accepted. The protective value of the venom, in making it unnecessary to hold onto and subdue a struggling creature that might injure the snake, was also early recognized.

Variability in the Effect of a Bite

Just as there is a great variability in the effects of snake bite on a human being, so also observers have noted differences in the results on prey. When we consider that the danger from a given quantity of snake venom is inversely proportional to the bulk or weight of the creature bitten, it might be supposed that rattlesnake bite would be quickly fatal to the small creatures on which they feed. While such is generally the case, there have been instances wherein stricken animals have been observed to cover considerable distances before succumbing, and others in which they have escaped and may have recovered.

This occasional ineffectiveness may be attributed to two particular causes affecting prey, among the many variables inherent in snake bite. First, rattlesnakes have full control over venom extrusion; they can, if they wish, bite without injecting venom at all. Their natural inclination is to conserve venom, since it is their means of food procurement, and it is possible that, in striking prey, they release much less than when under the stress and fear of defending themselves against some such large creature as man. Secondly, the very lack of bulk of the prey acts to the snake's disadvantage since the momentum of the strike carries the prey along, without the interposition of sufficient resistance to permit the fangs to be deeply imbedded. Watching rattlesnakes strike small mammals in captivity, one cannot but be impressed by what seems to be the lightness of the touch. Except for the effects of the venom one will frequently think that the snake has missed — as indeed it often does, either by misdirection or falling short — for the target may not be appreciably moved by the blow.

Differences in Dealing with Prey

Although exceptions have been noted, there seems no doubt that rattlesnakes usually release a mammal which has been bitten but retain hold of a bird, reptile, or amphibian. In explanation of these differences it may be presumed that the mammal is released to avoid the danger of a retaliatory bite; the snake can readily follow its scent to the place where it dies and thus will seldom lose its prey. But this would not be true of a bird that might fly a short distance and be lost; or of reptiles and amphibians, on which venom acts more slowly so that they might die too far away to find. Also, the retention of a hold permits a bite, rather than merely a stab, and hence a greater assurance of venom injection.

Rattlesnakes may occasionally be injured by their prey, despite the habit of releasing mammals. This has been shown by a study in which a rattlesnake was turned loose one day and recaptured the next. During the one-day interval between captures, this snake had caught and eaten a pocket gopher, and, in so doing, sustained bites on the anterior part of the body. In this case, the bites were not serious, for the wounds were healing when the snake was again captured a week later.

Constriction

Although several of my correspondents have written of seeing rattlesnakes constrict their prey, I am disposed to doubt the accuracy of these observations. For, in the thousands of times that rattlers have been observed in the course of being fed at the San Diego Zoo, no rattler has been seen to use this method. Constriction is, of course, a customary way of holding and subduing prey used by many kinds of snakes, but is not needed by those having the powerful alternative, venom. Rattlers, while swallowing prey, do not even steady or hold it down with a part of the body, as is the practice of some of the racers. They do, however, form a loop in the neck and drag the prey backward. This method of eating, in which the head and a short section of neck face backward along the major length of the snake's body (which slowly advances forward), is characteristic of many snakes, constrictors as well as nonconstrictors. Sometimes a rattler will take his prey to the shade of a bush before swallowing it.

One of my correspondents mentions that he found, in a rattler's stomach, a rabbit that he thought was both longer and slimmer than a normal rabbit, from which he judged it had been constricted before being swallowed. This is in reality a usual condition of prey, caused by the swallowing process as well as by partial digestion; it may be one source of the widespread belief that rattlesnakes constrict their prey.

Method of Swallowing

The size of prey that any snake — rattlesnakes included — can successfully engulf, is a constant source of astonishment, especially because a snake can neither chew nor dismember its food but must swallow it whole, and must accomplish this unaided by either hands or feet. It is made possible by the elastic nature of the jaws and throat, and particularly by the fact that the halves of the lower jaw are flexibly attached to each other at the front, rather than being rigidly united as in lizards and mammals. Also, the bones of the jaws are thin and pliable. This elasticity of the whole structure allows distension of the mouth aperture, while the partial independence of the jaw sections permits the advancement and retraction of each half separately. Each is equipped with fine, curved teeth, whose points are directed backward; they comprise, in effect, four rows of hooks, each of which may be independently advanced, imbedded in the prey, and then pulled back, dragging the food with it toward the gullet. One or two half-jaws are advanced while the others hold the prey to prevent it from slipping out of the mouth. Thus, by a properly synchronized series of ratchet-like motions, the snake pulls the prey in, past the stretched constriction of the mouth and throat. Once past the point of narrowest opening, a series of lateral undulations of the throat is used to force the food down the gullet into the stomach.

In swallowing large and difficult prey, rattlesnakes have one advantage over nonviperine snakes; they can and do use their fangs as hooks to aid the small, solid teeth in dragging the prey into the throat.

When not in use the fangs lie folded up against the roof of the mouth. When it is desired to use them, either for biting or to draw in prey, the maxillary bone to which each fang is attached is rotated so that the fang swings downward and forward until its base is perpendicular to

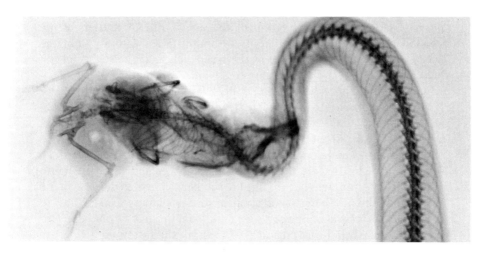

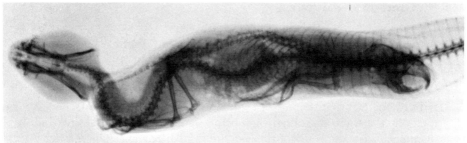

Fig. 24 X rays of a red diamond rattlesnake (*C. r. ruber*) swallowing a hamster. The jaw and throat distension is made possible in part by the lateral outward spread of the angle of the jaw. Note relative positions during (upper figure) and after swallowing (lower figure). (X rays through the courtesy of Charles M. Bogert, American Museum of Natural History, New York)

the roof of the mouth. The two fangs, one on each side, can be manipulated independently. By this means, first one and then the other is hooked into the animal being swallowed, and then each is both rotated and drawn back so as to pull the animal in. The fangs are more effective than the solid teeth because of their ability to rotate and also because their much greater length insures engagement through hair or feathers. Sometimes a rattler, if it has trouble in swallowing a large animal, will rub the prey against the ground or a rock; this helps to release a leg or wing that may have become caught in the angle of the snake's mouth.

The length of time required to swallow prey depends on the relative sizes of prey and snake. A small mammal may be swallowed in a minute or less, while a large one may require an hour or more. In the case of large and difficult victims, the snake will be seen to tire, taking lengthening rests between the rhythmical ratchet movements of the jaws.

Even when a rattler holds its mammal prey, instead of letting it run until it dies after a strike, the snake will often drop it and nose it carefully before beginning the swallowing process. This is evidently for the purpose of getting an advantageous hold of the creature's snout or head so as to facilitate swallowing, for prey is usually swallowed head first. The head-hold reduces the impediment to swallowing imposed by hair and limbs. Most often a side of the prey is upward, rather than its back. Sometimes a snake will take two or three tentative holds before securing one that seems satisfactory. Even so, owing either to the eagerness of hunger or to an inability to recognize the head, a mid-body hold is sometimes taken. This makes it necessary for the snake, either to double the body of the prey, or laboriously to transfer the hold, by lateral movements of the snake's jaws, to the head of the prey.

After a snake has swallowed its prey, it will usually yawn as if to relieve the stretched mouth parts and settle them again into a comfortable position. During this action it may alternately advance the fangs and refold them once or twice.

The mechanism whereby a snake can breathe, even though its mouth and throat be crowded with a large animal, is provided by the structure of the trachea or windpipe. This opens near the anterior tip of the lower jaw and is not collapsible. The glottis (the outer end of the trachea) is not fixed in position; it may be caused to extend out into the air below the prey to serve as an air intake while the swallowing proceeds.

Disgorging Prey

Rattlesnakes have no difficulty in disgorging prey that has been completely swallowed, as anyone knows who has taken the trouble to force-feed a snake, only to have the food shot out in a fraction of the time it took laboriously to work it in. But during the course of the swallowing process a rattler may sometimes have difficulty in getting rid of the food. For example, a writer reports finding a small southern Pacific rattlesnake with its mouth so completely gagged by a half-swallowed meadow mouse that it was unable to rid itself of the prey and therefore could be handled quite safely.

Fatal Meals

Occasionally one hears of a meal that has proved fatal, the rattlesnake being unable either to engulf the prey or rid itself of the half-swallowed encumbrance. One writer tells of a small young northern Pacific rattler that ate a pocket mouse, and on the following morning its movements were so handicapped by the large meal that it was unable to reach its accustomed shelter and was killed by the sun.

Horned toads are occasionally fatal to rattlesnakes, as they are to other snakes as well. A horned toad when held will swing its head from side to side, as if deliberately seeking to cause injury with the horns; also the horns can pierce the snake's throat as the snake tries to swallow it.

Reactions of Prey

While the fascination of birds and mammals by rattlesnakes is no longer believed possible by those who have given the closest study to the habits of the snakes, other reactions upon the part of animals that normally serve as the food of snakes are well authenticated. It is not unusual to find birds chattering noisily about a snake. Prairie rattlesnakes have been located by observing a characteristic neck-craning by meadow larks. The actions of turkeys, which, upon finding a snake in a field, circle around it with warning cries, are well known. This reaction has even been used to aid in the collection of snakes.

FOOD REQUIREMENTS

Frequency of Feeding

Studies of captive specimens indicate that mature rattlesnakes thrive on an adequate meal every fourteen to eighteen days. In the wild it may be supposed that the energy consumed in hunting for food and similar activities leads to a slightly greater intake requirement; necessarily the meals secured would be more irregular with respect to both frequency and size. Young rattlers, with a greater need for food during their stage of rapid growth, feed more often than adults, if they can find suitable prey.

Most statements regarding the frequency with which rattlers eat in the wild have been mere guesses. They have varied from every two days to three times per year. But recent field studies of the stomach contents of large series of rattlesnakes are beginning to give us data that, co-ordinated with observations on captive snakes, permit more reliable conclusions. These seem to indicate that two to three weeks approximates actuality, at least at times of major activity.

As to seasons, it is believed that rattlers are especially active in pursuit of food, first, in the spring to replenish the fat consumed during the long period of hibernation; and, second, in the fall to restore any fat depleted by inactivity during summer, in order to be prepared for the next winter's complete seclusion.* However, it is essential, when the food is

---

*I have collected some statistical data on loss of weight by rattlesnakes during hibernation. It was my conclusion that adult rattlers going into hibernation in autumn are about four percent heavier than when emerging in the spring. Juveniles lose somewhat over twenty percent of their initial weight during hibernation. There is virtually no increase in length during hibernation.

secured in the latter season, that there be sufficient warm weather to in-sure digestion before the onset of the cold stops this vital process.

Temperatures, whether seasonal or diurnal, are of great importance in the feeding schedules of snakes. First, they affect the metabolic rate and hence both the food requirements and the rate of digestion; and, sec-ondly, they control muscular activity and hence the ability to hunt suc-cessfully. These two effects are obviously interrelated. There are upper limiting conditions as well; and at both extremes they work together rather than in opposition. Below the optimum temperature the snake needs less food, and has less ability to secure it; above the optimum he will be driven below ground by temperatures he cannot withstand, but is fortunate in that his refuge—some mammal hole or rock crevice—may also be the refuge of his prey, to the latter's undoing.

It goes without saying that hunting must be arranged to coincide with the activities of the prey. Since mammals comprise the bulk of rat-tlesnake food, it is to the advantage of the snakes that rodents are also largely nocturnal or at least active at twilight.

Chance in securing prey entails wide discrepancies in the intervals between meals. A snake that finds a small creature will be satisfied only for a few days and then must search for another; or it may continue to hunt until it has eaten a bulk equal to a full meal. Should a rattler be fortunate in finding a single animal equal to its capacity, there will no doubt follow a maximum interval between feedings consistent with the seasonal food requirement.

Summarizing these effects of seasons, temperatures, bodily require-ments, and the availability of prey, I should express the opinion that adult rattlers of the larger species require a full meal every ten days to two weeks in the spring, every three weeks in the summer, and every two weeks in the fall, with the total meals per annum dependent on the length of the active season.

It is only natural that one should ask why a rattlesnake's schedule of meals is so completely at variance with what we should ordinarily expect from a knowledge of other animals. This results from several different, but interrelated, phases of a single general condition—the reptile's almost complete lack of internal temperature control. It can exert muscular ac-tivity and digest food only when external factors maintain a body tem-perature within a certain range. But, coincidentally, the lack of a tem-perature differential between its body and the surroundings results in lit-tle heat loss by radiation or convection; its digestion is geared to a lower temperature and is slower than that of birds and mammals, for otherwise it could eat only in midsummer; and its muscular activity is less. All of these factors lead to a lower food intake requirement than that of warm-blooded, or endothermic, creatures.

One other condition tends to stretch the average time between feedings. The snake's extensible jaws, and the physiology of its intestinal tract, permit the ingesting of food items that are relatively large compared with the bulk of the snake. The prey of northern Pacific rattlesnakes has been found to average forty percent of the weight of the captor, the range being from 3½ to 123 percent; two-fifths of the body weight would probably be an average adult meal. At the San Diego Zoo we have observed that a rat equal to one-fourth of the snake's weight constitutes a satisfying meal, but one that cannot be considered large as measured by its full swallowing capacity. But these ratios of meal to body weight are all relatively larger than the ratios represented by the meals of warm-blooded carnivores.

One writer has calculated the differences in the food requirements of rattlesnakes as compared with certain warm-blooded predators. He estimates that an adult rattlesnake — although weighing one-fifth as much as a red-tailed hawk — annually eats only as much as would satisfy a hawk for a few days; and a rattlesnake's yearly requirement would furnish but a single meal for a coyote. More frequent feeding is the penalty the warm-blooded creatures pay for their higher metabolic requirements and reduced capacity per meal.

Both the vicissitudes of the rattlesnake's method of securing prey, and the necessity for a long hibernation in most climates, render a considerable fat storage necessary for survival. Nearly all cold-blooded vertebrates are plentifully supplied with such storage. This allows the snake to survive irregularities in the feeding period that may multiply, by many times, the intervals we have assumed to be normal. In fact, it is not unusual, as I shall mention later, for captive rattlers to live over a year without food. There are lobes of fat deposited in rattlesnakes along both sides of the intestines from the stomach to the vent. This is nearly or completely exhausted at the end of the period of hibernation. (It is from this fat that rattlesnake oil is secured.)

Some curious ideas about the digestive processes of snakes have been recorded from time to time. It has been said that rattlers, in dealing with unusually large prey, may swallow it part way, and then wait several days for digestion to dispose of this portion before they complete the swallowing process. This is impossible since the anterior part of the prey would not reach the snake's stomach where digestion could proceed. The opinion has also been expressed that rattlesnakes have no stomachs, the food merely staying in the body until it decays.

Multiple Meals

A newspaper report appeared July 11, 1937, concerning a farmer at Jamestown, Tuolumne County, California, who found, in one of his rabbit hutches, a northern Pacific rattlesnake that had eaten a litter of ten young rabbits.

Several of my correspondents tell of experiences to the effect that a rattlesnake, coming upon several birds or mammals, may strike and kill a number before starting to eat one.

But most of the stomach contents of rattlesnakes containing several food items indicate that they were secured at different times—that the snake, not satisfied with the first animal, continued to hunt for more, sometimes immediately after the first was eaten, sometimes later, but at least before digestion of the first was complete.

Water Requirements

In captivity, when water is available in their cages, most rattlesnakes drink occasionally, and no doubt they do also in the wild.

It has been reported—particularly of the timber rattlesnakes (*C. h. horridus*)—that at times when it is dry in their usual rocky habitats they roam down into the lowlands seeking water. Rattlesnakes drink by putting their snouts under water, making no attempt to keep the nostrils out so they can breathe. As a rattlesnake drinks, a pulsating movement is evident, particularly at the angle of the jaw. The water is forced inward by the opening and closing of the lower jaw; however, the movement is slight, and it is necessary to have one's eye on a level with the snake's head in order to see that the mouth is actually opened when the snake is drinking. A thirsty rattlesnake has been observed to drink steadily for three minutes.

Rattlers have also been seen to drink from wet surfaces—a way of getting moisture from dew practiced by some lizards. A correspondent notes:

A rattler often gets its water off the surface of a rock, from dew on leaves, or the raindrops. I have watched them sip the moisture from a wet board, sucking it in like a vacuum cleaner.

I have watched *pricei* drink water caught in the depressions of lily pads, while the snake was swimming in a goldfish pond with water all about.

The importance of water to rattlesnakes has been a matter of some difference of opinion. Probably there are considerable differences between species, those accustomed to well-watered habitats having a greater need than the desert forms. No doubt a desert dweller, such as the sidewinder, encounters water only on rare occasions and therefore must be largely dependent on its prey for its moisture requirements. As far as captive snakes are concerned, much depends not only on the food supply, but on the temperature and humidity at which they are kept.

A rattlesnake's annual liquid intake, beyond that secured in its prey, has been estimated to be about equal to its body weight.

Rattlesnakes in captivity will almost never drink milk, even though water be denied them, thus exploding a myth that has long been current.

FEEDING RATTLESNAKES IN CAPTIVITY

The food given to rattlesnakes in captivity is likely to yield more information on their methods of killing and swallowing prey, and on the frequency with which they eat, than on the kind of prey taken under natural conditions. It is impossible to offer all of the species of prey occurring in a snake's natural habitat, and therefore stomach-content studies of wild snakes yield better data than observations on captives, which do not show any great discrimination as to the prey they will accept when hungry. This may be of species markedly different from anything available in their native haunts. Rattlers do, however, discriminate between animals as different as rats and toads.

The reptile department of the San Diego Zoo began operations in 1922. We knew nothing then of the feeding of snakes in captivity; and such further handicaps as the lack of an adequate food supply, together with the unfavorable conditions under which the snakes were kept, made for short-lived exhibits, even after we adopted suggestions courteously given by older institutions. With the completion of a new reptile house, and, more particularly, after the advent of C. B. Perkins to take charge, there was a complete reversal of conditions. By the exercise of intelligence and patience he achieved results in breeding and raising snakes in captivity, and in longevity records, of which our local institution is deservedly proud. It is upon Mr. Perkins's observations and results that the following notes are largely based.

Rattlesnakes are nervous creatures, and are thus among the most difficult snakes to feed successfully. There are both species and individual differences in their readiness to accept food; there are even differences between the members of a single brood. In such institutions as zoos, which usually have available many more snakes than can be exhibited because of space limitations, there is a natural tendency to eliminate the finical feeders in favor of those that will eat readily; the attendants have little time to waste on such fastidious tenants, whose poor condition through lack of food soon becomes evident. Only the rarest, irreplaceable species are retained if they fail to eat. But a regular feeder becomes a valued possession, since a good exhibit is thereby assured over a long period.

Effects of Conditions of Captivity

Rattlesnakes will feed more readily in fairly large cages—say with a floor area of four by four feet—than when more confined, although at the San Diego Zoo many of the specimens retained for study, rather than for exhibition, eat readily and thrive in cages only one and a half by two feet. But crowding is to be avoided, since, if the snakes are really hungry and in the mood to feed, it is difficult to see that each gets its share if they can interfere with each other.

Temperature conditions are especially important. At too low a temperature a snake is usually too lethargic to feed, and even if it does, will have difficulty in digesting the prey. One writer observed that at 50° F. or below rattlers that had fed would throw up half-digested food they had eaten about fifteen days before. Too high a temperature causes discomfort and nervousness. Probably a range of 77 to 85° F. is best, with an optimum for most species at about 80° F. This is the temperature at which the cages are maintained at San Diego.

Although a certain degree of stimulation, such as waving the prey before the snake, may be effective in causing it to seize the food, too much excitement, especially of an unaccustomed nature, may have an adverse result, frightening it into refusal. Although some rattlesnakes will feed readily enough in the exhibition cages, others are disturbed by the people crowding up to the window, for which reason a blind is generally placed over the glass when they are fed; this also avoids complaints from the tenderhearted, even though the snakes are not offered live prey.

Dead Food

The feeding of dead prey to rattlesnakes does not offer any particular difficulty and has many advantages, from the standpoint both of the availability of food and of the safety of the snakes. Rattlesnakes are occasionally found eating carrion in the wild so that the use of dead prey is not entirely unnatural. It has been the experience of Mr. Perkins that rattlesnakes which will feed at all, will almost as readily accept dead food as live. In one or two instances individuals had to be started on live food, but later became accustomed to eating dead prey.

When rattlesnakes are fed dead prey, it is much easier to see to it that, in their excitement, they do not strike the wall of the cage or each other. The danger that a snake may be bitten by the prey is avoided. With dead prey it is a relatively simple matter to be sure that each rattler gets its share, and this without interfering with its cage-mates. Even so, the attendant must be on the alert to see that one snake does not take hold of a mouse that another has started to swallow, for if this happens one of the contenders will continue blindly swallowing until it has engulfed the other, sometimes with fatal results to itself because the meal is too large to be digested. At the San Diego Zoo we have had many experiences with rattlers that have tried to swallow the same prey simultaneously, and in one instance the danger was not noticed in time to prevent a tragedy. In this case one massasauga *(S. c. catenatus)* ate another. Although forced to disgorge the meal almost immediately, the swallowed snake died on the following day, and the swallower two weeks later.

Methods of Feeding

A rattlesnake accustomed to captivity and anxious for food will come eagerly to the cage door when it is opened, and with raised head and darting tongue manifest a lively expectation. He will quickly seize the prey held before him with a pair of forceps and retire to a corner to swallow it. At such a time rattlers seem to have at least some of the alert intelligence with which they are credited in the popular mind. As previously mentioned, they even learn to distinguish between attendants, or probably the routine that they follow, showing more interest when the one who usually feeds them opens the door. In order to reduce the excitement when a cage is opened for cleaning rather than for feeding, the snakes are allowed to smell the shovel used for the purpose. They soon learn to recognize this and retire quietly to their corners as soon as they have sensed its presence.

When there are several rattlesnakes together in a cage, even though they are fed dead rats or mice presented to them individually with a pair of forceps, great care is necessary to prevent their striking each other. Such strikes are caused by excitement and primarily result from a desire to seize the food, rather than any wish to inflict injury on each other. Although rattlesnakes are virtually immune to their own or another rattler's venom, a fatality quickly follows should a fang accidentally penetrate the brain, spinal cord, or a vital organ of a cage mate.

It is seldom necessary or efficacious to leave the prey in the cage for more than an hour or so; if it is not accepted at once it seldom will be later. However, with some individuals, leaving the food overnight has proved effective, for most rattlesnakes are nocturnal, particularly in summer. Sidewinders *(C. cerastes),* the most consistently nocturnal of all rattlers, seem to show a preference for night feeding in captivity.

Every beginner must be cautioned never to leave a live rat overnight in a cage with a rattlesnake, unless there is plenty of food available for the rat, for otherwise the rat will kill or injure the snake. Many an amateur has been surprised next morning to find his rattlesnake minus a part of its head or tail; for a hungry rat will get food at any risk, and a rattlesnake, unless hungry and alert itself, seems no match for so active and intelligent a creature. We have had experiences wherein rats ate the tails of

otherwise uninjured rattlers, indicating that the rattler was without sufficient feeling or energy to protect itself.

Forced feeding of captive rattlesnakes is not recommended; it is a
dangerous and seldom-successful procedure that should be resorted to
only in the case of some rare specimen that cannot be replaced. When
forced feeding is employed, it is customary to push several mice down a
snake's throat by mounting them on a thin skewer or splint. The first
mouse is fixed to the end of the skewer so as to cover the point, thus
preventing injury to the snake; the rest are threaded on the stick behind,
like chickens on a spit. Once they are down, the rattler's neck is constricted with the fingers and the skewer withdrawn. Often the snake will
regurgitate the food. An attempt to prevent this by the application of
some form of tourniquet about the snake's neck is seldom successful; or it
may be too efficient and kill the snake. Sometimes a rattler may be
brought through an illness by the administration of an egg-milk batter
with a syringe, a type of forced feeding possibly justified.

Frequency of Feeding

At the San Diego Zoo young rattlers are fed once a week for their
first year, grown snakes every two weeks. If a young snake, when
changed to a biweekly diet, becomes thin, it is put back on a weekly
schedule. Old snakes that have become very fat, as they sometimes do, are
fed every four weeks. (The fat is not periodically consumed in a long
hibernation as is the case with snakes in the wild, and probably they
exercise less, since they have no need to search for food.) The standard
meal comprises one mouse or rat of suitable size, selected so as not to
stretch the snake's mouth unduly. Newly born mice are fed to the smallest snakes, and the largest rattlers are given the biggest rats available. For
feeding purposes, a mouse-rat farm, comprising albino mice and rats, such
as are raised extensively for laboratory use, is maintained.

Snakes born at the zoo are fed after their first shedding, which is
usually from five to ten days after birth. Adult rattlers, fresh from their
wild habitats, are tempted with food from two to twenty days after estab-
lishment in their permanent cages, the time depending on the degree of
nervousness shown by the snakes. Failure to take food the first time is
never accepted as final.

Seasonal Effects

In the case of captive snakes, shedding does not seem to interfere
seriously with the feeding schedule. Rattlesnakes have been observed to
eat within two days of shedding, and also one day after. Some will eat
when the eyes are in the semiopaque stage that ends about two days before the shedding. Thus, the widespread notion that snakes will not eat

during the shedding period is not borne out by observations of captive snakes, although in the wild, for protection, they probably do not roam abroad while "blue-eyed." After a snake has been fully surfeited with food it prefers to lie quietly in one corner of the cage.

The snakes at the San Diego Zoo are kept warm throughout the winter and consequently remain active, without an interval of hibernation such as they would experience in a natural habitat. They accept food as readily in the winter as in the spring or summer.

Individual and Species Differences

Some young in a brood will eat readily, whereas others refuse, for reasons that are not apparent. This has been found true at San Diego of a number of broods of several different species.

At San Diego, we have found some species differences among adult snakes. Speckled rattlesnakes (*C. mitchelli*) and sidewinders (*C. cerastes*) feed less readily than other species. This does not seem to be a matter of nervousness. Some snakes that manifest nervousness for months, or even years, by rattling whenever their cage is opened, eat as readily as those that have become quite tame.

Once a rattlesnake has become accustomed to feed in captivity, so that it shows expectation and interest when a cage is opened, it will eat almost any kind of small mammal or bird. Mice and rats are used at the San Diego Zoo because they are the easiest to raise. Rattlesnakes in captivity will eat white rodents as readily as they do the gray or brown creatures upon which they prey in the wild. In fact, at the zoo, we have tried them on mice artificially colored blue, green, yellow, or red, without any evidence that the color was noticed. At times of a shortage of rodents, defective baby chickens from the commercial hatcheries have been substituted with success. Almost any snake accustomed to feed on rats will eat chickens, although in one case — this was before dead food became the practice — a cage of prairie rattlesnakes showed fear of the baby chicks, and would not go near them until they had first been killed. Lizards comprise a good food supply for small rattlesnakes, but they are seldom available when needed. This problem has been solved by keeping a surplus frozen in a refrigerator.

Reactions of Prey

Much has been written about the intense fear shown by any animal that may be put in a cage with one or more live rattlesnakes. Our San Diego observations have been quite the reverse. Before we learned the advantages of feeding with dead animals, or found that patience and a proper cage temperature were more important than the simulation of conditions met in the wild, we were accustomed to place live creatures in the cages, including rats, mice, rabbits, birds, young chickens, and lizards. None of these showed any fear of the snakes, unless alarmed by some sudden movement of a rattler or the buzz of a rattle, should some snake be frightened into sounding off. On the contrary, they ran or hopped unconcernedly around or on the snakes, much more worried by their new surroundings than by the strange co-occupants of the cages. One writer saw a rattler about to strike a chicken. The chicken pecked the snake on the nose, whereupon it withdrew and left the chicken victorious. Another observed that birds repeatedly lit on the backs of captive snakes and paid no more attention to them than if they had been inanimate objects.

A rattler in captivity has been reported to play with a mouse for several hours before destroying it. This is a fantastic idea; unless a rattler has an immediate interest in feeding, or is afraid of a mammal or bird, he will ignore it.

Long Fasts

It is well known that rattlesnakes can live for long periods in captivity without food of any kind. They will last longer if they have water to drink; also they will survive longer in humid than under dry conditions, and at low temperatures rather than high, for under such circumstances life processes are slowed down and tissue and fat consumption are reduced.

There is little doubt that under the more favorable conditions of a lower temperature some individuals would survive for two years; and most specimens, if starting in good condition, would have no difficulty in fasting for one year. Of course, after a snake has refused food for a long time there is little chance that he will become a natural feeder, yet Perkins had a western diamond that, after a year of refusal, suddenly seized a rat intended for a cage-mate. After this it ate regularly.

CHAPTER EIGHT

Reproduction

SEXUAL CHARACTERISTICS

Some kinds of snakes are oviparous—that is, they lay eggs. Others are ovoviviparous—that is, the eggs are retained in the body of the mother until they are ready to hatch, so that the young are born alive. The rattlesnakes are among those that give birth to living young.

In the southern sections of the United States, rattlesnakes normally mate in the spring, after coming out of hibernation, and the young are born between late August and early October. It is probable that the females give birth to first broods at the time they reach the age of three years. In the more northerly areas, where there is a shorter season of activity and growth, the rattlers have a different life cycle, for it has recently been shown that they bear young only every two years, instead of annually, and probably not until they themselves are four or five years old. In these areas the young are born earlier than is the case farther south, in the late summer,

142

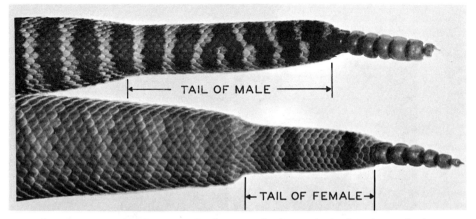

Fig. 25 Comparison between the longer tail of a male rattlesnake with the shorter, thinner tail of a female. The tale of the male is less sharply tapered from the body than that of the female. (The differences in pattern and color are not sex differences; the pictures are of snakes of different species but of almost equal body lengths.)

and mating may take place in the autumn.*

It will be noted that the additional time afforded in a biennial cycle is not consumed in an extension of the time between fertilization and delivery. It is required because of the long season of hibernation, which imposes a year and a half—in terms of calendar rather than active time—for egg growth prior to ovulation.

Male snakes also have a reproductive cycle and are incapable of producing fertilization at certain times of the year.

Studies of life cycles are complicated by the fact that female snakes may retain live sperm for a considerable time, and, in rare instances, a mating may not even be required for each brood.

Methods of Determining Sex—Tail-Length Differences

Since no sexual pattern or color differences are evident in most rattlesnake species, it becomes necessary to devise other means of distinguishing the sexes. As the anatomical differences cannot be ascertained with safety in live specimens, for handling would be required, one is usually forced to depend on a visual examination of the tail, which affords a dependable criterion in the case of adult specimens, especially after some

*Whether off-season births occur among rattlesnakes, as do off-season matings, I do not know. I have occasionally seen rattlesnakes (especially southern Pacifics) in the spring that were so small it seemed they must have just been born. Had they been born at the usual time in the autumn, one could not imagine their having survived so long without either food or growth.

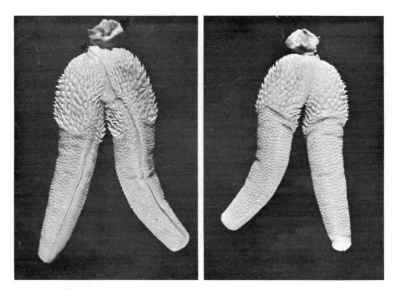

Fig. 26 Hemipenis of *C. adamanteus* (front and back view)

experience has been acquired. The tail of the male is relatively longer and thicker than that of the female; also it tapers gradually from the body, whereas in the female there is a sudden and rather marked narrowing at the vent, which sets off the tail from the body.

Male Organs

The intromittent male organs of snakes are normally retracted into their tails and thus are hidden. They are paired, there being one organ on each side, and, in addition, each organ may be more or less bifurcated. Only one organ is used at a time. There are rather remarkable generic and even species differences manifested in the degree to which each hemipenis is bifurcated, and in the character of the exterior ornamentation in the form of fringes and spines. The ornamentation referred to is visible only when the organs are extruded for use, for it is only then that the fringes and spines are on the exterior of the organs. The organs are hollow, and in the course of being extruded via the cloaca, they are turned inside out.

The male organs, since they are both paired and deeply bifurcate, so that superficially there appear to be four when fully everted, have been the basis of statements to the effect that rattlesnakes have hidden legs. These legs are most likely to be evident, so it is said, when a snake has been singed in a fire.

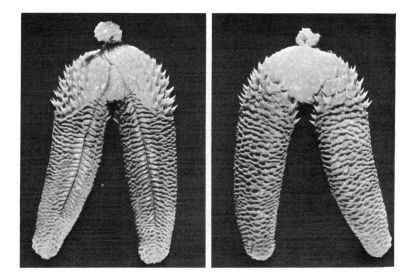

Fig. 27 Hemipenis of *C. mitchelli stephensi* (front and back view)

SEXUAL ACTIVITIES

Traveling in Pairs

The popularly accepted story that rattlesnakes travel in pairs is both true and false; it is a matter of degree, for sometimes they do, but more often they don't. Rarely a rattler will be found lying beside another recently killed, and this has led to the belief that they do this invariably, awaiting a chance to avenge the late lamented.

That rattlesnakes may be, and frequently are, found together during the mating season is unquestioned; but that they travel in pairs at other times requires more than verification by an occasional encounter. Only the accumulation of a considerable body of statistics can verify a belief of this kind. Unfortunately, a person with a preconceived notion respecting a habit of this nature will unconsciously remember the occasions when two rattlers were found near each other and will forget the times when only one was observed. And again, when two are found together — which is not unusual — are they a pair? After all, rattlesnakes are somewhat gregarious, especially near the denning season.

Any territory, favorable from the standpoint of food and hiding places, is likely to contain several snakes, yet they may be entirely without attachment for each other; in fact we may be sure that creatures with so low an order of intelligence will have no such attraction outside of a brief mating season. On several occasions I have found two red diamond rattlers *(C. r. ruber)* together and they have almost as often been of the same sex as of opposite sexes.

The myth that the survivor of a pair will untiringly and unerringly seek out and destroy the person who killed its mate has grown out of the fact that in the mating season a male occasionally has been seen to court a freshly killed female. From this small and unusual beginning have come the stories of the avenging rattler that follows the body of its mate over any devious course through which it may have been carried, until we come to the tale, said to be of Indian origin, of the human murderer who uses the vengeance of a rattlesnake to achieve his own by dragging the body of a dead rattler through the bedroom of an enemy, assured that the mate will follow.

The traveling-pair story has been cited as an example of Mississippi-Florida folklore, and as a belief of the Pennsylvania Germans. There is no doubt that it is believed in every section of our country. In the spring of 1954, a rattler was found in a downtown Los Angeles apartment. The snake was destroyed by an officer from the appropriate city department, but when a search of the premises failed to disclose the inevitable mate, the occupant left to spend the night elsewhere.

The Male Combat "Dance"

The combat "dance" of snakes — rattlesnakes among them — is an affair wherein the obvious isn't the truth, and the truth is stranger than the obvious. Even the title assigned to this weird performance is of doubtful applicability, since "combat" poorly describes a bout that nearly always ends in a draw, with neither participant injured; and "dance" is hardly appropriate to a sparring or wrestling match between creatures with neither arms nor legs. But the stylistic gyrations of the performers suggest a symbolic dance, and as the term "dance" has been applied to the exhibition by those who have most often observed it, dance it shall be, hereafter without the quotation marks.

The reason this interesting performance has been so infrequently mentioned in popular accounts of snake activities is because the dance was thought to be courtship or actual mating. This is why I use the word "obvious" in the introductory remark; snakes linked together or twisted about each other are obviously mating, and so it was assumed, beginning with Aristotle, down through the Arabian naturalist Damiri of the fourteenth century, almost to the present.

Some of the descriptions of the dance that I have received from my correspondents follow:

Fig. 28 Male combat dance of prairie rattlesnake (*C. v. viridis*). (Photographed by E. W. Cottam. Published through the courtesy of the U. S. Forest Service and Mr. Cottam)

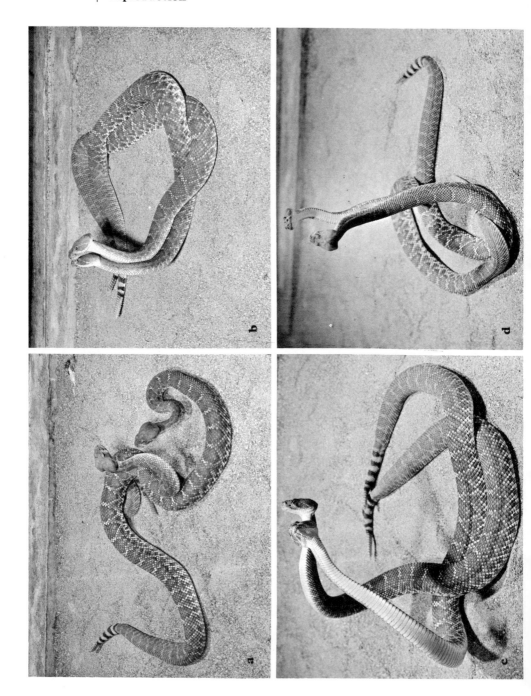

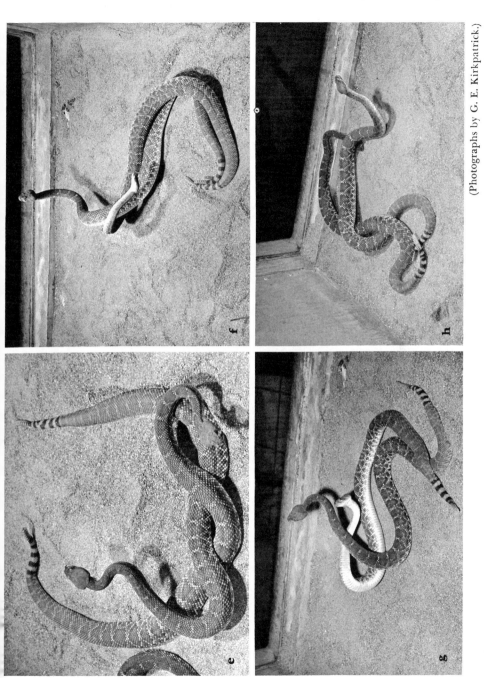

(Photographs by G. E. Kirkpatrick.)

Fig. 29 (a–h) Episodes in the male combat dance of two captive red diamond rattlesnakes (*C. r. ruber*)

☐ As often as twenty-five times or more, I have seen large specimens of the canebrake rattlesnake *(Crotalus horridus atricaudatus)* fighting, always in the fall of the year, and have watched these fights sometimes for as long as a half hour or more. Their method of fighting is to face each other, put the sides of the heads together, and gradually rear, still facing each other, with the necks in a half-turn around each other. When the snakes are as high as they can reach into the air, which in a large specimen will be as much as three feet, one will give a lurch upward and then bring its entire weight down across the neck of its opponent, slamming it on the ground with considerable force, evidently to injure the opponent. If the attempted damage fails, the snakes again rise in the same manner, and they will continue thumping each other against the ground until both become weary, whereupon they withdraw in different directions. In the many encounters which I have witnessed, I have never seen a snake injured or run away. The battle has always ended in a draw.

☐ In late July, 1928, while returning from an Arizona journey, I passed through the lower Laguna Mountains in San Diego County. It was late in the evening, sometime after sunset. One of the lads riding with me exclaimed, as we went past a roadside gully: "I just saw an octopus* lying there and it was moving its arms. However could it get here so far from the ocean?" "That's impossible," I said, but the lad insisted he had had no dream. Stopping the car, we walked back and there was, indeed, an octopus-like tangle of arms, but consisting of two rattlesnakes coiled about one another. They soon uncoiled and, now separated, began a most weird and gentle series of antics. The head and at least a third of the body of each was raised upright and, in a sort of rhythmic unison, they swayed back and forth before one another. Every few minutes, they would make passes at each other, embrace (neck and head separate) and then recede — only to repeat the odd performance. This we watched at close range with the serpents paying no attention whatever, either to our motions or spirited conversation. After about half an hour, it became so dark we could scarcely see them.

The dance and mating patterns are so different that there is no difficulty in distinguishing them, for in the one case the rattlesnake's bodies are raised well above the ground, while in the other they lie beside each other.

The dancing pattern seems to be the same for all species and subspecies of rattlers. The snakes do not rattle during the dance. Dancing is not restricted to a single season of the year.

There remains a difference of opinion as to the reasons for these stylized performances. They seem rarely to result in any injury. One writer is of the opinion that a homosexual behavior of one male toward another often initiates a dance. It may result from the lack of an adequate cue for sex recognition. The participants need not belong to the same species; in one case, over a period exceeding five months, a male southwestern speck-

---

*Once, as I was driving along a hilly back-country road watching for snakes, I saw, lying on the pavement what appeared to be a dead octopus. Of course I backed up to make an examination. It was an octopus.

led rattler *(C. m. pyrrhus)* repeatedly tried to mate with a male southern Pacific rattler *(C. v. helleri),* and whenever it did the southern Pacific would respond by taking the dance position, as if to convey a repellent cue. These rattlers had been caged together for several years before the dances took place.

Sometimes males have danced when there was or had been a female in the cage, but more often when no female had been present. Certainly the presence of a female is not necessary to stimulate males to dance. Once, in the case of three male prairie rattlers, dancing suddenly started after they had been cage mates for six years. There was no female present at any time. In another instance when two males had been together, the introduction of a third male caused the first two to dance.

An observation made on some hybrid rattlesnakes *(unicolor* x *scutulatus)* at the San Diego Zoo is of interest, since it may throw some light on the problem of the dance stimulus. There were four rattlers in the cage, three males and a female. One of the males was courting the female, which had just changed her skin. Whenever one of the noncourting males was held near the excited one, the latter immediately rose up to dance, to which the noncourting male would make no response. But whenever the excited male was given the shed skin of the female to smell, he would cease his attempts to dance.

A histological examination of two male sidewinders *(C. c. laterorepens)* that had engaged in a dance, found both to be in breeding condition. The examiners believe that aggressive behavior on the part of snakes may be an expression of territorial defense, social domination, or sexual domination; also sex discrimination may be involved. But they do not draw a final conclusion as to the urge that leads to the rattlesnake dance. For myself, upon the evidence thus far adduced, I believe it probable that the dance stems from some sexual impulse, rather than from one based either on territorial defense or social domination.

Courtship and Mating Pattern

In the course of the last thirty years, we have often had opportunities to observe the behavior of mating rattlesnakes at the San Diego Zoo. From notes on different species and subspecies, it is apparent that the pattern is quite uniform, both with respect to courting and copulation.

The following example summarizes the mating pattern of rattlesnakes of the genus *Crotalus* as described by C. B. Perkins: the male courts by crawling beside or over the female; he nudges the female's body with an alert head, and his tongue is frequently darted out. Meanwhile the posterior part of his body jerks convulsively as the tail is curled under that of the female first on one side, then on the other.

The forked hemipenis on one side is everted slightly until finally the female's tail is raised and a successful admission is gained. There is no evidence that both hemipenes are ever used successively in a single copulation, despite the suppositions of some that they are.

After copulation has begun, a considerable pulsating or pumping action by the tail of the male is to be noted. This may be fairly regular or intermittent. For long periods both snakes may lie quietly with their bodies separated except at the tails; at others the male may continue a nudging courtship. Eventually the female seems to become restless and drags the male around until separation is finally effected.

Our studies indicate that copulation in the rattlers rarely lasts less than an hour or two, and probably six to twelve hours duration is more normal. One *ruber* pair remained locked for at least 22½ hours, and another pair at least 19½ hours. Unfortunately it is seldom that the times of both the start and end are known.

It is evident that the mating pattern is frequently misunderstood. Although it has been stated that the anterior parts of the bodies may in rare cases be raised and intertwined, this position is certainly not normal in rattlesnakes. The male dance has often been inaccurately considered a nuptial performance. Also mere courtship has many times been reported as a successful mating.

The early travelers in America frequently supposed that the rattlesnakes found lying about dens were mating, since they were often found intertwined or even in such concentrations as to be referred to as "balls of snakes." Such ideas persist even to this day. For example, one of my correspondents writes: "I have seen rattlesnakes breed. They form a ball. I have seen a ball of snakes as large as a 30-gallon barrel. They also give off a terrible odor while breeding." Studies of large numbers of rattlesnakes at their dens have failed to verify any of these statements. At the San Diego Zoo we have had, in a single cage during the mating season, congregations of rattlers of from a dozen to several hundred. Although they frequently piled up in a corner of a cage until they reached a depth of a foot or more, no signs of sexual excitement were observed, much less a real mating.

It is presumed, although adequate experiments have yet to be made, that in the rattlesnakes either the postanal glands or skin odors, or both, are important in the location of mates. The glands are larger in female rattlers than in males. That the sense of smell—that is, the tongue–Jacobson's-organs combination—is involved, is evident by the amount of tongue-flicking indulged in by an ardent male.

BIRTH

Although some of my correspondents have expressed the belief that mother rattlers repair to the hibernating dens to have their broods, so that the young may readily locate the dens when the season for hibernation arrives, this is highly doubtful. Were all the mothers to have their young at the dens — this would be two to four weeks before denning time — it is obvious that the concentration of young rattlers thereabouts would be very high, which would reduce the possibility of their securing the food so necessary to their survival through their first winter.

The observation has been made that in the northern areas of the country, the females find suitable crevices in rocks, usually about a half mile from the dens, wherein to give birth to their young. Where no rocks are present and prairie-dog towns are used for denning, old holes on the outer edges of the towns are sought by the pregnant females. In our more southern areas, where the snakes do not gather at dens from wide areas, but rather seek individual retreats for the winter, so likewise the females, when about to deliver, seek the most convenient rock crevices, mammal holes, brush piles, or other places that afford some shelter.

Just as the discovery of snake embryos in mother snakes is the basis of many of the tales of snakes that swallow their young for protection, so the finding of eggs before the appearance of embryos has led to some of the mistaken statements that rattlesnakes lay eggs. Actually, the eggs are retained in the body of the mother until the young are ready to break out of the parchment-like flexible coverings.

There are a few myths concerning oviparity in rattlesnakes. One is that diamondback rattlers lay eggs under rocks — that is the wild ones do; those in captivity bear living young. Another is that timber rattlers lay eggs (always in odd-numbered clutches) in the sand on river banks. After they have been in the sand for a month, the mother snake returns, swallows them, and subsequently — or so it is to be presumed — gives birth to living young. And there is a story now going the rounds to the effect that rattlers east of the Rockies give birth to living young, whereas those to the west lay eggs.

There is a considerable variation in the intervals that separate the births of the successive members of a brood, although probably less in the wild than under the conditions of captivity, in which the spacing may vary from a minute or so up to hours or even days. When the brood comprises fully formed and live young, the intervals are likely to be shorter and more regular than where live young are interspersed with infertile eggs and defective young. In the latter case, the full delivery may take several months, whereas, under more natural conditions, the brood, often born at night, will be observed in its entirety on the following morning.

There are contractions in the posterior of a female before the birth of young, and as the young are born, rhythmic contractions of the body force the remaining embryos toward the cloaca.

In rattlesnakes, the point of attachment of the umbilical stalk is shown by a short longitudinal scar that crosses one or more ventral plates. It remains in evidence for some months.

C. B. Perkins has observed that nine out of ten young rattlers are born with the fetal membrane unruptured. They apparently have little trouble in breaking out, and this they do rather promptly if they are full-term young. They then crawl to the corner of the cage where they coil up and rest.

YOUNG RATTLESNAKES

Congenital Defects

One is impressed with the high proportion of defective young that appear in broods of snakes, rattlers and others, born or hatched at the San Diego Zoo—snakes with small heads, without eyes, with a fused loop of the body. Many of these defects are lethal, so that even if the young are born alive they cannot long survive.

It may be presumed that the proportion of defective young and infertile eggs in captive broods exceeds the proportion that occurs under normal conditions in the wild. Here are the records of two broods of red diamond rattlesnakes: first brood—two normal young, one live young but with a defective eye, four born dead, nine eggs (probably infertile); second brood—eight normal young, one alive but with a very small head, one alive but without a left eye, one born dead, nine eggs (probably infertile).

Young Per Brood

Generalities with respect to the numbers of young rattlesnakes in broods are rather unsatisfactory, since there are considerable differences among species. From data including published figures, San Diego Zoo records, those of my correspondents, and counts of eggs and embryos made on preserved specimens in my laboratory, I conclude that our commonest and most important species of rattlesnakes found in the United States probably average eight to ten young per brood.

The smaller kinds of rattlesnakes have proportionately larger young than the larger species. In the smallest species the young, when born, measure thirty percent or more of the ultimate size reached by the adult males, whereas in the largest species the young at birth measure only about eighteen to twenty percent of the full-grown males.

Within each species or subspecies of rattlesnake the larger mothers tend to have more young than the smaller.

Maternal Protection and Association

Though there have been many accounts to the contrary, it is generally accepted that young rattlesnakes remain with their mother for only a few hours after birth, or a day or so at most. Their propinquity, such as it is, does not result from any maternal solicitude; rather it is only because the refuge sought by the mother is also used as a hiding place by the young.

One source of confusion in this area is that the sizes of young rattlesnakes are consistently underestimated by observers, just as adult lengths are exaggerated. Hence one correspondent mentioned snakes as short as 2½ to 3 inches, and another thought the young must have been with the mother for some time as they had attained a length of eight inches. But, as a matter of fact, the young of even the smallest species of rattlesnakes are about 6 inches long at birth, and such common forms as the subspecies of the western rattlesnake are about ten inches, and of the diamondbacks twelve inches or more.

Only the finding of a mother rattler with a brood, the members of which had not shed for the first time, would prove that the young had remained with the mother upward of a week or so, which is the average interval between birth and the first shedding. But unfortunately, even then, there would be no proof that the young were actually hers.

In captivity mother rattlesnakes evince no solicitude for their recently born young. True, the young often lie close to or even on the mother, but this is only because of the natural gregariousness of snakes in or out of captivity; they seem just as likely to do this with some other rattler, not the parent.

One writer tried to excite a mother rattler (*C. v. viridis*) by tantalizing her young, but she seemed to have no regard for them.

Some presumptions in a fantastic vein have been voiced, such as the idea that young rattlers might be taken into the mother's stomach so that they could be fed; or that young diamondbacks are not weaned until they reach a length of a foot, at which time the mother leads them along a trail, where she leaves them to fend for themselves, at suitably spaced intervals. One author found a mother rattler crooning (by rattling mildly) to her newly born young.

Activities

Soon after birth, the individuals of the brood scatter in search of food. Since snakes swallow their prey whole, having no mechanism, either in their teeth or otherwise, for dismembering prey, stories of mother rattlers sharing food with the young are quite contrary to fact. Each must obtain its own live prey.

There is a considerable reduction in the proportion of a population constituted by the young-of-the-year, during the time from their births in the fall, to the end of their first hibernation in the spring. Lack of food, freezing, and vulnerability to predators take their toll. The food problem is a serious one for the little rattlesnakes, for they are limited to lizards and very small rodents. Although we have observed at the Zoo that young can survive for some months after birth without food, because they are fat with absorbed yolk sacs when born, the growth and vigor of those that have fed, compared with those that have not, show how serious the lack of food must be to those in the wild that are unsuccessful in securing any prey.

The second serious cause of mortality among the young is that of freezing during the first winter, particularly because many of the young fail to reach havens previously tested and found safe by their forebears. Thus they may take refuge in holes or crevices that are not below the frost line, or that are subject to fatal drafts, and never awaken to see the spring.

Finally there is predation, for the young snakes are inexperienced, careless of concealment when searching for food, and are preyed upon by creatures that could not successfully cope with their more dangerous parents.

CHAPTER NINE

Poison Apparatus

THE BITING MECHANISM

The biting mechanism of a venomous snake, such as a rattlesnake, comprises provisions for venom generation and storage, and means for injecting the venom into the victim; these means, in their highest development, consist of hollow fangs having the mechanical perfection of dual hypodermic syringes. Snakes have their venom primarily to subdue, secure, and digest their prey; any value that it may have as a protection against larger enemies is incidental to these primary purposes.

Although the head bones of the rattlesnake, like those of other snakes, are thin and delicate, they are beautifully formed to withstand the stresses of both biting and feeding, the latter, because of the relatively large size of the prey engulfed, being particularly important.

The biting mechanism is most highly developed in the viperine snakes, in that the fangs can be rotated from a resting to an acting position. In the resting position,

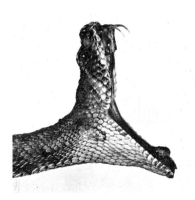

Fig. 30 Fig. 31 Fig. 32 Poses of a freshly killed specimen of *Crotalus atrox*, with mouth opened to show the fangs. In fig. 30, the fangs are folded against the roof of the mouth, and are covered by their sheaths. In fig. 31, the fangs are slightly advanced, and the sheaths have been cut away. In fig. 32, the mouth has been widely opened and the fangs advanced, as at the end of the forward drive of a strike.

they are folded back against the upper jaw, with base and point at about the same level, and with the bulge of the fang-curve fitting into a hollow in the lower jaw. To assume their active — their striking or biting — position, the fangs are rotated downward, the points describing a forward arc of about ninety degrees. It is this duality of position that allows the snake to have such long fangs, which, were they fixed in an erect position, would pose a serious handicap to the snake in all activities except biting.

The fangs are periodically lost and replaced, but little seems to be known concerning the normal frequency of fang change. Judging from available data, I would guess that the normal active life of each fang in an adult rattlesnake is from six to ten weeks, and that each change involves a double-fang overlap of four or five days.

It has been suggested that fang changes are synchronous with skin-changing. This is certainly not true, since the fang replacements on the opposite sides of the head are not synchronous with each other.

The maximum size of fangs attained in the rattlesnakes may be of interest. I have not had the opportunity to measure the fangs of any very large specimens of the eastern diamondback *(C. adamanteus),* the largest

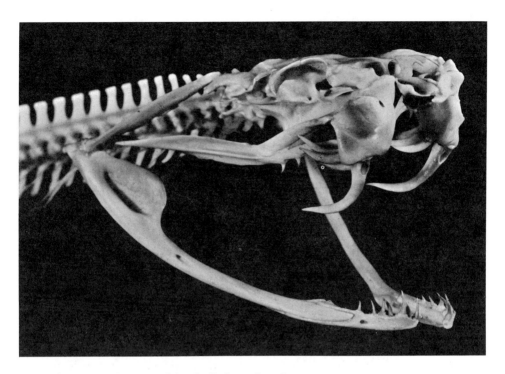

Fig. 33 Angular view of the skull of a rattlesnake

of the rattlers, and the heaviest of all venomous snakes. The largest available to me have been a little over five feet long, with fangs having a length of 11/16 inches. Calculations lead one to expect that an eight-foot specimen would have fangs about ⅞ inch in length.

Rattlesnakes do not by any means have the largest fangs found among the venomous snakes. The African Gaboon viper, an extremely heavy-bodied snake, is said to have fangs approaching two inches in length.

I have mentioned elsewhere the popular supposition that rattlers threaten their enemies with open mouth. This is certainly not true of uninjured snakes. Another somewhat related action, often attributed to rattlers, is that of spitting venom at their enemies. Certainly the rattlesnakes, unlike the spitters among the cobras, do not possess fangs designed for spitting. The lower orifice in the fang directs the venom downward, not forward. I have seen many rattlesnakes strike, but have never seen an uninjured rattler spit, with the ejection of either venom or saliva in the course of the strike.

There is an idea as old as Pliny, and probably much older, that the fangs of a snake are barbed, and are lost when it bites.

VENOM

As previously mentioned, the primary purposes of the venom of poisonous snakes are to secure food and to aid in its digestion, the latter a particularly important function where teeth are too thin and pointed to masticate food. The prey is killed by venom injection, so that it need not be held while it struggles, which avoids the necessity of holding a creature, such as a rat, that might injure the snake.

In most poisonous snakes, including the rattlesnakes, the venom glands are located on either side of the head toward the outer edge of the upper jaw. Their extent and size depend not only on the size of the snake but on its genus as well. Experiments have shown that snakes, in biting, have control of the venom discharge; they can, at will, discharge venom from either fang, from both, or from neither.

In the early days it was thought that the act of biting caused the piercing of the venom gland by the fang itself, with a consequent release of venom into the wound. But later, the presence of a more positive means of delivery was recognized—a duct between the gland and the fang, serving to carry the venom directly to the upper entrance of the venom canal in the hollow tooth.

Enzymes are now considered the toxic principles of venoms, and, though the knowledge of the activity of enzymes in snake venoms is incomplete, it is known that they have an important role in the production of shock, hemorrhage, and blood clotting. The enzymes also affect the spreading or diffusing factor of the venom.

For a time, snake venoms were divided into two broad categories, the neurotoxic and hemorrhagic, depending on whether their destructive actions were largely on the nerves or on the blood and tissues. But it was soon recognized that few, if any, venoms were purely of the one type or the other. Rather, it was often the case that one effect was more serious than the other, so that the presence of one type was masked by the more important symptoms of the other.

Death from venom is believed by some to be the result of the neurotoxic factors, even though the early symptoms may be predominantly hemorrhagic.

For a long time, the source of the danger from snakes was obscured by myth and misunderstanding. Pliny more often mentioned the stings of snakes than their bites; and the possession of a dangerous sting is still erroneously attributed to some snakes. Actually, no snake of any kind has a sting. The ideas that the venom of the viper is contained in the tongue, not in the fangs, or in the tail or breath, are as old as English literature.

From these early ideas have come a number of misconceptions, some of which were applied to rattlesnakes after their discovery in America.

The most persistent of these was the presumption that the virulence of the venom was generated by, or was proportional to, the snake's anger.

Another ancient belief, dating back to Pliny or before, was that the venom was nothing but the gall of the serpent conveyed to the mouth by certain veins.

Manual Venom Extraction

Sometimes it is necessary to extract venom from captive snakes for scientific purposes. The methods used should be designed with two objectives in view: first, to safeguard the operator; and, second, to secure a maximum quantity of natural or unmodified venom, properly segregated for scientific use. The details of handling must depend on the kind of snake to be milked — the danger from its bite, the character of its fangs, and its strength and agility. Also, there are differences in procedure depending on the rarity of the snake and whether it is to be milked repeatedly or only once.

In the procedure developed for rattlesnakes in San Diego, an assistant catches the snake immediately behind the head by means of a noose-stick and holds it with the head resting on the edge of a table. When the snake is so caught and held, it has no opportunity to reach any object with its fangs and thus to waste venom. The operator, by means of a short metal bar with a hook at the end, catches the snake's upper jaw under the rostral and tips the head back. Then the rim of a porcelain cup, or other suitable container, is introduced below the fang points, and the fangs are drawn downward and forward into the erect position, with the sheaths pushed upward to bare the fangs. Because the head is tipped back and steadied by the hook while the cup approaches, the snake can neither see the cup nor slash at it until it is in place, pressing against the fangs and ready to catch any venom expelled.

As the fangs are drawn forward and held erect, the edge of the cup is pushed steadily against them; this tends to hold the head firmly and gives the snake a feeling of something yielding on which to bite. The hook is now withdrawn, and the operator, further forcing the head against the cup with his index finger, presses on the venom glands with the thumb and middle finger. The snake will usually eject some venom in an attempt to bite when it feels the steady pressure of the cup against the fangs, but in all cases the flow is increased by the mechanical manipulation of the glands. The glands are pressed several times at intervals of about ten seconds.

Undoubtedly a strong pressure exerted on the venom glands frequently causes some injury; and if the snake is more important as an exhibit than its venom is for scientific purposes, or if additional venom is expected from subsequent milkings after an appropriate period of rest

Fig. 34 The mouth of the snake is opened, and the jaw is tilted back by the use of a metal hook.

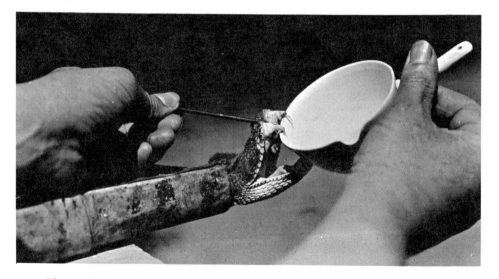

Fig. 35 The venom cup is slipped under the fangs, while the head is steadied with the hook.

(about a month), then the pressure on the glands must be gentle, or the snake should be allowed to bite through a diaphragm without any external pressure on the glands.

At the San Diego Zoo, using the methods I have described, we handled rattlesnakes at a maximum rate of about forty-five per hour.

Liquid rattlesnake venom is usually bright yellow in color and is virtually odorless when fresh and unaffected by bacterial action. In recent tests at the San Diego Zoo, one out of three observers thought that a very faint sweetish odor could be detected, the others believed there was none. We have noted a slight taste, astringent at first and then turning sweetish, when held on the tongue. But the taste is by no means strong. There is a slight tingling effect on the lips.

Toxicity

It would be of general popular interest if it were possible to rate poisonous snakes in the order of their danger to human beings. Some kinds are undoubtedly much more dangerous than others; but an accurate rating is quite impossible because of the many uncertainties involved in snake-bite cases. It is difficult to rate, with any accuracy, even so limited a group as the rattlesnakes.

It might be assumed that one criterion—the venom—should permit a fair evaluation, and it does, to a limited degree.

The South American rattlesnake *(C. d. terrificus)* is surely the most dangerous of all the rattlers because of its combination of large size and extremely powerful venom.

Of the species found in the United States, I should give first place to the eastern diamondback *(C. adamanteus)* because of its extreme size and high venom delivery, although its venom is not a strong one. Next would come the western diamond *(C. atrox),* which results in more serious cases

Fig. 36 The hook is removed, and the glands are pressed to expell the venom.

in the United States than any other venomous snake, rattler or otherwise; and may, in fact, cause more serious cases than all the other kinds of rattlers in our country put together. This is because it ranges over a large area, is extremely plentiful in some places, including farming districts, and is nervous, easily aroused, and, in rare cases, even aggressive. It has a more powerful venom than *adamanteus,* but the quantity is smaller. Following *atrox,* I should place *C. viridis* and *C. horridus,* particularly their larger subspecies, and *C. m. molossus.* All of these are thoroughly dangerous snakes. But my not having mentioned the other kinds of rattlers should not suggest that they are to be trifled with.

Despite all the doubts surrounding the data on venom yields and strength, we can make some fair guesses concerning relative hazards. For example, there is no indication that the sidewinder *(C. cerastes)* fulfills its fearful reputation of being the most deadly rattler that crawls. Its venom yield is certainly low and the venom is not especially powerful; and as a dangerous snake, it is not to be compared with some of the larger rattlers that I have mentioned.

Venom Utilization

Rattlesnake venom is in occasional demand for two purposes: for the immunization of animals (especially horses) in the production of antivenin; and for therapeutic uses. The latter is still in the experimental stage and there is no certainty, or even probability, that rattler venom will ever have any recognized place in pharmacology.

Production of Antivenin. From time to time, people in areas where rattlesnakes abound—particularly in the Southwest—have been encouraged by the high prices reputedly paid for venom, to consider going into rattlesnake farming, with the expectation of a continuous and valuable crop. Actually, rattlesnakes are difficult and expensive to raise in captivity, and repeated milkings are not fruitful; therefore, a steady venom supply must come from the continuous availability of freshly caught snakes. Even this might be profitable, as an off-season side line for farmers and stock raisers in a few places, were there a steady market for dry venom, but such a market does not now exist, and probably never did. It was widely reported during World War II that "the government" would buy all the rattlesnake venom available at some fantastic price, and as a result I received from hopeful farmers many requests for instructions in the art of rattlesnake culture. I do not know how the rumor started. It was probably true that large supplies of antivenin were purchased for Southern army training camps, and therefore there was an increased demand for venom for immunological purposes by the manufacturers of antivenin.

Use of Venom in Therapeutics. There is something peculiarly fascinating in the conversion of a dangerous creature such as a rattlesnake—and the essence of that danger, the venom—into a cure for disease, and so it is that experimental work looking to such a use has often resulted in sensational and premature accounts of the curative properties of venom.

Rattlesnake venom, either taken orally or injected subcutaneously, has been suggested for a number of maladies, and no doubt it has been tried and unreported for many more. A letter written in 1824 states that pills made of pounded venom glands and cheese were alleged to relieve palsy, rheumatism, and typhus fever. They gave a general feeling of well-being and, indeed, of "ethereal delights." The writer of the letter—no doubt under the influence of his remedy—hazarded the prophecy that rattlesnakes would one day be raised for medicinal purposes. Another writer said that rattlesnake venom was an excellent tonic for anyone with the courage to swallow it.

Rattlesnake Venom in Homeopathy. Snake venom, including rattlesnake venom, has long had a place in the homeopathic pharmacopeia. Its usage has been in accordance with the homeopathic theory that a medicine is efficacious for any disease whose symptoms approximate those resulting from the medicine alone.

Among the afflictions or conditions for which rattlesnake venom was recommended were: delirium tremens and alcoholism, melancholia, vertigo, convulsions, meningitis, apoplexy, deafness (of certain kinds), bronchitis, neuralgia, mumps, diphtheria, laryngitis, gangrene, peritonitis, acute jaundice, dysentery, cholera, asthma, pneumonia, hemophilia, pernicious anemia, rheumatism, syphilis, elephantiasis, typhus, yellow fever, plague, epilepsy, and rabies.

Rattlesnake Venom in Standard Medical Therapeutics. Although rattlesnake venom has never achieved a place in the regular (allopathic) pharmacopeia, it has been used experimentally in the treatment of a number of afflictions, occasionally with some appearance of success. One of the most famous early cases was that of a youth in Rio de Janeiro, who was treated with *terrificus* venom for leprosy or elephantiasis, or both, in 1831. This case ended fatally within twenty-four hours.

Rattlesnake venom has also been recommended for alcoholism. It is not difficult to trace the source of this idea. In the days when alcohol was the standard remedy for rattlesnake bite, it was believed that a patient would never become inebriated, regardless of the amount of whiskey or brandy he consumed, as long as there remained any unneutralized venom in his system. The recommendation for venom as a cure for alcoholism is an obvious corollary.

Rattlesnake Venom Treatment for Epilepsy. Probably the widest use of rattler venom for curative purposes, and the nearest that it has come to adoption by any appreciable element of the medical profession for any purpose, was in the treatment of epilepsy. The theory of the beneficial effect of the venom was that it decreases blood coagulability and that epileptic blood is above normal in this respect.

The treatment was apparently initiated through a case report in 1908 of an epileptic thirty-five years old who had been subject to seizures at intervals of a month for over fifteen years. He was accidentally bitten by a moccasin, and suffered no further attacks within the next two years. When this was reported in the medical press, it was picked up by a researcher who gradually developed an elaborate technique for treating epileptics with rattlesnake venom, and, as he thought, with considerable success.

Miscellaneous Therapeutic Uses. Several writers, in addition to those associated with homeopathic practices, have presented extended lists of diseases claimed to have been successfully treated by the use of rattlesnake venom. These lists include epilepsy, asthma, neuralgia, neuritis, laryngitis, pleurisy, nerve exhaustion, insomnia, whooping cough, hysteria, hay fever, acute homicidal mania, arthritis, and eye inflammation.

I have mentioned elsewhere the wide differences that exist between the venoms of the many kinds of rattlesnakes—differences not only of relative potency but of physiological effects. It is notable that the physicians who used rattlesnake venom therapeutically at the turn of the century were seldom definite as to the species of rattler from which the venom was derived. We may assume the drug manufacturers to have been equally uncertain of their products, for there was then little appreciation of the differences in the venoms of the many species of rattlesnake. And there are even differences—maybe from sexual, climatic, alimentary, or seasonal influence—in different batches of venom from the same species of rattlesnake. These and other differences may lead to toxicity variations as great as five or even ten to one. The accurate identification of the various kinds of rattlesnakes requires some experience. With all of these obstacles to the standardization of venom as a drug, it is evident that the preparation of a remedy derived from rattlesnake venom, or one of the constituents of that venom, would not be a simple task.

In current therapeutics, rattlesnake venom has no recognized place in regular medical practice; and the values attributed to it from time to time by enthusiasts hoping to turn this ordinarily dangerous substance to man's advantage have failed to be substantiated.

CHAPTER TEN

The Bite and Its Effects

THE NATURE OF THE BITE

I shall now consider the rattlesnake's bite, whether that bite is the terminal feature of a strike (the forward lunge of the head and body), or is the mere process of opening the mouth and sinking the fangs in the prey or in an enemy close enough to be reached without a lunge.

It is not difficult to observe how a rattlesnake bites when no strike is involved—such a bite, for example, as is likely to occur when a snake is held in restraint or stepped upon. Accidents of this kind are frequent when a rattlesnake is carelessly handled, with a sufficient length of neck left free to allow the snake to turn and bite the hand that holds it. Under these circumstances the snake opens its mouth wide enough to advance the fangs and to sink them into the victim, an action which the jaw muscles are amply strong enough to execute. This is a simple bite, the effectiveness of which depends

167

somewhat on the radius of curvature of the surface bitten, as well as on the penetrability of the substance beneath. Objects of small radius—a wrist, for example—can be bitten more effectively than flat surfaces, since the lower jaw of the snake affords a viselike action that permits full use of the jaw muscles. When biting in this way, a rattler can, if one or both fangs meet obstructions, raise each fang separately to seek a new point of entrance. The curve of the fang aids in penetration.

Bite or Stab?

The nature of the bite that terminates a strike is less easy to analyze, for the forward motion of the head is too rapid to permit seeing what the snake does. However, since the advent of high-speed photography, many instructive pictures of striking rattlesnakes have been taken, and further research is under way to disclose certain details of the sequence.

A rattlesnake starts its forward lunge with the mouth closed. During the advance, the mouth is opened very wide, the angle between the upper and lower jaws attaining almost a 180° opening. The fangs may either be driven in by the force of the forward thrust, or they may be imbedded by a biting action produced by a hingelike closing of the jaws. It is upon this detail—stab or bite—that there has been a difference of opinion among herpetologists.

To some extent, the shape of the rattlesnake's fang favors the bite, rather than the stab theory, for the pronounced central curve in the fang assists subcutaneous penetration by continued pressure—a sort of folding in—after the fang has once entered.

It may be suggested that rattlesnakes either stab or bite, depending on the conditions involved, such as whether they are biting prey or acting in defense, and depending also on the size, contour, and penetrability of the object struck. There is likewise the matter of the relative sequence of contact of the fangs and the lower jaw, an uncertain feature since rattlesnakes are rather poor shots. They often completely miss a small object such as a rat; thus it is that they sometimes strike their target too high or too low, so that either the lower jaw or the fangs may make the first contact.

Probably the best pictures illuminating this question comprise a series taken by Walker Van Riper, of the Museum of Natural History, Denver, showing a number of strikes by prairie rattlesnakes, *C. v. viridis.* The target in the most informative pictures was a three-inch latex bulb, a soft and flexible object. These photographs may be judged to favor either the stab or bite adherents, depending on interpretation. In some, the indentation of the bulb appears only under the fangs; although the lower jaw is touching the target, it seems to be exerting no upward pressure, and thus the snake is stabbing. But, in other pictures, there are indentations both

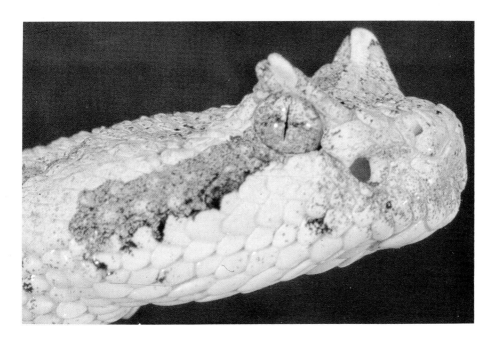

1. Desert sidewinder (*Crotalus cerastes*), showing "horns," vertical pupil, pit (larger orifice), and nostril. Nathan W. Cohen

2. Sidewinder sidewinding, leaving typical tracks. Nathan W. Cohen

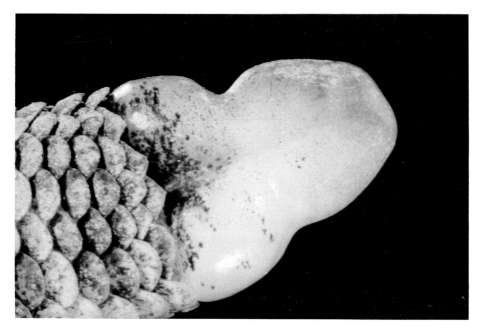

3. Button of a juvenile rattlesnake. Nathan W. Cohen

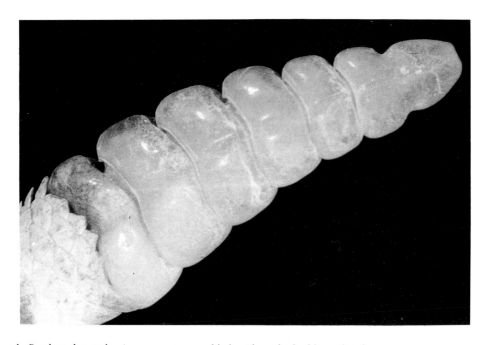

4. Rattlesnake rattle. As segments are added with each shedding, distal segments become brittle and eventually break off (note fractures). Nathan W. Cohen

5. Western diamondback rattlesnake (*Crotalus atrox*), extremely dangerous because it is the most irritable and aggressive of rattlesnakes, inflicts more bites than other species in the United States and ranges over a large area in the Southwest. Nathan W. Cohen

6. Albinistic Western diamondback. San Diego Zoo

7. Mojave rattlesnake (*Crotalus scutulatus*), the rattlesnake with the most potent venom in the United States. Nathan W. Cohen

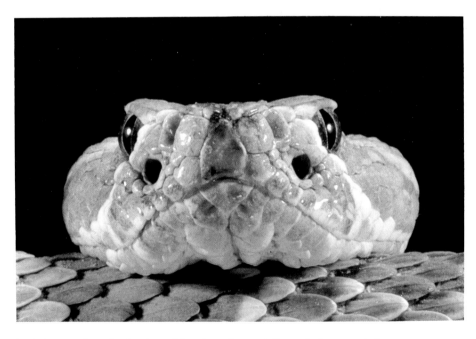

8. Mojave rattlesnake, head-on. (Note large forward-directed pit organs below nostrils.) Nathan W. Cohen

9. Mojave rattlesnake. Nathan W. Cohen

10. *Crotalus enyo.* Nathan W. Cohen

11. Hemipenes, paired copulatory organs, of male Western diamondback rattlesnake. Nathan W. Cohen

12. Diamondback rattlesnake skull showing fangs and replacement fangs. Nathan W. Cohen

13. Red diamond rattlesnake (*Crotalus ruber*), the rattlesnake with the most docile temperament, the least likely to strike — but dangerous nonetheless. San Diego Zoo

14. Grand Canyon rattlesnake (*Crotalus viridis abyssus*). San Diego Zoo

15. Northern Pacific rattlesnake (*Crotalus viridis oreganus*). Nathan W. Cohen

16. Northern Pacific rattlesnake. Nathan W. Cohen

17. Southern Pacific rattlesnake (*Crotalus viridis helleri*). Nathan W. Cohen

18. Speckled rattlesnake (*Crotalus mitchellii*), the rattlesnake most highly variable in color and pattern. San Diego Zoo

19. Northern black-tailed rattlesnake (*Crotalus molossus*). Nathan W. Cohen

20. Black-tailed rattlesnake showing forked tongue (a feature common to all snakes). San Diego Zoo

21. Aruba Island rattlesnake (*Crotalus unicolor*). San Diego Zoo

22. Mexican west-coast rattlesnake (*Crotalus basiliscus*). Nathan W. Cohen

23. Uracoan rattlesnake (*Crotalus vegrandis*). San Diego Zoo

24. Timber rattlesnake (*Crotalus horridus*). Nathan W. Cohen

25. Pigmy rattlesnake (*Sistrurus miliarius barbouri*), the rattlesnake with the smallest rattle, therefore capable at most of producing a little buzz. Nathan W. Cohen

26. Tiger rattlesnake (*Crotalus tigris*). Nathan W. Cohen

27. Central American rattlesnake (*Crotalus durissus*). San Diego Zoo

28. Banded rock rattlesnake (*Crotalus lepidus*). Nathan W. Cohen

29. Speckled rattlesnake, unfed and skin not stretched. Nathan W. Cohen

30. Speckled rattlesnake, skin expanded by a large meal. Nathan W. Cohen

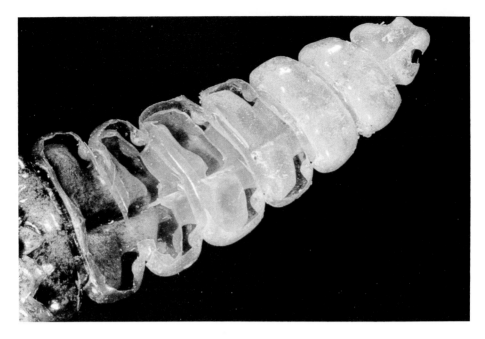

31. Rattle partly dissected to show internal structure. Nathan W. Cohen

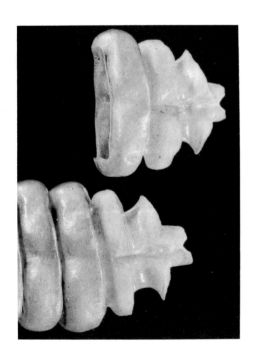

32. One segment of a rattle disarticulated from the rattle to show interlocking structure. Nathan W. Cohen

under the fangs and under the anterior mandibular teeth; clearly the snake is exerting a pinching action—a bite—on the fold of latex between its jaws. So both theories are in some degree verified.

In a normal strike, as shown by the high-speed pictures, the mouth is opened to its widest soon after the forward lunge begins, and some distance before the target is reached. The head withdrawal is slower than the forward lunge. The snake's mouth is closed when it returns to the coil from which it struck, and it is now ready to strike again if necessary.

Bite Patterns

The question of stab versus bite is of more than academic interest, since it may have a bearing on the treatment to be rendered, and the treatment of rattlesnake bite is too serious and painful a procedure to be taken lightly. Persons in the fright that follows a snake bite may be quite incapable of judging whether or not the snake was venomous, or even of noting whether it had rattles, in which case its venomous character would be a foregone conclusion. In some accidents the snake is never seen. Under such circumstances the nature of the marks left by the snake's teeth may be of critical importance. If rattlesnakes only stabbed and never bit, then in most cases it would be quite easy to distinguish a rattler (or other pit-viper) bite by the two punctures made by the fangs, as differentiated from the rows of smaller punctures made by the short, solid teeth of harmless snakes. (I am speaking of conditions met in the United States.) But, on the other hand, if a rattlesnake bites and the harmless teeth also make puncture marks, then to distinguish between rattler and harmless bites would not be so easy or certain, particularly with the complications of intervening clothing, partially ineffective bites on irregular surfaces, teeth scratches made when the victim jerks away, and other confusing effects.

The bite pattern may also be of some importance in judging the size of the snake and the depth of penetration, if the fang marks are clear. This is of interest to a physician contemplating incisions for drainage.

Actions after a Strike

A rattlesnake, when striking an intruder in defense, usually withdraws its head and re-coils in preparation for a second lunge. However, in some instances rattlers have been observed to hang on. It is probable that this is not a part of a normal sequence, but results from an inability to withdraw the fangs because of some condition involved in the bite, entanglement in clothing, for example. One writer makes it appear that a rattlesnake's hanging on is to be expected; he gives directions for pulling it away from the victim to whom it might otherwise attach itself for twenty to thirty minutes. Although this is an exaggeration, there have

Fig. 37 A striking prairie rattlesnake (*C. v. viridis*), showing the position of the jaws as the head reaches the target

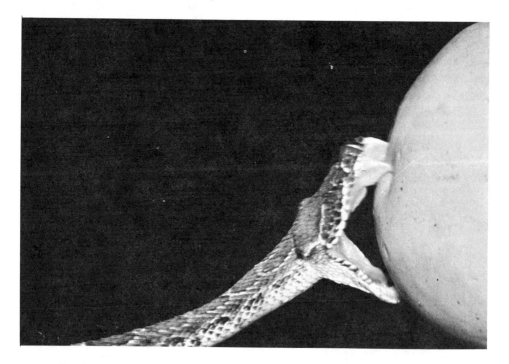

Fig. 38 A striking prairie rattlesnake, showing a stabbing action as the fangs reach the target

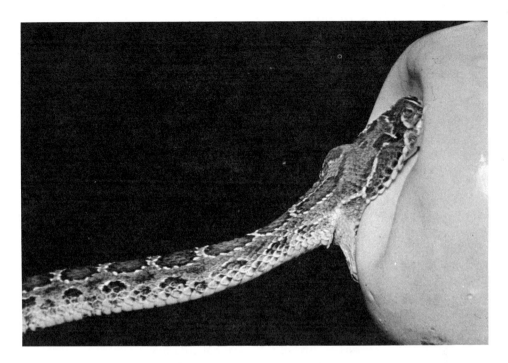

Fig. 39 A striking prairie rattlesnake, showing the fangs imbedded in the target with little pressure exerted by the lower jaw, 1/50 second after the fangs penetrated the latex target

Fig. 40 Another strike of a prairie rattlesnake, in which there was a biting action immediately on contact of the fangs with the target (Figs. 37–40 are published by courtesy of Walker Van Riper, who took the photographs.)

been instances in which rattlers have hung on; for example, there was the case noted in 1857 in which a girl bitten in the ankle dragged a snake several yards before it was shaken loose.

While there is still much to be learned about the method of biting used by rattlesnakes, we need no longer believe, as did a writer in 1655 that a rattlesnake has a head like a dog, and can bite a man's leg off as if hewn with an ax.

The Snake-Bite Hazard

The relative dangers from various kinds of venomous snakes are dependent on many factors other than the toxicity of the snake's venom and the perfection of its biting mechanism, subjects that have already been discussed. Some of these additional factors have to do with the snake itself — its disposition, and its prevalence, habits, and habitats, and how these impinge on the human residents of the same area. And then there is the density of the human population, and its habits — whether people seek to decimate the snakes, and whether they roam about unshod and with legs unprotected. Hence we may have snakes that are quite dangerous from the standpoint of the virulence of their bites and the high percentage of the fatalities resulting therefrom; yet they may be quite unimportant hazards because of their particular natures.

Other snakes may be dangerous but cause no problem since they inhabit remote places where few human beings ever go. Finally, there is the adequacy of the treatment available if an accident should befall. For all of these reasons, neither statistics of the number of bites in any area, nor the ratio of the fatalities to the total bites from any kind of snake, are as important as the total fatalities, or the ratio of the fatalities to the total population.

In *U.S. Army Air Forces Informational Bulletin* of April, 1945, as an indication of the relative unimportance of the snake-bite risk, the statement was made that there are more deaths in the United States each year from lightning than from snake bite; and that to a soldier, with shoes and leggings, mosquitoes are a thousand times more dangerous than snakes. One writer says that more people are killed and injured annually in their bathtubs in this country than by snake bite. In a way this is a reassuring statistic, but it must be admitted that a good many more people encounter bathtubs than snakes and, we hope, oftener.

The danger from rattlesnake bite has always been exaggerated, from the standpoints of both frequency and severity. It is doubtful whether bites by venomous snakes in this country ever exceeded two per one hundred thousand of population per annum, or that the fatalities exceeded fifteen to twenty percent of those bitten, even when the population was largely rural, and despite the fact that the treatments in those bygone days were often as dangerous as the bites they sought to cure. The corresponding figures for today are much lower.

SNAKE-BITE STATISTICS

Fatalities from Rattlesnake Bite in the United States

How serious is rattlesnake bite? How many people are bitten by rattlesnakes each year in the United States, and how many of these cases prove fatal? It is difficult to ascertain accurate information on fatalities from rattlesnakes because the category for reporting deaths from snake bite in the *International List of Causes of Death* includes other venomous creatures such as centipedes, scorpions, spiders, and insects; but some statistics are available.

One set of data concerning deaths in Arizona from the bites and stings of venomous animals covers the twenty-two years from 1929 to 1951, inclusive, but without 1937. The fatality causes that could be allocated were segregated thus: scorpions sixty-nine, rattlesnakes eighteen, insects (unspecified) five, spiders (unspecified) four, black widow spiders three, Gila monster one, bee one, centipede (secondary infection) one; total 102. Thus rattlesnakes caused only eighteen percent of the fatalities. Of course, Arizona is not to be considered typical of the usual distribution of deaths in this category, because of the presence in that state of the dangerous scorpions. Arizona contains more different subspecies of rattlesnakes (sixteen) than any other state, and, as to numbers of individuals, they are plentiful in many areas; yet despite these facts, rattlesnakes caused less than one death per year in the twenty-two years enumerated.

In California, in the fourteen years 1931 to 1944, inclusive, there were nineteen deaths from rattlesnake bite. During the same years, the total deaths from venomous animals were seventy. Rattlesnakes therefore accounted for twenty-seven percent of the total. It is believed that the black widow spider was the principal offender in this state. I am under the impression that today (1955) in California rattlesnakes probably cause at least fifty percent of the fatalities from venomous animals (rattlesnakes are the only dangerous snakes in the state), or two or three per year in a population of thirteen million.

Although I have no statistics to prove it, I should assume that in Texas the percentage of snake-bite fatalities, in relation to the total venomous-animal deaths, would be higher than in Arizona and California, owing to the prevalence there of the large and dangerous western diamondback, *Crotalus atrox*. In Florida, in the years 1940 to 1951, inclusive, of the forty-six deaths caused by venomous animals, thirty-eight deaths or eighty-three percent resulted from snake bite, most of them no doubt caused by the eastern diamondback, *Crotalus adamanteus,* the largest of all rattlesnakes.

Incomplete as the snake-bite statistics in the United States may be, it is nevertheless apparent that snake bite in this country is of relatively minor importance. Deaths in the United States now occur at the rate of about 1.5 million per annum, and of these only about fifty are to be attributed to venomous animals, including snakes. Even among accidental deaths, snake bite is rather unimportant. In 1948, one accidental death out of 2,450 in this country was caused by the bite or sting of a venomous creature; in 1949 the figure was one in 1,917 accidents.

Death rates are usually expressed in terms of deaths per annum per hundred thousand of population. At present the rate for the United States as a whole is about 0.033 per hundred thousand per annum for all venomous creatures, and maybe 0.015 to 0.02 for snakes alone. In California, in the years 1931 to 1944, inclusive, the death rate from rattlesnake bite was 0.02 per hundred thousand per year. With a present population (1955) of about 13 million, it is believed that rattlesnake-bite deaths in California will average about two per annum but might reach five in an exceptional year, or a top figure of 0.038 per hundred thousand. The rate in Florida may reach 0.2 per hundred thousand per annum.

There has been a recent decline in the number of deaths from the bites and stings of venomous animals. A peak in numbers was reached in 1935 with 211; the latest statistics indicate an annual toll of less than one-fifth of this number, despite the increased population. A number of facts may be cited as contributing to this decline: The advent of antivenin in 1927 and its subsequent improvement in quality and distribution; better methods of incision and suction treatment, and the prevention of shock; improved remedies for scorpion stings and black widow bites; the use of antibiotics and other drugs to prevent infection; more adequate information on treatments in the hands of hospital staffs; faster means of transportation from field to hospital; the mechanization of agriculture; and the urbanization of the population. I do not suggest that these factors have been given in the order of their relative importance. I also wish to call attention to the fact that all of these statistics include so-called illegitimate bites, meaning those that result from handling snakes, a subject to which I shall return.

Proportion of Rattlesnake Cases Ending Fatally

The statistics previously given have involved the numbers of fatalities in this country and their relation to population; I shall now touch on the fatalities as related to the numbers of bites or cases.

Many of the early accounts of rattlesnakes stated that the bite was invariably fatal. Others, however, were equally sure that the bite could be cured if some plant remedy known to the Indians were applied.

Gradually the extent of the danger, first cited by travelers as being one of the greatest hazards encountered in the wilds of America, has been more soundly evaluated. In 1890 it was stated that there never had been an authenticated case of death from rattlesnake bite in the United States. This led to an investigation which disclosed some, but surprisingly few, fatalities; but there seems to have been a tendency, in the survey that was made, to attribute most of the fatalities disclosed either to infection or to excess alcohol used in the treatment, and not to death by rattlesnake bite. However, there can be no doubt that the number of fatalities has usually been exaggerated.

To the questions—How many people are bitten each year by venomous snakes in the United States? What proportion of these bites is by rattlesnakes? What percentage proves fatal?—only the most sketchy answers can be given.

First, as to the number of bites by poisonous snakes: in recent years these have usually been estimated at about 1500 per year for the entire country. I would say that of the bites by venomous snakes, seventy percent, or say about one thousand per year, are inflicted by rattlesnakes.

Rattlesnakes are the most important venomous snakes occurring in the United States. They range over the largest areas, and in those places that they share with other venomous snakes they are often the most plentiful (especially is this true west of the Mississippi River) and probably cause the most fatalities, although not always the most snake-bite cases.

In many areas in the United States rattlesnakes are the only venomous snakes present and therefore cause 100 percent of the snake-bite cases.

As to the proportion of rattlesnake bites that prove fatal, again the conclusions are largely guesses; in fact, even with perfect records only average mortality figures could be given, not an answer applicable to every bite, so much depends on the kind and size of rattlesnake, its condition when biting, the effectiveness of the bite (the depth of fang penetration and the quantity of venom injected), the size and health of the victim, and, finally, the availability and character of the treatment. Also, in judging mortality statistics, it is well to remember that the serious cases are the ones likely to reach hospitals and thus to be recorded.

Keeping these factors in mind and extrapolating from the records available, I think we may make the following conclusions with regard to rattlesnake bite in the United States: total bites 1,000 per annum; fatal bites 30 or 3 percent of those bitten; mortality rate 0.02 per 100,000 of population per year.

VARIABLE FACTORS IN SNAKE-BITE CASES

The gravity of a rattlesnake bite is something that cannot be closely defined or predicted, any more than one can predict the seriousness of a fall, without knowing the exact circumstances of the accident, such as the height of the fall, the character of the surface struck, the age of the victim, and similar pertinent details. And in a snake-bite case the conditions are even more obscure, since there are important factors that cannot be ascertained, even after the accident has occurred. Those having the responsibility for its treatment can only presume the factors to be adverse, and take appropriate steps upon this supposition.

As one studies the reports of snake-bite cases, in an attempt to become informed upon the symptoms—for a knowledge of symptoms should be valuable in judging the gravity of a particular case and hence the treatment to be adopted—it is invariably found that the symptoms of the bite are partly masked, not only by the psychological impact on the victim, but by the effects of the treatment. This uncertainty begins with the moment when any initial first aid in the field may have been given, and continues with the remedial measures taken by the doctor. The gravity of the bite is thus obscured and confused, and there is a resulting difficulty in assessing the effectiveness of the cure. As treatment is always necessary, this confusion of malady and treatment is a difficulty that cannot be surmounted, except to whatever extent the symptoms may be determined by tests on experimental animals. And here another source of uncertainty arises, since different animals react differently to the venoms of the several kinds of snakes.

The very fact that rattlesnake bite is an accident of infrequent occurrence means that few doctors become experienced in its treatment. Many cases reach the physician only after the patient's condition has been materially modified by field or home treatments. These may vary from the currently approved methods recommended by the public health agencies to folklore remedies, some of which may be as dangerous as the bite itself. For these reasons, although remedial measures are fairly well standardized in recent medical texts, the treatment does not afford the physician that confidence which would accompany his handling of a more frequent and less variable malady.

The variability involved in the gravity of the rattlesnake bite makes accurate prognosis impossible and treatment difficult. The patient may presume he has been bitten by a venomous snake when actually he has been bitten by one completely innocuous. This is an occurrence not at all infrequent, for people are so badly frightened by snakes that the all-important factor of identification may be mistaken or neglected. Or, for a variety of reasons, even though the bite may have been that of a venomous snake, no venom may have been injected. Snake bite, even when a venomous snake has been the culprit, is not a simple affliction with uniform results; it varies from a condition involving no danger and little discomfort to one of extreme gravity. It is this wide variability, including, particularly, the many cases that would recover without any treatment, that has appeared to validate one after another of the folklore remedies that have been used, and are still being used down to the present day. It has seemed to confirm the efficacy of the many other remedies, having at least some semblance of scientific justification, that physicians of the past employed with apparently beneficial effects, when in reality they were completely useless or even harmful. So important are these variable factors in the effects of snake bite, and in the history of its treatment, that I shall begin the discussion by listing the more important ones:

☐ 1. The age, size, sex, vigor, and health of the victim, which are important in determining his absorptive power and systemic resistance to the venom. There are more snake-bite cases involving men than women because men are more often in the woods or fields, either working or hunting. Bites among children are proportionately more frequent than in the population as a whole. Rattlesnake bite is particularly dangerous in the case of children because of their lower neutralizing capacity, of the way in which the younger ones crawl about, and because, with their short stature, they are more likely to be bitten in the head or body. And sometimes they do not immediately tell anyone they have been bitten, so that treatment is delayed.

Since the earliest studies of the effects of snake venoms on animals have been made, it has been known that the smaller the animal the more quickly it succumbs to a given quantity of venom, each unit of body weight undertaking its share to dispose of a certain quantity of venom.* And it is the same with humans; the young and the infirm are more seriously affected than adults and those in good health.

☐ 2. The allergy complex of the victim; his susceptibility to protein poisoning; the sensitization (anaphylaxis), or partial immunity imposed by previous bites or treatment.

☐ 3. The emotional condition and nature of the victim. Extreme fear and apprehension will affect his heart action and therefore the rapidity of venom absorption; and there may be more direct reactions from fear alone.

☐ 4. The location of a bite on the body. The depth of penetration may be reduced if the fangs should strike a bone or tough tissue. If, by unfortunate chance, the venom should be injected directly into a vein or artery, the outlook will be grave indeed, since the absorption will be rapid and without the opportunity to delay or interrupt it. On the other hand,

---

*This consideration of capacity also governs the amount of antivenin that should be used. Ordinarily we think it natural to give a child less medicine than an adult, but the reverse is necessary with respect to antivenin. For the purpose of antivenin is to neutralize that part of a venom dose that the body cannot neutralize for itself, so the antivenin has a greater task in the child and more is required.

injection into fat leads to slow absorption. Bites in the extremities are less serious than in the body, for absorption is usually slower and the opportunity to delay or prevent venom diffusion by a constriction band, or other curative measures, is enhanced.*

There is a well-known rattlesnake story in the Southwest concerning a cowboy who was struck in the neck while crawling through a barbed-wire fence and died instantly. This is told of many times and places; it may have had its genesis in some actual occurrence wherein a man was found dead in a fence, having been unable to untangle himself after being struck.

☐ 5. The nature of the bite, whether a direct stroke with both fangs fully imbedded, or a glancing blow or scratch. The movement of the victim (a jump backward, for instance) may cause a partially ineffective bite; or a bone may be struck, thus causing imperfect penetration. Only the point of the fang may penetrate the skin, in which case there will be no venom injection, for the discharge orifice of the fang is well above the tip. The snake may misjudge its distance and have the fangs only partially erected at contact, thus effecting only a slight penetration; or it may, for the same reason, eject and lose venom before the fangs are imbedded.

☐ 6. The protection afforded by clothing, which, by interposing thickness, will permit less depth of fang penetration, and will cause the external and harmless absorption, in the cloth, of part or all of the venom.

☐ 7. The number of bites. Occasionally an accident involves two or more distinct bites.

☐ 8. The length of time the snake holds on, for it may withdraw or be torn loose before injection takes place.

☐ 9. The extent of the anger or fear that motivates the snake. The muscles that wring the venom glands and thus eject venom are separately controlled from the biting mechanism. The snake's natural tendency is to withhold venom, since this is its means of securing prey; but if hurt or violently excited it is likely to inject a large part of the venom contained in the glands.

☐ 10. The species and size of the snake, affecting the venom toxicity and physiological effects, the venom quantity, and, by reason of the

---

*The locations of the bites in a total of over two thousand cases in the United States were tabulated in a study done in 1930. The majority of these, but not all, were rattlesnake bites. Where the hands were differentiated, it was found that the bites in the right hand exceeded those in the left by about two to one, a not unexpected result. Bites in the forearm exceeded those in the upper arm about twenty to one, and bites in the shin or calf were about twenty times as numerous as those in the thigh. Bites in the hand were about seven times as frequent as bites in the wrist; and in the front part of the foot about twice as prevalent as in the heel. But, of course, the most important conclusion from these data is that only about one-and-a-half bites out of a hundred are in the head or body, compared with the balance in the arms or legs, for head and body bites are likely to prove more serious.

length and strength of the fangs, the depth of injection. The age of the snake is likewise important; as is characteristic of young snakes of many kinds, both venomous and harmless, juvenile rattlers are more "on the prod"—ready to bite—than adults. But the fact remains that young rattlers, with their short, delicate fangs and limited venom supply, are very much less dangerous than adults of the same species. However, they are not to be trifled with. They have both fangs and venom ready for use immediately upon birth and can inflict a very painful injury. One correspondent wrote me concerning the bite of a canebrake rattler *(C. h. atricaudatus)* less than two hours old, that almost proved fatal. I have had the experience of watching the progress of two cases of bites by juvenile southern Pacific rattlers *(C. v. helleri),* both of which were painful and even serious. Some of the neurotoxic symptoms were quite alarming. Snakes that have passed their prime may also secrete less venom, and this of a reduced virulence.

☐ 11. The condition of the venom glands, whether full or partially depleted or evacuated by reason of recent feeding, defense, ill-health, or captivity. The season of the year (proximity to hibernation) may also cause a variation, but this is not definitely known.

☐ 12. The condition of the fangs, whether entire or broken, lately renewed or ready for shedding.

☐ 13. The presence, in the mouth of the snake, of various microorganisms, some of which, gaining access to the wound, may, abetted by an antibactericidal action of the venom, entail serious sequelae.

☐ 14. The nature of the instinctive first-aid treatment, if any, such as suction, or circulation stoppage by pressure.

☐ 15. Probably the most important, and until recently, least appreciated variable: the wide difference in the virulence of the venoms of the several kinds of rattlesnakes. Drop for drop, some of these have been shown to be sixty times as powerful as others. In the early days, even in scientific research, a rattlesnake was a rattlesnake, and it was generally assumed that the venoms of the several species were quite similar. Now it is known that they differ greatly both in strength and in the nature of their physiological effects. Some kinds of rattlers are a hundred times more dangerous than others; some, indeed, may be incapable of causing a human fatality.

Illegitimate Bites

One item of interest in connection with snake-bite statistics is the question of what proportion of the bites or fatalities results from the handling of snakes, as compared with those that happen to persons who have no intention of indulging in so unnecessary a risk. The latter might be referred to as legitimate bites.

Probably the most accurate statistics on this question now available for the United States were compiled in 1935. They showed that a surprisingly large number of bites were inflicted on persons who were intentionally handling poisonous snakes. Of the 2,342 bites, 163, or more than one in every fifteen, were received in this way. Of these, forty-seven were in ignorant persons, often children, who picked up the snakes, not realizing their danger. Forty-eight were in professional snake catchers, about half being received while capturing the snake and half in handling recent captives. Thirty-one were showmen at fairs or carnivals, and twenty-three were scientists studying snakes in the laboratory or extracting venom. Snakes supposed to be dead inflicted fourteen bites.

Temporal Variables in Snake-Bite Cases

Snake-bite cases are most frequent when there is a coincidence of snake and human activities. In most areas of the United States, the snakes, rattlers among them, are most active in the spring, May especially. (In some areas of the Southwest, where summer rains are prevalent, the snakes may reach a peak of activity later, in July or August, although their sorties then are largely nocturnal.) But the human activities are later—the vacation, hunting, fishing, and harvesting seasons— so that the snake-bite season reaches its peak in July.

Like the seasonal incidence of snake-bite cases, the hourly incidence depends on the correlation of snake and human activities, but much more on the latter, if the snakes are above ground at all; for the majority of snake bites probably involve hidden and resting, rather than active, rattlers—snakes that are not even seen by the human victim before the accident. Thus it is no surprise to learn from a brief set of statistics compiled in 1950, that most accidents occur between six and nine in the morning, and between noon and six in the afternoon.

Seasonal and Climatic Effects on the Gravity of Snake Bite

There has long been a belief that venomous snakes are more dangerous in the summer and in hot weather than in the winter. (Although one writer said that dogs bitten in April almost always died, whereas those bitten in late summer recovered.) Certainly there are seasonal differences in snake activity, and it is possible that the theory may have gained credence from the fact that a cold snake strikes slower and with less effectiveness than a warm one; and if the snake is cold enough it cannot strike at all, in fact, it cannot even bite. It is true, as a matter of practical interest to hunters and fishermen, that the rattlesnake hazard is reduced or completely eliminated in cold weather, but this is more the effect on rattlesnake habits than on a change in the gravity of the bite.

Related to the idea of seasonal variations in venom virulence is the idea that southern snakes are more dangerous than those in the north (in the Northern Hemisphere), which is well founded if the effect of snake size be taken into consideration. A report of rattlesnake bite in which the phase of the moon reputedly affected the bite is evidently a survival of an ancient belief.

Rattler bites have been said to be more dangerous in wet weather; female rattlers more venomous when pregnant. In 1832 the belief was expressed that rattlers were blind in summer, not because they were shedding their skins, but because they are so full of venom and absorb so much of it into their own systems that blindness results.

Pausanias, who lived in the second century A.D., said that venom virulence depends on the kind of food eaten; vipers that are vegetarians — there are, in fact, no vegetarian vipers — are particularly dangerous.

There is a myth that it is safe to handle a rattler if it be deprived of water for some time.

Is there any truth in these ideas concerning the importance of seasonal, climatic, and other effects on the comparative gravity of rattlesnake bite? There is some basic truth in many of them, particularly such as may influence the venom supply. Obviously, snakes that have recently expended some venom in feeding are less dangerous than those that have conserved the supply. Snakes are more dangerous in warm weather because they are more active and alert. It is probable that, at favorable temperatures, they restore their venom supplies more rapidly after partial depletion than when they are cold and lethargic. But whether the unit toxicity of venom in any species is influenced by these factors remains to be proved. Any difference is probably of negligible importance compared with such an obvious factor as the size of the snake.

SYMPTOMS AND AFTEREFFECTS

In a discussion of the symptoms of rattlesnake bite, we have the difficulty of sifting out the variables that affect the gravity of the bite. But the source of more confusion than all of these is the superimposed effect of the treatment and the difficulty of segregating its symptoms from those of the venom alone. It is unfortunate that there is no single symptom sufficiently dependable and distinctive to leave no doubt that the bite was inflicted by a venomous snake, and that sufficient venom was injected to warrant the radical, painful, and even dangerous remedial measures that are necessary in serious cases; and, indeed, are necessary before later symptoms can have verified the early diagnosis.

In an outline of the symptoms of rattlesnake bite, it is first necessary to make a distinction between the South American rattlesnake, *C. d. terrificus,* and the species inhabiting the United States, of which such Nearctic forms as *C. adamanteus, C. atrox, C. viridis,* and *C. horridus* are typical. There is no doubt that *terrificus* venom, primarily neurotoxic, differs in its effects to a major degree from the others, which are largely hemor-rhagic. There are many snakes that are not rattlesnakes, and not even pit vipers, whose venoms are more like that of *terrificus* or, alternately, like those of the Nearctic rattlers, than these two are like each other.

Discussion of Symptoms by Categories

All of the following remarks, based on case reports of rattlesnake bite in the medical literature, apply to the bites of the kinds of rattlers found in the United States.

Swelling and Discoloration. Almost all case reports dwell on the presence and importance of this manifestation. A swelling, in most cases, is pronounced at the site of the injury. Besides the swelling, there is much surface discoloration. It may be red, blue, or black; it tends to darken with time. The discoloration is one of the most evident of the hemorrhagic effects of the venom, as the breaking down of the lining of the smaller blood vessels permits the blood to diffuse through the tissues. The swell-ing gradually spreads toward the heart. It sometimes becomes of very large and alarming proportions; it may be severe enough to rupture the skin. From the extemities, if the site of the bite is in one of these, as it usually is, the swelling may reach the trunk within twenty-four hours or less; the rate will depend on the quantity of venom, and on the use of a constriction band, if any. Sometimes red streaks can be seen extending proximally from the site of the bite. The swelling appears quite soon after the accident, and, if not masked by remedial measures, most of which also produce swelling, is a fairly good diagnostic indicator of the presence of venom.

Pain. Instantaneous pain—as differentiated from subsequent pain result-ing from the swelling, or from incisions, or other treatment—is men-tioned in nearly all accounts of rattlesnake bites. Were its presence invariable, it would constitute the best single confirmation of the pres-ence of venom in the bite, thus justifying treatment. Unfortunately, there have been a few cases, some of a dangerous character, wherein no pain was felt; but even so, pain remains the most characteristic early symptom of rattlesnake bite.

The symptom of immediate pain is advantageous to a snake in that it should tend to draw the attention of an enemy away from the snake be-fore the victim could injure it.

Weakness and Giddiness. Accounts of rattlesnake bite frequently mention various degrees of weakness, faintness, and giddiness. Some of the other terms met in these reports are shock, collapse, pallid countenance, etc. No doubt these symptoms are strongly affected by the fright and mental shock that accompanies a bite, and by the pain incident to the treatment. Great excitement, confusion of mind, and extreme nervousness are occasionally mentioned. The victim may become comatose or unconscious in the early stages, as distinguished from the coma that may be expected in the final stages of fatal cases. The various manifestations of weakness may be very serious if they affect people when alone in the field and far from help.

Respiratory Difficulty. This characteristic neurotoxic symptom is not infrequently met in cases of Nearctic rattler bite, particularly in the later stages.

Nausea and Vomiting. These are occasional symptoms, as are also digestive disturbances such as diarrhea or constipation. Blood may sometimes be passed. Nausea may be a relatively early symptom.

Hemorrhage. There may be considerable oozing from the wound, owing to an anticoagulant action of the venom, and this may take place even though no drainage incisions have been made. There may also be bleeding from the mouth, nose, intestines, or kidneys.

Circulatory Disturbance. The pulse is often rapid (double normal), feeble, and fluttery or thready. There is a concomitant lowering of blood pressure, which may have serious secondary reactions. The temperature is generally subnormal, except in cases where an infection becomes important. An anemia of considerable severity and duration may follow.

Miscellaneous Effects. Gangrene is sometimes a direct venom effect, but more often follows the improper use of a tourniquet — that is, one that is too tight and permitted to remain too long with the pressure unrelieved.

Various nervous or systemic disturbances are occasionally observed, including convulsions. There may be local paralysis involving difficulty in swallowing and numbness of the face, especially of the lips and tongue. There may be excessive thirst. Cold sweats are not infrequently mentioned in case reports.

Although more typical of the Neotropical rattler, cases have been reported in which the bites seriously affected the vision, resulting in the inability to focus, retinal hemorrhage, or temporary blindness.

In view of this great variability in the symptoms of bites by Nearctic rattlesnakes, it is to be regretted that uniformity is not evident in at least one symptom upon which the diagnosing physician might place dependence. Studies have been made in the hope that a blood-smear test might be devised to determine accurately whether crotaline venom had entered the blood stream of the victim of a bite, but so far without success.

Symptomatic Case Reports by Species

There are available some case reports, either in the medical literature or among those that I have collected, wherein the species of the offending snake is known beyond question. I list a few of them below. The treatments inserted in parentheses are introduced because they often modify the subsequent symptoms.

Crotalus adamanteus. The details of a very serious bite by an eastern diamond rattler of large size have been given by several writers. The essential symptoms were instant pain "like two hot hypodermic needles," spontaneous bleeding from the wound, intense internal pain, hemorrhage from the mouth, low blood pressure, weak pulse, great swelling and black discoloration of the leg (the site of the bite) with accompanying severe pain.

The following symptoms are quoted from a recent (1952) account of an *adamanteus* bite in the side of the leg, near the knee cap:

This was the worst snake bite in the history of the Reptile Institute. Due to the great quantity of venom injected, different and heretofore undescribed symptoms were experienced. Twenty-five minutes after the bite [the victim] could not walk and it was difficult to breathe. Every muscle in his body jumped and twitched spasmodically, due to the neurotoxic effect of the venom. This continued for five days and was the most dreadful and exhausting experience of any of his many injuries. The hemolytic effect of the venom caused his right leg to swell and turn black from the ankle to hip. During the fourth and fifth days, which the doctor said were the most critical, he was too weak to talk. The hemolytic effect also caused an anemic condition by the fourth day, in spite of four blood transfusions. On the sixth day, like a miracle, a marked improvement was evident.

The symptoms evident in this serious case have been set forth in detail, partly in the words of the sufferer himself. Tingling of the hands, chest, and face, with numbness of the upper lip, were experienced soon after the bite. These are neurotoxic effects not usually so evident in bites by *adamanteus.* Some of these symptoms were still present when the patient left the hospital twenty-two days later. The pain, normally prominent in bites by this species, was apparently deadened by the large dose of venom, so that initially no pain was felt. Muscle twitching began five minutes after the accident. Paralysis of the legs prevented walking within half an hour. One peculiar symptom was that everything appeared yellow to the victim. This symptom has been observed in other bites by the eastern diamondback.

Crotalus atrox. The following are the symptoms of a *crotalus atrox* case, a bite in the right index finger: there was an instant stinging pain at the bite; one minute later the finger was stiff and black at the tip. (A tourniquet, incision, and suction were applied at two minutes.) At four minutes the stinging continued; at eleven minutes the finger was black from the end to the bite, and swelling had commenced. (The tourniquet was loosened at fifteen minutes. At twenty-four minutes blisters near the wound were punctured; at thirty-seven minutes, as the knuckle was swollen, the tourniquet was moved to the hand. At forty-two minutes the swelling continued and the tourniquet was moved to the wrist.) Discoloration and pain were now evident at the bend of the elbow. (More blisters around the bite were punctured.) At forty-seven minutes discoloration was evident from the wrist almost to the elbow; there was some sweating; the arm was discolored. (The tourniquet was again moved upward.) There was a tremulous feeling. At two hours the soreness was less below the elbow, but was worse at the inside of the elbow; there were many large blisters below the wrist, and much discoloration on the inside of the upper and lower arm. Later symptoms were obscured by the treatment. Eventually the bite caused the loss of the end of the finger.

Crotalus viridis viridis. This summary of the symptoms of *viridis* bite (a composite of several experiences) was given in 1946: there is an immediate sensation like the sting of a wasp; there is little or no bleeding; a spreading feeling of numbness, especially at the tongue and lips; the tips of the fingers and toes tingle; there is a tendency to faint, and thereafter to remain in a coma for some time; nausea and vomiting are usually present; swelling begins about ten minutes after the bite, with discoloration, and continues with severe pain for thirty-six to forty-eight hours.

In a bite by a prairie rattlesnake involving a child, there were marked neurotoxic symptoms in an arched back, hypertonic muscles, difficult respiration, vomiting, regular convulsive movements, and an inability to recognize people. A duodenal ulcer became evident four days after this bite and was judged to be of neurotoxic origin.

Crotalus viridis oreganus. In a case of a bite by a juvenile of this subspecies, the snake sank its fangs in the middle joint of the left forefinger. There was tingling of the scalp and the soles of the feet. There was nausea within twenty minutes; this lasted for the rest of the day. The swelling was rapid and severe, with the arm swollen to twice its normal size. The hemorrhagic discoloration covered the entire arm and extended onto the body. Aching and throbbing at the site of the bite increased during the first twelve hours and became almost unbearable. The patient collapsed once while trying to stand. The swelling was evident for several weeks.

Crotalus viridis helleri. In a case involving this snake, there was pain at the point of the injury; a difficulty in speaking because of stiffness of the tongue and lips; a state of shock; profuse perspiration; a pallid face; the pulse rapid and thready; and drooling at the mouth.

Case Duration

In rattlesnake-bite cases, with their many variable factors affecting severity, there is a corresponding variability in the duration of the case and the period requiring treatment or precautionary observation. Naturally much depends on the treatment and its efficacy. Complications, such as infections or gangrene, sometimes caused by the treatment rather than the bite, may also prove important.

In 1642 it was written:

There are Rattle Snakes (in New England), which sometimes doe some harme, not much; He that is stung with any of them, or bitten, he turns of the colour of the Snake, all over his body, blew, white, and greene spotted; and swelling, dyes, unlesse he timely get some Snake-weed; which if he eats, and rub on the wound, he may haply recover, but feele it a long while in his bones and body.

It has been stated that in uncomplicated cases patients should be discharged from the hospital in from five to seven days after the accident; that recovery from constitutional symptoms should be complete in two or three days, often a few hours, but there would be unhealed lesions of longer duration; that, with good treatment, half the cases should be completely recovered in ten to twelve days.

Some more specific case durations are shown by three cases reported to me: a man bitten in the arm by a northern Pacific rattler, *C. v. oreganus,* was able to return to work in six days; a man bitten in the ankle by a rattler of the same subspecies, was in the hospital for three days, and was away from work for twenty days; a boy bitten in the thumb by a red diamond rattler, *C. r. ruber,* was hospitalized for six days. In another bite by the same species, a man passed the crisis in seventy-two hours, although he did not recover for several weeks.

Some data are available on the duration of fatal cases of rattlesnake bite. The presentation of these figures should not be taken as an indication of a high mortality rate, for fatalities involve only a small proportion of all cases, as has been pointed out elsewhere.

From the statistical summaries on this subject, I conclude that the first day, and especially the last part of it, is the most dangerous period. I have had six reports from correspondents of deaths in forty-five minutes, one hour, three hours, four hours, seven hours, and fourteen hours.

As a matter of reassurance, in closing this discussion of the rapidity of death from rattlesnake bite, I wish to repeat that the fatalities from this cause for the entire United States, with a population of over 160 million, seldom exceed thirty per year. Of every one hundred people bitten by rattlesnakes, only about three will die.

Causes of Fatalities

With all of the variable symptoms of Nearctic rattlesnake bite, what physiological effect, then, leads to an occasional fatality? It is difficult to say; with so many complicated interrelations of symptoms, it is usually impossible to select the fatal one or the organ that fails. Many of the fatalities of the past have been secondary in character—from wound infections, or gangrene resulting from such infections, or from the improper use of a tourniquet. There has been a feeling on the part of some physicians that neurotoxins are more important in Nearctic rattler bites than generally supposed, although their more insidious effects are masked by the much more apparent and dangerous-looking hemorrhagic symptoms, and that the neurotoxic effects reach a crisis later.

It has also been stated that in quickly fatal cases of rattlesnake bite, the hemorrhagins of the venom were the causal factors, but deaths after the first day—assuming no infection or similar complication—were likely to have been caused by the neurotoxins.

With better methods of treatment—antivenin, scientific drainage, the prevention of infection, supportive transfusions, and a closer watch for any sudden onset of adverse symptoms—such unfortunate terminations should be largely avoided, where it is possible to get the victim to a hospital soon after the accident.

The Symptoms of *Crotalus durissus* Bite

With respect to the Neotropical rattlesnake, *Crotalus durissus,* we find the symptoms rather sharply differentiated from those of the Nearctic rattlers with which we are more familiar, and which are the source of our rattlesnake-bite problems in the United States.

The symptoms of the bite of the Brazilian rattlers in cases involving human beings have been summarized by one writer as follows: impairment of vision or complete blindness; paralysis of the eyelids and eyeballs; and paralysis of the peripheral muscles, especially of the neck, which becomes so limp as to appear broken. He also comments on the lack of local pain, such as is almost invariably and immediately evident in Nearctic cases. Another writer mentions paralysis of the respiratory system, with death by suffocation. Hemorrhagins may be present in *terrificus* venom but they are completely overshadowed by the startling and serious neurotoxic effects.

Infection in Rattlesnake Bite

Infection often follows rattlesnake bite—as well as the bites of other venomous snakes—for the venom contains a principle that inhibits the effectiveness of the normal antibacterial property of the victim's blood. As the snake's mouth is always contaminated with bacteria, and field remedies are seldom applied with due regard for asepsis, an infection of a serious nature may complicate and delay recovery, even leading to a fatal outcome. It is to be hoped that modern drugs will greatly reduce the danger from this by-product of snake bite.

Permanent Aftereffects

Any permanent aftereffects of rattlesnake bite depend on the seriousness of the bite, and, still more, on the treatment. As in any wound treatments, necrosis may result, or nerves, tendons, or muscles may suffer permanent damage, especially in the case of inexpert field incisions. Gangrene resulting from overtight or overprolonged tourniquets may necessitate limb amputations. But permanent systemic ill effects are probably rare, notwithstanding the statement made in 1805 that those who survive rattlesnake bite are always "sickly and sensible to changes of the atmosphere," or in 1872 that snake bite might stunt a child.

Effect of Venom on the Eyes and Other Organs

The fact that rattlers are not natural venom spitters has been touched on elsewhere. However, there may be cases in which a snake, falling short in a strike, or having a damaged mouth, may expel venom that reaches the eyes of the person at whom the strike is aimed. It is a relief to know that in such circumstances no serious effects are to be expected. The eyes do become irritated and inflamed, but if washed promptly should clear up in three hours or so. (Some snake venoms, especially those of the spitting cobras, may have serious effects, leading to blindness, if the venom is not quickly washed out.)

According to researchers, various organs are susceptible to the action of rattlesnake venom, which exercises a destructive affect on the functional cells. There may be degenerative changes in the liver; yellow eyes and skin from a mild jaundice are common in severe cases. The kidneys, also, are affected by the destructive action of the venom. The spleen can become enlarged and congested, and the duodenum and lower colon filled with blood.

The Recurrence of Snake-Bite Symptoms

Throughout the literature of snake bite, there is frequent mention of the recurrence of symptoms, usually on anniversaries of the accident, but sometimes at other intervals.

One colonial writer records the case of a man who was badly frightened (but not bitten) by a rattler and had recurring symptoms every day for fourteen days; and of a dog that had to be shot, so acute were the annual symptoms of the bite.

There is a case report from army records in Texas, of a new and painful swelling that necessitated amputation of a finger some three years after a rattler bite. We may hazard the guess that a totally unrelated infection was involved. Another report tells of a boy who became speechless at annual intervals following a rattlesnake bite.

It need hardly be said that the medical profession gives no credence to these recurrence beliefs.

SUSCEPTIBILITY, RESISTANCE, AND IMMUNITY TO RATTLESNAKE VENOM

Not only are some snake venoms much more toxic than others, but some venoms are more toxic to one kind of animal, whereas other venoms are more effective against another animal. So, in all questions of susceptibility and immunization, both the kind of snake and the kind of victim are important. In general, venoms are likely to be particularly effective on the natural prey of any species of snake.

It has long been known that cold-blooded animals are less susceptible to snake venoms than birds and mammals.

Immunity to Snake Bite in Man

Famed among the ancients were the Psylli, the Marsi, and the Ophiogenes. These tribes had a way with snakes and, best of all, they were immune to snake bite. The Psylli lived in Asia Minor near Rhodes, or, possibly, in north Africa; the Marsi inhabited central Italy, and the Ophiogenes the region of the Hellespont. Some of the same powers were ascribed to the Syrians on the Euphrates.

To summarize the reports of Pliny and others: these peoples were not only immune to snake bite, but they could cure those bitten, by the touch of a hand or the use of saliva; they were fatal to any snake that bit them, and their odor drove snakes away; they exposed their children to snake bite to test their legitimacy.

It is probable that the stories of these snake people originated with travelers' tales of the exploits of the snake charmers in the countries visited. And, it has been pointed out, none of the snakes found in the countries of the Psylli and Marsi were very dangerous, which accounted for the belief in their immunity. In ancient days it was the general belief that all snakes were deadly.

Sometimes certain elements of a population were thought to be immune. Some writers of colonial days thought the Indians immune to rattlesnake bite. There is a fairly widespread myth in the South and West to the effect that blacks are immune to rattlesnake bite. Occasionally a belief in natural immunity among men is still voiced, but there is no present acceptance of the idea that any people are immune, although some are more susceptible than others, as is true of individual susceptibility to other protein poisons.

Acquired Immunity. We turn now from beliefs in natural to those of acquired immunity, acquired either by eating some substance — usually a snake-bite remedy, such as some plant, a piece of snake, or some snake venom — or by the injection of an immunizing substance into the body. The advantage of taking the cure before the accident is an obvious one. We must not censure the ancients too severely for their theories, remembering that until rather recently it was generally believed that because whiskey was a sure cure for rattlesnake bite — in truth, it is worse than useless — anyone who had imbibed too freely was immune to snake bite.

There has been a limited modern belief in the venom-imbibing method of acquiring at least a slight immunity, but one recent experimenter fed venom from several kinds of rattlesnakes to animals and found no resultant immunity.

Pioneers in the Pacific Northwest believed that to eat a rattler's heart would produce immunity. There is a belief along the Rio Grande in Texas that huaco bulbs in whiskey or mescal will afford immunity against rattler bite.

That immunity may be acquired by snake-show operators through their having been bitten repeatedly is a common belief, and may, in accordance with modern theories of immunization, have some slight basis in fact. A showman, who said he had been bitten 288 times, claimed complete immunity. Yet another wrote me that his latest bite was the most serious he had ever experienced. Peter Gruber (Rattlesnake Pete), the keeper of a museum, was said to have been bitten thirty times in forty years; no mention was made of his having acquired any immunity.

From immunity acquired through one or more accidental bites, it is a natural step to immunity secured through deliberate injections. One writer said she had heard that the natives near Tampico, Mexico, were in the habit of inoculating themselves with rattlesnake venom as a protection against all venomous snakes. The venom was introduced into the tongue, both arms, and other parts of the body, by the use of a snake fang. There followed a rash lasting several days, after which there was complete immunity. Collaterally there was acquired the power to call snakes, when wanted, to cure other people; or, alternatively, to cause the snakes to kill people by biting them.

A friend of mine, an experimental pathologist, once immunized himself against rattlesnake venom until his blood had a high titer; unfortunately, he did not test for the duration of the acquired immunity.

With the problem of immunization arises also that of anaphylaxis, or hypersensitivity.* As early as 1892, it was pointed out that repeated doses of venom are more dangerous than the same total amount in a single dose. A sufficient time must intervene between immunizing doses or the cumulative effect is marked.

Is the snake-bite risk to any person sufficiently grave to justify immunization? Certainly not, except in the case of operators of snake pits in carnivals or similar places of amusement. And if these operators are frequently bitten, one can only suggest that they had better engage in some other occupation where carelessness is not so strongly penalized.

Susceptibility of Domestic Animals

Elsewhere, in discussing stock losses from rattlesnake bite, I present a number of experiences of my correspondents who mention, in a general way, the degree to which domestic animals are affected by rattlesnake bite.

There seem to be no marked differences in the effects on cattle, as compared to horses. Although hogs were, for a long time, believed to be immune to rattlesnake bite it was early evident that the supposed immunity was only the protection afforded by the layer of fat that prevented the venom from entering the circulation. In 1892 the queer idea was reported that the degree of susceptibility of hogs to rattlesnake bite depends on the color of the hog.

The belief is occasionally expressed that animals, especially dogs, if repeatedly bitten, become immune just as snake-pit operators are believed to acquire immunity. One writer even permitted his dog to be bitten, an exceedingly cruel expedient, in the hope that he would then have a "rattlesnake dog," that is, a rattlesnake killer unafraid of the snakes. Several writers mention instances of immunity in dogs, and my correspondents have done likewise; but tests have shown that dogs have an average susceptibility.

Immunity in Mammals That Prey on Snakes

In the United States, it is not known whether any of the mammals that eat rattlers have any conspicuous resistance. There is no mammal of which rattlesnakes comprise more than a minor or incidental part of the diet, although rattlers are occasionally eaten by coyotes, foxes, badgers, peccaries, opossums, wildcats, racoons, and skunks. So far as I know, comparative immunity tests have not been made on any of these.

---

*A person can develop hypersensitivity to either venom or antivenin. In either case, future envenomation or treatment with antivenin can cause a fatal anaphylaxis.

Susceptibility of Reptiles

Said Pliny: "The sting of the serpent is not aimed at the serpent." Gradually the truth of this statement has been verified — snakes are relatively immune to snake bite. The immunity is not a complete one, and some snake species are less affected than others, but compared with that of mammals and birds the susceptibility of snakes is low.

At the San Diego Zoo we have seen a number of harmless snakes accidentally bitten by rattlers. Ill effects are sometimes observed. The most apparent symptoms, when there are any, comprise a swelling at the bite and sluggishness. Fatalities are rare, but they do occur.

Immunity of King Snakes. King snakes, the presumed deadly enemies of rattlesnakes, have long been thought to be immune to rattlesnake bite. I say "presumed enemies of rattlers" because a good deal of exaggeration has grown up concerning this relationship. There is a widely dispersed legend to the effect that king snakes spend their lives trying to clear the world of rattlers; that whenever one comes upon a rattler, usually after trailing it for weeks, he kills it by constriction, and then, taking a bow from the cheering audience, starts in quest of another.

But the truth isn't quite so felicitous. Actually, king snakes feed on snakes, as well as on mammals, birds, and lizards, and when, if hungry, one comes upon a rattler small enough to be eaten, it is swallowed just as it would have been had it been a gopher snake or some other harmless species. The story of the king snake that kills a rattler much too large to engulf is to be doubted, even if told you by one of your favorite and most trustworthy relatives. Hundreds of such stories have appeared in print; other hundreds have been told me by eyewitnesses with whom I didn't argue, but still I am unconvinced.

However, there is one feature of the rattler–king-snake relationship that may be credited, namely, the relative immunity of the king snake to rattlesnake bite, for the resistance is much greater than in mammals of similar size. I describe elsewhere in some detail the stereotyped pattern followed by the king snake, in swallowing a young rattler. In four out of six cases observed, the rattler was not seized by the snout, and thus was free to bite the king snake, which it did, usually several times. These bites were ignored by the king snake, and no subsequent harm was suffered from them. Hence, although it cannot be said that king snakes are completely immune to rattler bite, we have evidence that they are unaffected by a dose of venom that would make a man, weighing about three hundred times as much, seriously — and even dangerously — ill, if left without treatment as the king snake was.

Immunity of Rattlesnakes to Their Own and Other Venoms.

Since colonial days it has been quite generally believed that rattlesnakes, if cornered or injured, will bite themselves with suicidal intent, and that the venom is quickly fatal, that is, within seconds or minutes.

That rattlesnakes do sometimes bite themselves when attacked or injured is unquestioned, but certainly they have no suicidal objective. They strike wildly at anything that moves, or they even lunge or bite without aiming at any particular object. Quite by accident they may hit some part of their own bodies, and, having done so, and feeling a yielding substance, they may inject venom. They are particularly likely to bite themselves if they are held down with a stick. There the likeness to the story stops. They do not die instantly; in fact, they would rarely die at all were it not for the injuries that were originally the cause of the wild strike.

The literature discussing the degree of resistance that rattlesnakes have to their own venoms, to the venoms of rattlesnakes of other species, and to the venoms of snakes other than rattlesnakes, is quite extensive. Some immunity tests have been well conducted and are quite convincing. In general, the conclusions are: rattlesnakes are quite resistant to their own venoms, slightly less so to rattlesnakes of other species, and still less immune to the bites of snakes of other genera. Occasionally a fatality may be explained by the possibility that the snake's fang punctured the spinal cord or some vital organ, or that an unusually large amount of venom was ejected.

It should be understood that all of the really informative tests on immunity have been carried out by the injection of measured quantities of venom. Tests in which one snake is allowed to bite another rarely prove anything, since the quantity of venom injected is never known.

Effect of Rattlesnake Venoms on Plants

Some colonial stories, undoubtedly descended from Old World myths, tell how rattlesnake venom will permeate a plant and cause its destruction. The affected plant sometimes splits, and sometimes withers and dies.

Real tests on the effects of rattlesnake venom on plants began in 1852, when venom on the point of a pen knife was inserted into the shoots of young plants. The plants withered above the cut, but no controls were used, and subsequent investigators have thought that the damage was done by the knife and not by the venom. One researcher inoculated vegetables with venom on the point of a lancet; the next day they were withered and dead.

More recently a number of experiments were made on plants, first on yeast, then on higher plants. This time controls were used, and the findings were negative; there were no resultant differences between the poisoned and nonpoisoned plants.

In 1932 experiments on plant seedlings showed rattlesnake venom toxic to plant protoplasm. This, of course, is not equivalent to saying that the beliefs in the destruction of large plants are now vindicated.

Treatment and Prevention of the Bite

THE TREATMENTS PROPOSED: THEIR HISTORY AND PRESENT STATUS

When I was a youngster, spending my summer vacations hunting and hiking in the mountains back of San Diego—this was in the 1890s—a new and certain cure for rattlesnake bite was announced. This was ammonia; and as soon as I, like many of my fellows, had secured a bottle to carry wherever I might go, I felt a great relief, for I knew that I had a sure protection against a danger that had previously caused some concern. So thereafter I tramped without worry in the granite and chaparral, where rattlers were moderately common.

This might be taken as an example of the way that snake-bite cures come and go, become popular and are discarded, only to be revived anew. For ammonia was not a new remedy in the 1890s; it had been advocated as early as 1738, and had been tested and found without merit in 1765. It was used again from time to time in the

195

early 1800s. My experience was the result of only one of several rediscoveries or revivals of ammonia as an unfailing cure; and this example of my unwarranted confidence is typical of the undeserved reputations of remedies for snake bite through the ages.

As one studies the case reports of snake bites, including rattlesnake bites, one cannot but be amazed at the number of remedies that have come into prominence, that have been adopted with enthusiasm, only to be discarded eventually as worthless. Whence did these cures emanate and how were they authenticated?

They have come from a variety of sources—from pre-Columbian Indian medicine men; from European importations tracing back to ancient Greece and Rome; from the alchemists of the Middle Ages; and from the biologists and physicians of colonial and modern times. And even after they have been discarded by reputable practitioners, or have been shown by experiment to be valueless, they still persist as folklore cures in the more remote areas.

Some will wonder how these cures ever gained the confidence of so many people, and why this trust continued even after they were shown by test to have no value. As to the ancient cures, this question is not hard to answer, for in the old days all snakes, and many harmless creatures such as lizards, salamanders, and frogs, were thought to be dangerously venomous. So when an application of radishes or goat's-milk cheese—two of the hundreds of remedies recorded by Pliny—cured the bite of a reputedly fatal but actually harmless snake, they became accredited remedies.

In the case of the rattlesnakes, the explanation is not quite so simple, yet it is basically the same. All rattlesnakes are venomous, most of them to a serious degree, but many persons are unable to distinguish a rattler from harmless species, despite the infallible clue of the rattle. Furthermore, rattlesnake bite is not as serious as sometimes supposed; often it is stated to be invariably fatal, which is far from the truth. But most important is the fact that many variable conditions involved in every snake-bite case often reduce the gravity of the bite to such a degree that there is no real danger. So the remedies gain their reputations by saving people from dangers that never existed; and these reputations persist despite their shaky foundations.

In discussing remedies, I have characterized some of them as folklore cures. It might be assumed that the application of folklore remedies can still be justified on the theory that every possible source of relief is warranted in the case of so serious a contingency as rattlesnake bite. No delusion could be more unfortunate. These remedies may in themselves be dangerous; they delay the application of the truly effective medicines known today; and they confuse the course and treatment of the malady itself. It is unfortunate that under the stress of such an accident as a snake bite, people are likely to revert to methods that, in less excited moments, they would ridicule. At a loss what to do, they apply some such fantastic remedy as the split-chicken treatment because years ago there was a family tradition that it once cured some second cousin.

Scarification, Incision, and Suction

The method of snake-bite treatment, in which the wound is enlarged and suction is applied to remove venom, is so old and so natural that its derivation is lost in the ancient rites of medicine men and witch doctors. Undoubtedly it was invented independently in many times and places. Whether it was thought that the danger from a bite lay in the liquid venom injected, or, as was believed by some American Indian tribes, in the introduction by the snake of some solid object into the wound, it was logical to try to remove the poisonous substance at the place of entry. So the instinctive sucking of an injury, as one claps an insect sting to the mouth, developed into more elaborate, and sometimes more successful, methods of withdrawing a part of the venom. This is one of the few of the aboriginal treatments still considered of value.

The extent of the scarification or incision used varied considerably at different times and places. Sometimes the scarification was more cere-monial than effective; it might be used rather to facilitate the introduction of remedies than to withdraw venom. But often relatively deep incisions were made, and mechanical devices were used to increase the flow of blood. Cupping glasses were well known to the ancients, and leeches were also employed. More recently various rubber or plastic cupping, or syringe-type devices have been employed; they are usually included as an important component of modern field-use, snake-bite kits. The most primitive of all methods—suction by mouth—I shall discuss later.

There has always been a difference of opinion as to the effectiveness of the suction treatment; it is admitted that mere scarification or deeper incision can have little value without subsequent suction to withdraw venom. It is certain that the effectiveness depends considerably on the kind of snake involved and the site of the bite, since the rate of absorption of the venom is affected by the character of the venom and the nature of the tissues in which it is deposited. Suction is most likely to be efficacious in cases that involve the rattlesnakes, most pit vipers, and the smaller vipers.

There has been some dispute as to the danger to anyone sucking the bite of a venomous snake with his mouth. Such danger as there may be could come as a consequence of the venom's entrance into a mouth sore. It has been known for a long time that most snake venoms are relatively innocuous when taken internally, in other than massive doses.

As to rattlesnake venom, a writer in 1854 reported having swallowed the contents of a rattlesnake's gland without ill effects. Another in 1943 swallowed, by accident, the venom from fifteen western diamondbacks (*C. atrox*) without suffering any difficulty. A showman was said to regularly drink rattler venom. But another writer took some internally and was nauseated.

If it is presumed that anyone sucking a rattlesnake bite will spit out the exudate, there is certainly no danger from the small amount of venom that might be swallowed, so that this hazard, from a practical standpoint, can be ignored. The chance of venom getting into a mouth sore or defective tooth is probably worthy of somewhat more attention; at least it has been mentioned rather frequently as involving an appreciable hazard.

Rather logically, but uselessly, it has been suggested that various presumed antidotes, such as olive oil, wine, salt and garlic, or milk should be used when sucking a bite.

Some stories, probably mythical, of the dire effects of sucking a snake bite have been noted. One tells how a black slave sucked a rattler bite and saved his master's life, but his own head swelled to "a frightful degree" and he died.

It is to be remembered that the sucking procedure also has the objectionable feature of possibly introducing pathogenic bacteria into the wound. However, it may be doubted whether this would seriously increase the danger of contamination already resulting from the pathogenic condition of the snake's fangs and the sepsis involved in a field incision.

Several items of folklore in this connection are of interest. It is a folklore belief of the blacks of Mississippi that a person with red gums can draw out venom, but if a black with blue gums should attempt it, disaster will follow. There is an Illinois folklore belief that it is safe to suck a bite through a silk handkerchief. Nebraska folklore has it that if one sucks a rattler bite, his teeth will fall out.

Some folklore has also crept into the making of incisions. One writer said that the bite should be enlarged with lizard fangs, because they contain a counter poison—a fantastic idea. This was in Baja California. According to Nebraska folklore, the skin must be punctured with soapweed points to let the venom run out. In New Mexico and Texas, yucca spines are used, as in the treatment of bites suffered by livestock.

Excision or Amputation

The treatment of snake bite by removal of the bitten part—by cutting it out or off—is an ancient remedy. It has been used for rattlesnake bite since colonial days, although now with decreasing frequency.

A few cases of amputation for rattler bite may be cited: a woman who had defended her child against a rattler was bitten in the finger, and a man saved her life—so it was claimed—by cutting the finger off. In 1866 a man chopped off his big toe when bitten by a rattler. A tannery barker cut off his thumb when he was bitten by a small rattler. The story is told of Three-fingered Smith of Idaho, who had seized an ax and chopped off two fingers when bitten by a rattler. One of my correspondents writes that the chief-of-party of a government surveying crew in Arizona cut off the finger of one of his men when bitten. Under date of December 26, 1948, there appeared in the newspapers the account of a young man in Mississippi who shot off the index finger of his right hand when he was struck there by a rattler.

I think it may be stated that the amputation of a finger or toe in the field is never justified in the case of a bite by a Nearctic rattler, because a constriction band, with incision and suction, will provide as beneficial a result with much less damage.

Cauterization

Cauterizing a snake-bite wound, by means of either heat or chemicals, was a drastic, painful, and damaging snake-bite remedy that had its advocates through many years, although today it is only a folklore cure restricted to remote districts. The lack of benefit from cauterization as from excision or amputation, is the result of the rapid absorption of the venom into the bodily tissues—its failure to remain in a localized pocket, where it might be burned out and destroyed.

Cauterization is very old, and the adoption of cauterization for rattlesnake bite was a transplant from the Old World. Cauterization with a hot iron was recommended for rattlesnake bite as early as 1648 as a second line of defense, if a certain herb was not to be had. The cowboys, as a result of their frequent use of branding irons, often treated wounds with a hot iron. Hot coals were also used.

Another method of cauterization used on the Western frontier was to sear the wound by lighting a patch of black powder placed over it. A small brass or iron ring was customarily carried by miners, stockmen, herders, and Indians in frontier days, to hold powder on a snake bite in order to facilitate cauterization. Unexploded gunpowder, moistened and mixed with salt, was used extensively in the form of a poultice.

A homeopath, writing in 1853, favored the use of a hot coal or hot iron, but held far enough from the skin or wound so as not to cause a burn. Even a lighted cigar would suffice since, he believed, only a mild, dry heat was needed to destroy the venom.

Of the chemical caustics that were used in the treatment of rattlesnake bite, the one most often mentioned is silver nitrate. Others are nitric acid, potassium hydrate, and carbolic acid.

From the first, animal experiments with cauterization have proved disappointing, although occasionally there were sufficient indications of partial relief to confuse the situation. At least, it was shown that even a slight benefit could be obtained by the application of the cautery only if it were applied within a few minutes, at most, after the bite. Since this speed was rarely attained in the treatment of rattlesnake bite, it is probable that in almost every case so treated, the unfortunate victim suffered from this painful remedy in vain; if he recovered he would have done so without the cauterization.

The Tourniquet, Ligature, or Constriction Band

The tourniquet or constriction band is a means of delaying venom absorption. It is often combined with incision and suction, since it may hold venom in tissues close to the bite where it may be withdrawn, at least in part, by suction. But, more than that, it may impede venom absorption to such an extent that the system defenses are not fatally affected.

The use of a tourniquet for impeding circulation is very old, and was recommended for snake bite by Celsus, an encyclopedist of the Augustan era in Rome. Some Indian tribes used either actual or ceremonial ligatures before the coming of Europeans. The ligature was a standard element of treatment among the colonists and pioneers, particularly as an adjunct of suction. A newspaper report of May 13, 1953 told of a Mexican in San Diego County who made a tourniquet of the skin of the rattler that bit him.

It might be natural to assume that so long as the venom from a bite could be restricted to some local area — a finger or a hand, for example — the patient's life would be safeguarded; therefore it would only be necessary to effect such a restriction with a tourniquet or constriction band until the venom could be neutralized by some appropriate antidote. But the danger from gangrene was early recognized. Unfortunately, even today this danger is not given due weight, especially by laymen.

As a matter of fact, part of the venom is diffused through the body not via the blood vessels, but is carried by the lymphatics, as was recognized as early as 1793. Only moderate pressure, as provided by a constriction band, is required to restrict lymph flow, and thus a tight tourniquet is not only dangerous but unnecessary.

Freezing and Cold Packs

Freezing with an ethyl-chloride spray or ice has been suggested, on the theory that a low temperature will delay venom absorption and reduce the chemical activity of the venom. Although freezing or long-term cooling is no longer advised by the medical profession, a few physicians suggest the local application of ice as a first-aid measure.

Some rattlesnake-bite patients treated by freezing have suffered permanent limb damage.

Cures Derived from the Snake

One of the more persistent tenets of folklore medicine is that the cause of an injury will also furnish the cure, if the method of application can only be discovered. Thus, the use of snake parts to cure snake bite is both ancient and widespread.

In general, the ancient cures involved the use either of all of the snake that caused the injury (or one of the same kind) or of some part such as its head or liver, prepared in various ways, often mixed with other ingredients. The applications were sometimes external, sometimes internal.

Whether these cures were derived from European sources or from the Indians with whom they came in contact, the early explorers and emigrants used snake parts for the cure of rattlesnake bites, and these prescriptions still persist as folklore cures in many sections of the United States and elsewhere in the Americas.

In Brazil, rattler's head, mashed and applied to the wound was prescribed in 1648. In New England in 1672 it was recommended that the rattler's heart be swallowed while fresh, or be dried and pulverized and taken in wine or beer; and that its liver be bruised and applied to the bite. In the 1700s rattlesnake oil was reported to be beneficial, as was rattler fat rubbed into the bite.

As to present-day folklore remedies in the United States, we have various methods reported. Kentucky: bind the liver and intestines of the guilty snake to the bite. Mississippi black folklore: apply to the bite a piece of the snake's body, on the theory that it will reabsorb the venom. Oklahoma: apply the warm flesh of the rattler, or oil prepared from its fat, to the bite. Texas, Mexican folk-cure: let the victim bite off the head of the offending snake. Ozark Mountains: bind a piece of the rattler's flesh to the wound; but always kill the snake and burn it before applying any treatment. Nebraska: cut the rattler into three-inch pieces, split and apply to the bite. Tennessee: drink rattler oil. New Mexico, Spanish-speaking people: grab the snake by the head and tail, and bite it in the middle, whereupon the poison in the person's mouth will kill the snake and cure the bite.

One of my correspondents reports the following:

I saw one man that was bitten, a sheepherder. He killed the snake and used the carcass for a poultice, then used mashed raw potato and tobacco for another poultice. He stayed in camp and cured himself. I hope nobody else tries this, though.

The Split-Chicken Treatment

One of the most widespread and uniform of the folklore cures for snake bite is that which may be termed the split-chicken remedy, for chickens were most often used, although other animals have had some acceptance. In this treatment, a live chicken is split or slit and the bleeding flesh is immediately applied to the snake bite as a poultice. According to popular belief, the chicken's flesh should turn green, or its comb blue, from the venom that it draws from the wound, whereupon another chicken should be applied until the abnormal color no longer appears, after which the patient is assumed to be cured. In some forms of treatment the chicken, or other animal, is killed in preparing it for application; in others the flesh is bared without a fatal injury. Needless to say, not the slightest benefit has ever been shown to result from this cruel method, yet it is still practiced in some backwoods areas.

Other animals reportedly used are toads or frogs, rabbits, and a turkey buzzard crop. In the cattle country, a cow might be killed, the abdominal cavity split, and the injured arm or leg of the human sufferer plunged in and left until the body of the cow became cold.

In an early folklore cure in the Middle West the bloody skin of a freshly killed skunk or black cat was applied to a bite.

Some variations of the split-chicken remedy have been reported: a chicken wing might be used; the chicken must be black; the chicken must be carefully buried after the treatment; a decapitated black rooster must be used, and if the treatment is successful all its feathers fall out.

The Mexicans around Santa Fe, New Mexico, believed that several chickens (usually three) had to be used, and the cure was complete when a chicken survived. It is said that in Nebraska as many as twelve chickens have been used in a single case.

Milk

Milk is an ancient remedy for snake bite; it dates back to Pliny. Both cow's milk and goat's milk were used; the application was either external, in the form of a poultice, or as a drink. It seems never to have been particularly popular in the treatment of rattlesnake bite, although it was said to be in use in the Western settlements in 1793. Snakeweed boiled in sweet milk is one of the Ozark folklore cures. In a recent West Virginia case reported in the press, the victim was given milk to drink. As liquids are sometimes recommended in the modern therapeutics of rattlesnake bite, milk taken internally might be beneficial, but there is nothing to recommend its external application.

Miscellaneous Animal Products

Several folklore treatments that apply animal products to rattlesnake bite are of Old World derivation. Among these is the saliva of a fasting man, a poultice of melted cheese, and eggs in various forms.

The scrapings of a crocodile's tooth were used for rattlesnake bite both in Baja California in 1772 and in Paraguay in 1784. Although Pliny suggested powdered human teeth and fails to mention this use of crocodile teeth—but he does describe nineteen other remedies derived from crocodiles—we may be sure that so unusual a cure, noted in such widely separated places as Paraguay and Baja California (where there are no crocodiles), must have had a common European origin. Powdered crawfish were used near Niagara Falls in the late 1700s, and mosquito bites were suggested in 1844.

Treatments that involve various kinds of excrement are of ancient origin. They are mentioned in some of the earliest accounts of the rattlesnake (1615, 1628) as well as in some current folk cures.

Botanical Cures and Vegetable Products

In colonial days and for a time thereafter, various plant cures for rattlesnake bite were the most popular and generally used of all treatments. Sometimes extracts or infusions of the leaves or roots were given internally; sometimes poultices or fomentations were applied; often both methods were used together. Although many of these cures were attributed to the Indians, who were supposed to have that deep knowledge of nature often attributed to primitive people, others were derived from European and Asiatic lore, for plant cures for snake bite had an extensive place in ancient and medieval pharmacology. Nicander, in about 150 B.C. listed many of these plants. Pliny, in his great *Natural History*, records no less than 196 ways in which plant preparations could be used to cure snake bite; probably at least a hundred different plants were involved. Often they were to be mixed with such substances as wine, vinegar, honey, or grease.

In the eighteenth century it was reported that rattlesnake bite was extremely dangerous, but fortunately various herbs wrought certain and immediate cures. These were known to the Indians, who sometimes would, but at other times would not, divulge them to the white man. Through a kindly Providence the particular plant needed was always available where rattlesnakes were found.

When the Indian cures failed, the cause was usually attributed by them to some infringement of ceremonial detail, for their treatments often had a mystical basis. The whites were more likely to cite some botanical misidentification.

In 1741 a writer stated that he regretted that the root for curing rattler bite lost its effectiveness in crossing the ocean. Evidently the skeptical Britishers had tested its reputed effectiveness.

In 1768 it was claimed that one form of rattlesnake root would cure even if the patient had been bitten several days before, provided the weather had been moderate in the meantime.

In 1833 it was reported that two women almost lost their lives by taking the wrong plant by mistake, the remedy being worse than the bite.

The value of plants fell into disrepute as methods of animal experimentation were perfected.

Caesar's Cure. One of the most famous and romantic of the early plant cures for rattlesnake bite was Caesar's cure. Caesar was a black slave in Carolina in 1750. Sometime prior to that time he had devised an antidote for poisons and another for rattlesnake bite. These he agreed to divulge for his freedom and a monetary reward. Tests also were agreed upon. Evidently the tests proved satisfactory, for, in the *Carolina Gazette* of May 9, 1750, the details of the remedy were published, as commanded by the Commons House of Assembly, which had bought Caesar's freedom and awarded him a pension of £100 annually for life.

Plantain roots and horehound were the bases of both the poison antidote and the rattlesnake remedy. For the bite, an infusion of these plants was to be taken internally, and a leaf of tobacco moistened with rum was applied to the bite. Caesar's antidote, which is quite worthless for both purposes for which it was devised, does not differ greatly from other botanical cures popular in the eighteenth century.

The test, by which Caesar proved the efficacy of his cure to the legislators, seems to have been made by him and some of his fellow slaves. They jumped into a tub of large rattlers and allowed themselves to be bitten. They had no fear because of their confidence in Caesar's cure. We may guess that the rattlers had been defanged or milked before the test.

Onions and Garlic. Onions and garlic were important remedies for snake bite in the ancient and medieval worlds; they were also employed as repellents. Sometimes the onions were applied in poultices, sometimes they were eaten. The spread of their use to the United States might well have been expected, and they were reported to have been tried as early as 1753 in Louisiana. They later became a regular folklore cure. Sometimes a fresh onion was cut and held against the wound. It was the belief that when the onion turned green it was actively abstracting the venom, and a fresh one should be applied from time to time. Onions were used by cowboys in early Pennsylvania, in New Mexico, by the Pennsylvania Germans, by Midwestern pioneers, and by California miners in the gold-rush days.

Tobacco. An early and widespread snake-bite cure in the United States was tobacco dating back to 1615. In the days when the chewing of tobacco was a much more prevalent habit than today, freshly chewed tobacco became a standard first-aid poultice for cuts and bruises, and the extension to snake bite was inevitable. Occasionally it was mixed with gunpowder. Sometimes strong tobacco juice was taken internally for rattlesnake bite, and if the patient failed to get sick from so nauseating a drink, this was taken as proof of its effectiveness, just as a failure to get drunk proved the efficacy of whiskey.

In early California, the sheepherders used a poultice of equal parts of tobacco, salt, and onion. In 1784 it was said that the natives in Paraguay blew tobacco smoke into a snake-bite wound, put on a tobacco poultice, and took a drink of the juice. A Florida case is reported in which a man bitten by a rattler may have died from the remedy of four cups of water in which tobacco had been steeped.

Indigo. Indigo, a preparation derived from plants of the genus *Indigofera,* was sometimes used for rattlesnake bite a century or so ago. It was often one of the ingredients of complex poultices, as, for example, mixed with rattler flesh and salt, with iodine, with camphor gum and grain alcohol, and with sorrel, the last a folklore cure of the Pennsylvania Germans. It appears that powdered indigo, in the form of bluing, mixed with salt, was a favorite remedy of the cowboys in certain areas of the Southwest.

Vinegar. Vinegar, usually in combination with some plant, was often used in ancient and medieval medicine. It seems never to have been popular for the cure of rattlesnake bite, although a salt-vinegar combination was reported in 1848. A Louisiana folklore mixture taken internally, included lard, vinegar, boiled tobacco juice, quinine, and whiskey.

Turpentine. Turpentine is still a folklore cure in the country districts where it is so often used in cuts. One may assume that the burning pain that turpentine causes may give the victim—or, what is more important, his relatives—the feeling that something powerful is working for his good. It has recently been reported as a folklore cure for rattlesnake bite, generally mixed with camphor or gunpowder.

Alcohol. Alcohol, especially whiskey, was for a long time the most generally accepted remedy for rattlesnake bite. It differed from many of the other remedies that are no longer recognized as having value, in that it was not only accepted by laymen, but by the medical fraternity as well.

The use of whiskey for rattlesnake bite was the source of innumerable stories and vaudeville gags, particularly during the prohibition era. But the effect of its use was anything but a joke. Without doubt many a fatality was caused by the large quantities of whiskey given to the victim under the mistaken belief that no amount of alcohol could cause injury to a person who had been bitten by a rattler. There were several correlated theories basic to the treatment. For alcohol was not given primarily as a stimulant; on the contrary, it was believed to be a specific antidote for snake bite and particularly rattlesnake bite. It sought out the venom in the body and destroyed or neutralized it wherever it might be, whether in the blood or tissues. Thus it was thought that a person exhibiting any snake-bite symptoms was insufficiently supplied with alcohol. And, as a corollary, a person could not become drunk, regardless of the quantity of whiskey imbibed, until that quantity was more than the amount required to neutralize the venom in his system.

Alcohol did not gain much popularity for rattlesnake bite until the middle of the nineteenth century. But by then the alcohol treatment was well established and had supplanted all others.

How the theory of the direct antidotal effect of alcohol gained credence among doctors and laymen alike is not known, since there were no tests to substantiate it. But the general proposition that alcohol cures snake bite probably had its validation in the uncertainties that have authenticated all the folklore remedies of the past—the cure of cases that required no treatment, that recovered despite the treatment rather than with its help.

A modern twist to the venom-alcohol relationship is thus expressed by one of my correspondents: "The main danger from snake bites is hysteria. There have been cases of drunken people bitten by rattlesnakes, and they didn't know about it or didn't worry about it and survived. I believe if a person is bitten and relaxes, and doesn't worry, he will get all right."

Some of the quantities of liquor given (or taken) for rattlesnake bite were formidable, as for example, a quart in ten to twelve hours; two quarts of corn whiskey in twelve hours, until, no doubt amid rejoicing, the patient finally showed signs of inebriation, and the doctor desisted; one-and-a-half quarts of whiskey for a young girl; seven quarts of brandy and whiskey in four days; a quart of brandy in the first hour, and another quart within two hours; one quart of brandy and one-and-a-half gallons of whiskey in thirty-six hours;* one-half pint of bourbon every five minutes until a quart had been taken.** Two ounces of whiskey every two hours was the army treatment in 1869. A newspaper story of 1867 said that

---

*The doctor reported (seriously) that the patient was seen next day looking for another rattlesnake to bite him. Thus was born one of the sure-fire vaudeville gags of a later time. Then there was the tale of a dry town where a tired rattlesnake had ten men on its waiting list.

**Surprisingly enough this patient got drunk; the rattler that bit him must have been out of practice.

persons suffering from rattlesnake bite, including nondrinkers, had taken up to a gallon of whiskey without showing signs of inebriation.

One writer tells of an experience in the Confederate army in which a man suffering from snake bite was given a gallon of whiskey when the current price was $450 a gallon. An officer protested, saying the cure was worth more than the man.

There were objectors to the use of alcohol in large quantities almost from the time when it first became popular. Some of these doubted the value of alcohol, others thought it definitely injurious. As early as 1831, a fatality from rattlesnake bite was reported in the case of a man who was drunk when bitten. This should have led to doubt as to the specific antidotal quality of alcohol. One writer mentions a death from an overdose of whiskey following the bite of a harmless snake.

In the days of its popularity for rattlesnake bite, whiskey was sometimes injected or applied externally as well as internally. Frequently it was mixed with other substances, particularly red pepper; and although it was the most trusted treatment, other remedies such as a tourniquet and suction, or various poultices were not inhibited by its use. However, it is true that the complete confidence in whiskey often stopped the use of more effective treatments.

The decline of whiskey as a cure began in the early part of the present century, and it was practically eliminated from the methods approved by reputable physicians in the 1920s, when antivenin and an effective incision-suction program became available. A survey of snake bite in the United States in 1908 concluded that alcohol in large doses only added alcohol poisoning to snake poisoning; and that acute alcoholism may have resulted in five percent of the fatalities.

Today alcohol is not recommended in snake-bite cases except, rarely, in small quantities as a stimulant. It has the adverse effect of speeding circulation and thus increases the rapidity of venom absorption. Most first-aid manuals and doctors experienced in the treatment of rattlesnake bite oppose its use. It has been virtually relegated to a folklore status, where it is still popular.

Oils and Fats

Oils and fats, whether of animal or vegetable origin, have been popular as snakebite remedies since the earliest days, and even now have a place in folklore medicine. The application was sometimes internal, sometimes external, often both.

Hog lard was recommended for rattlesnake bite in 1752. It was tested in 1824 without success, but hog and other animal fats continued to be recommended from time to time. Olive oil was a European remedy for snake bite dating back at least to 1200 A.D., and was probably used much earlier. It was recommended for rattlesnake bite in the eighteenth century. It is doubtful whether olive oil is extensively used as a folklore cure anywhere in the United States today.

Kerosene

Kerosene has long been a favorite cure for rattlesnake bite among farmers and others in remote districts. They have great faith in it as a specific antidote that will reach the venom and neutralize it. The venom is presumed to turn the kerosene green, thus affording a visual proof of its effectiveness. Usually the bitten limb is immersed in kerosene and kept there, but sometimes the oil is used in a poultice with onion or garlic. Despite this belief, there is no evidence from tests that kerosene is in any way effective against the bite of the rattlesnake or any other venomous snake. The cure is probably a direct descendant of the early folklore remedies of various oils, fats, and greases.

Inorganic Chemicals

Iodine and Iodides. Iodine, in the form of the tincture, and various iodides, particularly potassium iodide, were often used for rattlesnake bite in the middle of the last century. The application was usually by painting the wound, and much importance was attributed to the treatment by some of the medical practitioners of those days who considered it a specific antidote. The only use of iodine today in snake-bite treatment is as an antiseptic.

Potassium Permanganate. Of the many inorganic chemicals that have been proposed and used as antidotes for snake bites, rattlesnake bites included, potassium permanganate probably attained the widest popularity, and this with the sanction of and use by the medical fraternity itself. For a time, from about 1900 to 1930, most of the rattlesnake-bite first-aid kits included potassium permanganate as the most essential item.

One of the earliest treatments of rattlesnake bite with potassium permanganate was given in May, 1883, by a medical student who was employed by the Northern Pacific Railroad. He had read of the use of the remedy and tried it out on a patient four hours after the latter had suffered a bite in the calf of the leg. Not only was the remedy injected into the wound, but also elsewhere in the leg, and it was likewise given internally. Despite this treatment, the patient recovered in four days.

Even from the first, and notwithstanding its subsequent wide use, doubts were cast on the efficacy of permanganate. Nevertheless, the use of potassium permanganate continued to spread in this country. However, with the availability of an anticrotalic antivenin in the late 1920s, the snake-bite treatment manuals no longer took a neutral attitude respecting the use of permanganate, but definitely recommended against it. Since that time this chemical is seldom met with in snake-bite cases, although it is sometimes employed as an antiseptic.

The rise and fall of potassium permanganate is an example of the difference between theory and practice. This chemical — in common with a number of others — will, when mixed with venom, completely destroy its toxic properties, as demonstrated by many test-tube experiments. But venom is so closely akin to bodily tissues, and its absorption by them is so rapid, that the permanganate cannot reach and destroy it in the body.

Salt. Salt has been used as a snake-bite cure since the days of ancient Rome, often mixed with some plant. It was used in the treatment of rattlesnake bite as early as 1765, and thereafter received fairly frequent mention. Salt is without direct effect on venom, as was shown as early as 1787. It is currently listed as being among the folklore remedies of Illinois, Nebraska, Kentucky (with fresh pork), Tennessee, Arkansas (with soap), Louisiana (with garlic), and New Mexico. It was used by the pioneers in the "old" Northwest.

In modern therapeutics a saline solution may be used to irrigate the fang punctures and to aid in washing out venom.

Ammonia. Ammonia has had a sustained life as a snake-bite remedy, although it was shown long ago to be ineffective. It seems to have been mentioned first in 1738 for the cure of rattlesnake bite in Louisiana. Its use in those days was premised on the theory that venom was an acid that could be neutralized by the alkali ammonia. It was also tried internally as early as 1792.

Miscellaneous Inorganic Chemicals. Although no other chemical ever reached the widespread use attained by potassium permanganate, several inorganic chemicals that were found to render snake venom harmless when mixed with it in a test tube attained some popularity. Of these probably the most important were gold chloride, chromic acid, and calcium hypochlorite. They failed for the same reason as did potassium permanganate — an inability to reach and neutralize venom already mixed with body tissues.

Many other inorganic compounds have been tried for the bites of rattlers and other snakes from time to time, most of them in the hope that the venom might be neutralized, others for their systemic effects. Because of the acid character of venom, many were either bases or the salts of the alkali elements such as sodium, potassium, calcium, or magnesium, or of the ammonium radical. Alum water (potassium aluminum sulfate) was a folklore remedy in the West. Other cures included compounds of arsenic, iron, antimony, or mercury.

Mud or Earth Applications

The application of mud or absorbent earth to the bite of a snake is probably of ancient origin. It is given as one of the cures for rattlesnake bite in the first printed account of the treatment of its bite (1615).

Whether the mud cure was of Indian origin, or came from an observation of dogs treating themselves, is not known. A few Indian tribes used the mud method, but it does not seem to have been popular among the colonists in the United States.

There were some variations in the method of application. Sometimes the limb was buried in the mud of a swamp or creek, despite the discomfort of the patient; sometimes a more convenient mud pack or poultice was used. On occasion the victim was buried to his neck in a manure pile. Sometimes absorbent earth or clay was mixed with milk and applied to the arm or leg, or the wounded part might be buried in dry earth and milk poured around it. The application of a paste of clay, soot, and vinegar was reported to be a folklore remedy in Oklahoma.

A researcher in 1787 experimented with absorbent earths applied to viper-bitten pigeons. The results were inconclusive. Certainly today the drawing power of absorbent earth is not considered as effective as the suction methods recommended.

Snake Stones

The snake stone, an ancient device designed to draw venom from a snake bite, has never been as important in the folklore treatment of rattlesnake bite, as it was in the treatment of snake bites in the Old World. There snake stones were often used, and great importance was attributed to them, and no doubt still is, in many backward places. They might be of animal, vegetable, or mineral origin; usually their derivation was clothed in mystery. Often they appeared to be partially calcined and porous pieces of deer horn; some comprised charred vegetable matter removed from the stomachs of ruminants, or calculi. In operation they were believed, when affixed to a snake bite, to adhere until all venom had been sucked out, whereupon they dropped off.

In 1805 a Virginian advertised a snake stone that would cure rattlesnake bite or rabies, and offered to sell it to the people of four or five adjacent counties for $2,000. Shares were $10 each. Evidently the sale was made, despite efforts to discourage it. Another account of a community snake stone, this time owned by twenty farmers in North Carolina, was given in 1905. The author claimed to have witnessed successful treatments with this stone, a small porous object about the size of a silver dollar.

Narcotics and Stimulants

In the early days of snake-bite treatment, opium, sometimes in the form of laudanum, was given as an antidote for snake bite. More recently, morphine, aspirin, or barbiturates, have been recommended to relieve the pain of the venom so that the patient might obtain comfort, rest, and improved morale.

Among the stimulants recommended and used from time to time are nitroglycerine, digitalis, and tea or coffee. One of the popular first-aid kits before the days of antivenin contained caffeine tablets. Oxygen inhalation has also been used. This might be of some value in cases wherein respiratory distress becomes an important sign, although positive pressure breathing may be more effective.

Stimulants are not now indicated following rattlesnake bites, and probably rarely following the bites of other snakes.

Emetics and Purgatives

Among the few snake-bite remedies of the ancients that have some basis in logic, and are still advised, were the use of emetics and purgatives, since some venom is excreted through the intestinal tract. Stomach washes and diuretics have also been used.

In 1706 a root was described that cured rattlesnake bite by causing profuse sweating, but that had no effect if taken by a person who had not been bitten. Herbs have been used in Paraguay to promote sweating in snake-bite cases.

Some New Drugs

In recent years, some of the newer drugs, such as the steroids and the antihistamines have been used in cases of snake bite in allaying some of the troublesome allergic effects of antivenin or venom. They do not appear to be indicated in acute poisoning.

The antimicrobials, on the other hand, have become increasingly important in the treatment of snake bite, rattlesnake bite included, to avoid infection.

Veterinary Treatments

In general it is found, as would be expected, that most of the remedies that have been applied to human beings, whether of a folklore or scientific nature, have also been used for domestic animals.

Below are a number of remedies used in Texas by Americans and Mexicans for cattle and horses. Some of them have been reported used in other sections of the country.

Apply hot lard.

Incise and apply fine salt.

If the swelling is serious, apply cold mud or clay as a poultice.

Apply cow manure, but be sure to keep it cool.

Pound up the inside of a prickly-pear (cactus) leaf, and apply as a poultice an inch thick.

Hold a hot coal at the bite, but not close enough to scorch the hair.

Stand the animal in mud and rub the wound with kerosene.

After puncturing around the wound with Spanish dagger spines, apply kerosene and tallow.

Apply soda and vinegar.

Tie on a tourniquet and jab in Spanish daggers.

Cauterizing with gunpowder was used in early Wyoming days on animals as well as humans. One writer claimed to have used the split-chicken cure with success on a horse.

The following are some of the treatments that have been given dogs for rattlesnake bite: salad oil; ammonia, nitrous salts, and a tobacco rub; a tea made from a weed; a mud bath.

Present-day therapy closely parallels the recommendations for human beings, that is, a constriction band, incision, suction, and antivenin.

It has often been stated that dogs, when snake-bitten, seek a mud hole or swamp, and immerse themselves in mud until the worst reaction is past. Presumably this is done to relieve the pain. I cannot say, from my own experience, whether this is fact or folklore, but certainly it is a widely circulated bit of natural history.

ANTIVENIN

Curiously enough, although snake-bite antivenins have not been so effective as some other antitoxins—such as those used against diphtheria and tetanus, for example—the first important work on this type of artificial immunity was that of a researcher in 1887, who was seeking to produce antibodies against rattlesnake venom. He discovered preventive inoculation, for he was able to produce, in test animals by repeated sub-lethal doses, a gradually increased resistance, and this without ill effects on general health; but the resultant immunity did not last long.

The next important step was taken in 1894, when it was found that the blood of a guinea pig immunized by the method above, when injected into a second guinea pig, protected the second against venom. Thus was evolved the idea of the remedial substance, antivenin.

Ever since the development of antivenin, there has been confusion in the popular mind between it—the curative serum from the blood of an immunized animal, usually a horse—and venom. It has become almost a folklore belief that venom itself is used in the treatment of snake bite. Southwestern folklore even has it that a second bite will cure the first, so that a rattler should be caused to bite twice.

Preparation of Antivenin

Antivenin is prepared from the serum of an immunized animal, usually a horse or donkey, whose resistance to venom has been built up by the injection of successively larger doses of venom, until it can withstand venom in quantities that would have been quickly fatal if given initially. After the immunization course has been completed, as shown by blood tests, blood is drawn off from time to time in quantities not large enough to injure the animal; and the serum, which contains the venom-counteracting antibodies, is segregated from the rest of the blood and processed for preservation and distribution.

Some horses react differently to venom than others; in some it produces maladies of the liver and kidneys. A horse that has withstood the immunization course successfully, and has proved to be a good antivenin producer, represents a considerable investment and is carefully tended. There is a limit to the neutralizing power attained by the serum; after the serum antibody-content reaches a certain concentration, larger doses of venom produce no further increase, and, in fact, there might be a decline in antidotal power.

The early method of inoculating a horse with increasing doses of crude venom had serious defects; the immunizing process was slow, and the health of the horse suffered, particularly, in the case of certain venoms, from local hemorrhagic reactions at the sites of the injections. Treatments were given every three to five days and the process required eight months or more to complete. Subsequent improvements have entailed the use of detoxicated venoms, or venom fraction mixtures.

Specificity of Antivenins

When the efficacy of antivenomous sera was first discovered, it was thought that, regardless of the venom used in the immunization program, the resulting antivenin might be equally effective against bites by all kinds of snakes. This was soon disproved, whereupon an exactly opposite assumption was made: that monovalent antivenins — that is, those resulting when the venom of a single species of snake has been used for immunization — would be most effective for the bite of the same kind of snake and partly effective for the bite of the nearest relatives.

Extensive experimentation has verified this theory in certain fundamentals, but not in detail. It is true that monovalent antivenins are almost never effective for venoms of a quite different type; for example, antivenins based on venoms that are largely neurotoxic are usually of little or no value in counteracting hemorrhagic venoms. But there are many important exceptions to the applicability of the rest of the theory, including the surprising result that a monovalent antivenin is sometimes less effective against the venom used in its production than against the venom of some other snake. So the venom-antivenin correlation is by no means a simple one. Polyvalency can be secured, at least to some degree, by treating the producing animal with a mixture of snake venoms.

Use of Antivenin

Antivenin should be administered by a physician. It should not be given uncontrolled in the field. In an emergency, when it is available in a remote camp and the case is serious, it might be used by a properly trained and equipped person familiar with serum reactions.

In the United States, only one kind of rattlesnake antivenin is available. This is Antivenin (Crotalidae) Polyvalent, Wyeth. It should be used only for the bite of a rattlesnake or the water moccasin (cottonmouth), and only if there is evidence that some venom has been injected.

When antivenin for our American crotalids was first made available in the United States—this was in the late 1920s—the idea was widespread that antivenin was a cure-all; that no matter how serious the bite, the victim need only take one 10-cc. shot and go about his business as if he had not been bitten. This was not the result of any statements made by the firm then manufacturing antivenin, but was, rather, caused by the uncritical newspaper publicity that greeted its availability in this country. Subsequently when patients died or lost an extremity despite the use of antivenin, there was a considerable reaction, and other forms of treatment were suggested. Today there is a better appreciation of the status of all methods of treatment, and of the value of the improved antivenin now available—its value is as the basis of treatment, but it is no cure-all.

There is no doubt that the use of antivenin has reduced the death rate from the bites of poisonous snakes in the United States, and has speeded the recovery of nonfatal cases. But it is difficult to derive authentic differences in mortality rates, because of the variations in the severity of envenomation, the amount of antivenin given, and the time elapsed before antivenin is administered.

Antivenins cannot do the impossible; some dangerous snakes can inject venoms in such quantities that quite impracticable quantities of antivenins would be required to counteract them.

RECOMMENDED TREATMENT FOR RATTLESNAKE BITE

It is to be kept in mind that such instructions for the treatment of rattlesnake bite as offered here may be superseded by improved treatments, whereas other parts of this book should retain their usefulness. But as someone in an emergency may look here for suggested methods of treatment, I shall offer a program that represents the present consensus of expert opinion. The procedures recommended are selected particularly with the treatment of the bites of Nearctic (United States) rattlesnakes in mind.

As a matter of reassurance to the victim of a bite, it can be said that even without any treatment, rattlesnake bite would probably not be fatal in more than five percent of the cases, although greater with some especially dangerous species. With proper treatment the mortality from rattlesnake bite should be less than one percent. As previously mentioned, snake bite is likely to be more serious in children and the elderly.

In an accident of this kind, be sure that the snake involved is a venomous snake before specific treatment is initiated. If there is any doubt as to the identification of the snake that did the biting, kill it and bring it in, or at least bring in the head and tail. Use a stick to push the dead snake into a box or can. However, do not consume time catching the culprit if this will delay treatment of the victim. But a sure identification may save the victim from painful and unnecessary treatment. You can have your snake identified by a zoo, a natural history museum, a park naturalist, or by the biology department of any high school or college.

The injection of rattlesnake venom into a wound is usually followed by some local pain, and this should be used as one of the criteria in determining whether the bite is that of a crotalid, and if venom has actually been injected. With most species swelling is also evident within a very short time. Also, a prickling sensation may be felt about the face, especially the lips. A minty or metallic taste is a common complaint following some rattlesnake bites.

Assuming that a person has actually been bitten by a rattlesnake, and that some venom has been injected, the following procedure should be adopted by the victim and his companions, if any are present:

1. **Rest and immobilization.**
   a. Keep patient at rest and give reassurance.
   b. Immobilize affected part in a functional position.
2. **Constriction band, incisions, and suction.**
   a. If patient is within twenty minutes of definitive medical care (including antivenin therapy), place a wide constriction band immediately proximal to the wound. The band should be tight enough to occlude lymph flow but not venous or arterial circulation. The patient should then be taken to the nearest medical facility.
   b. If patient is more than twenty minutes from medical care, and is seen within five minutes of the bite, place a constriction

band, and then incise through the fang marks. Incisions should
be no more than one centimeter long and no deeper than three
millimeters.* Apply suction directly over both incisions. This
procedure is of no value if delayed more than fifteen minutes
following the bite.

3. **Attempt to identify the snake.**
4. **Arrange for transportation or travel to the nearest hospital.**
5. **Under no conditions should the injured part be placed in ice, the bite area excised or surgical measures performed unless a physician is present and deems such procedures necessary.**

These suggestions will be affected by a number of variables. If the
victim is within twenty minutes of medical care, he should carry out 2.a.,
and proceed immediately to the hospital. Before leaving the scene, he
should make an attempt to identify the snake, or kill it and take it along
on a stick. Again, no time should be wasted on this act. If the victim is
more than twenty minutes from medical care, he should carry out 2.b.
and 3. and, if alone, walk to the nearest point from which transportation
to a hospital can be obtained.

If there is someone with the victim, the same procedures should be
carried out and a decision made as to whether or not to allow the victim
to walk, or to carry him, or to go for help. This decision will need to rest
on a number of factors: the severity of the envenomation, the general
condition of the victim, and the distance and time to the nearest medical
facility, among other things. It is certainly not wise to leave a severely
envenomated victim alone while going for help, but every factor of each
case must be carefully weighed in making a decision.

Whenever the victim has to walk out from the scene of the accident
he should walk slowly and rest every three to five minutes. He must
avoid over-exertion but should proceed as rapidly as possible to the nearest hospital or physician's office. If the bite is on the upper extremity, the
area of injury should be kept at slightly below heart level. If the bite is on
the foot, the foot should be immobilized in such a way as to minimize
motion, while still permitting walking. It is not necessary to loosen a constriction band, but if a tourniquet has been used, which is not advised,
this must be loosened for one of every ten minutes. Before loosening a
tourniquet, the victim should be at rest for at least three minutes, or until
heart rate is normal.

Contrary to popular opinion, most snake bites in the United States
occur within a short distance from medical care. In California, for in-

---

*Since the rapidity of absorption makes it improbable that any components of the venom remain concentrated at the points of injection for an appreciable time, deep incisions are unwarranted—and hazardous.

stance, over ninety percent of the rattlesnake bites occur within city limits, or within two miles of city limits in the foothill areas. The average time from bite to hospital in southern California was thirty-four minutes, in a recent review of two-hundred randomly chosen cases. In some parts of the country most bites occur in rural areas. Nevertheless, of two-hundred reports recently supplied by physicians, the average time from bite to hospital in their areas was fifty-four minutes. Few bites occur in backpackers, serious hunters or fishermen. In the past twenty years there has been only one report of a backpacker in the Sierras of California bitten by a rattlesnake, and he was bitten while changing a tire at the end of his hike.

THE PREVENTION OF RATTLESNAKE BITE

How may a person minimize the chances of encountering a rattlesnake, or of being bitten if he meets one? The danger from rattlesnake bite is rarely sufficient to justify anyone in changing his course or plans. Only during the spring or fall, when one's path might take him past a rattlesnake den, would he be warranted in detouring to avoid the snakes. But there are certain precautions that may be taken to reduce the chances of being bitten if a rattler should be encountered. These include alertness and care—how one should conduct himself in rattlesnake country to avoid being bitten, and the selection of protective clothing.

How People Get Bitten; Example Incidents

Before suggesting the precautions that people may adopt to avoid being bitten, I shall give brief descriptions of a number of actual occurrences. These will illustrate, better than any statistics, how accidents happen, and the kinds of actions to be avoided.

Citing illustrative examples of how people get bitten almost inevitably will convey an exaggerated idea of the danger; so before doing so I would like to point out that there are many instances of people living and working in country where rattlesnakes are quite numerous where no one is ever bitten.

Picking Berries or Flowers. A woman was struck in the lower calf of the leg as she crawled along the ground picking strawberries. A seventeen-year-old girl, walking in dim twilight, was bitten in the thumb by a timber rattler *(C. h. horridus)* as she picked a violet.

Picking Up Kindling Wood or Timber. A woman in her own yard was bitten as she reached down for some firewood. A two-year-old, helping his mother pick up sticks for a campfire, was bitten by a northern

Pacific rattler *(C. v. oreganus)* that he picked up, evidently mistaking it for a stick.

Picking Up Game. A man was bitten as he reached into a bush for a rabbit he had shot.

Reaching into Holes or Hollow Trees. A hunter, seeking to recover small game he had shot, was bitten by a rattler when he reached into an armadillo hole. A youngster in Arizona was bitten when he tracked a wounded cottontail to a hole and reached in to get it.

Picking Up Miscellaneous Objects. A laborer on an electric line-construction crew reached into tall grass for a rope and was bitten by a hidden rattler *(C. v. oreganus)* that gave no warning. A man repairing a fence dropped his hammer in a clump of grass and was bitten when he stooped to pick it up. A caretaker was bitten as he picked up scraps of paper in a school yard.

Turning Rocks. A man tearing down a rock retaining wall was bitten in the middle finger of the right hand by a hidden rattler. A farmer digging a hole, cut a rattler in two with his shovel without knowing it. As he stooped to remove a rock, he was struck in the hand.

Climbing among Rocks. A trout fisherman in the Sierra Nevada was using his hands in climbing out of a granite gorge. He was struck in the hand by an unseen rattler above him. A man sat down on some rocks at the foot of a bank. He leaned back against the bank where there was a hole and a Great Basin rattler *(C. v. lutosus)* concealed in the hole, bit him in the back.

Hunters Looking for Game. A man in the Sierra in California, while walking around a tree and looking upward for a hiding squirrel, stepped on a rattler and was bitten in the leg. A man took off his shoes to wade out for snipe and was bitten. A hunter, aiming at a deer, heard the rattle of a snake. Without waiting to ascertain the direction of the sound, he stepped backward onto the snake, and was bitten in the leg. This case ended fatally.

Walking and Hiking. A boy breaking his way through heavy chaparral was struck in the leg. A man walking on a trail slipped and put his hands out to break his fall, whereupon he was bitten in the hand through a buckskin glove.

A woman, walking toward a fishing camp along the margin of a lake, heard a rattler sound off but was too frightened to move. She was bitten in the ankle.

Drinking from a Spring. Two men, one in Wisconsin, the other in Kentucky, were struck when they knelt to drink at springs.

Stepping out of a Car. A man on a public road stopped his car, stepped out, and was bitten by a large southern Pacific rattler *(C. v. helleri)* under the car, which he had missed seeing as he drove up.

Stepping over or on Logs. A man stepped off a log directly on a rattler that was stretched along its underside. A seven-year-old boy was struck in the palm of the hand while climbing onto a log.

Brush Cutting. A forest-fire fighter was bitten in the thigh by a rattler he was trying to drive back into burning brush.

Agricultural Activities. Bites when people are tending crops or domestic animals, or while gardening, comprise an important proportion of all bites, although somewhat reduced by the increased mechanization of agriculture.

A woman was bitten while feeling among some vines for melons, and a boy was struck while picking watermelons. A 16-year-old girl picking grapes was bitten in the finger by a rattler that was in the vine, coiled above ground. One gardener was struck in the chin as he stooped to examine a plant.

Two men were struck by rattlers hidden in haystacks, as they pulled down hay to feed cattle. Another went out barefooted to feed his cows and was struck in the heel. A man was bitten as he crawled under his house to investigate a hen's stolen nest. A woman was bitten when she went into her poultry house to see what the chickens were cackling about.

Accidents in the Darkness. A man in the desert in darkness stooped to pick up what he thought was a rope. It proved to be a western diamondback *(C. atrox)*, which bit him, one fang taking effect through a glove. A boy catching insects after dark in front of his home was bitten in the heel.

Bites around the Home. A girl was bitten while rummaging in her garage. A rancher was bitten on the left hand by a rattler coiled in an empty bucket tipped on its side. A woman was bitten by a pigmy rattler that had crawled into a dresser drawer in her home. A man swung out of a cot in his home and stepped on a rattler. A woman in Texas was bitten as she slept on her porch.

Accidents to Children. A boy of four was bitten when he ran into some weeds while chasing a bird. A boy playing in his own garden overturned a board and was bitten by a juvenile rattler hidden beneath. A boy of seven on a bicycle rode over what he thought was a stick. He stopped to investigate and was bitten in the right foot. A nineteen-month-old child lifted a saw lying on the ground. She was bitten by a small rattler under it. The saw dropped back and covered the rattler so that the mother, who came to see why the child cried, could find nothing wrong. When the child's arm began to swell, the rattler was found, and appropriate measures were taken that saved the child's life. This is an example of the occasional cases in which young children cannot tell, or fail to tell, what has happened, which leads to delay and increased danger.

An Unexpected Second Snake. Occasionally an accident happens when a person avoids one snake, only to fall afoul of another. A man stooped to pick up a stone to kill a rattler and was bitten by a second that he had not seen, as it lay close to his intended weapon.

Miscellaneous. Of course, many accidents occur under strange and unlikely conditions. People have been bitten by rattlers in the course of such activities as:

> Changing a tire by the roadside in the evening
> Replacing a runway light at a city airport
> Picking up a golf ball in the rough
> Crawling under a barbed-wire fence
> Repairing a leaky water pipe
> Searching for supplies under a counter

Two boys were bitten fifteen minutes apart by a single small rattler coiled on the porch steps at a Halloween party. A man in a boat saw a snake swimming in a river. He knew there were no water moccasins in that area, so he seized the snake. It proved to be a rattler.

Lessons to Be Learned

Much may be learned from these cases that I have enumerated. A few of the accidents were unavoidable; a few involved children from whom better judgment could not be expected. But a majority fall into two categories: either people put their hands or feet into places where a rattler might lie concealed, or their attention was absorbed in other activities, such as looking for game, so that they were not alert to their immediate surroundings. Certainly most of these accidents could have been avoided by application of this simple rule: *watch where you put your hands and feet; don't put them in places without looking, and don't put them in places where you can't look.*

I have compiled the following safety-first suggestions for people in places where rattlesnakes may be found:

Don't lift a stone, plank, log, or any other object under which a rattler might be hidden, by placing your hand or fingers under it. First move it with a stick, a hook, or a pry-bar of some kind; or with your foot, if properly protected with heavy boots or puttees.

Don't gather firewood in the dark; gather all you need in the daytime, and employ the usual precautions of alertness as you pick it up.

Don't reach into a hole of any kind, whether in the ground or in a tree, for escaping game or anything else. Don't reach into holes or crevices in rock piles or under rocks.

As you walk in grass, brush, cactus, or rocks, stay in cleared spots as much as possible, and keep a sharp lookout for snakes that may lie concealed in, or beside, your path. Remember that rattlers, like most snakes, are protectively colored; it takes a sharp eye, not a cursory glance, to discover one in its natural surroundings.

Don't believe the stories you hear that rattlers are out only in hot weather; they are always hidden in the shade when it is hot. They are most active when the temperature is moderate.

Don't take too literally the tales of their shunning high altitudes. They are found higher in the mountains than most people suppose—up to 11,000 ft. or more in the Southwest, and up to 14,500 ft. in central Mexico. Rattlers can climb trees and bushes, and they swim readily.

While hunting, if you are keeping an eye out for game to run or fly, learn to glance frequently at your path to avoid stepping on a rattler. And remember that the leader of a file is not the one most exposed to snake bite; he may disturb a hidden snake that will strike one of the followers.

Don't walk around your camp in darkness, for rattlers are nocturnal much of the year. If you must move about, use a flashlight to light your way, and don't remove the boots and puttees that you wore in the daytime.

Step on a log, never over it, so that you can first see whether there is a rattler concealed below the curve on the far side.

Avoid walking close to rock ledges; give them ample clearance if the path is wide enough.

Don't reach above your head for a handhold in climbing amid rocks, and avoid placing your hands near crevices into which you can't see.

When you crawl under a fence, try to do so in a cleared spot; if there is grass or brush, beat it with a stick so that a rattler will escape or make its presence known.

First examine the surroundings before sitting down to rest on a log or rock.

Learn to recognize the venomous snakes of your area, and avoid killing the harmless ones that are competitors of the rattlers for food. If you go into a new area to work, hunt, or fish, learn something about the snakes to be expected there.

When a rattler sounds off suddenly, don't move until you know whence the sound is coming; you may step onto a rattler or into its range instead of away from it. A rattler seen in time is not a dangerous snake, provided you and the other members of your party, including your dog, avoid it.

Rattlers will usually strike only at a moving object. When you back away from one rattler, make sure all is clear behind you, so that you won't trip or back into another snake.

Don't get excited if a snake gets in your boat. He's looking for a rest, not for you. Lift him out with an oar or a paddle.

Don't lay out your bedroll near brush piles, rock piles, or rubbish. Don't leave your clothes and boots lying around where rattlers can crawl into them. Pitch your camp in a clearing, if there is one.

Don't handle an injured or dead rattlesnake. Don't touch the head of a decapitated rattlesnake. Dispose of it where other members of the party can't accidentally come in contact with it. Never play practical jokes on others with dead rattlesnakes. If you must examine the head of a recently killed rattler, use sticks to handle it, not your fingers. If you must have the rattles of a dead snake for a trophy, put your foot on the snake's head before you apply your knife to its tail. The head of a rattlesnake has been known to bite half an hour or more after it was completely severed from the body.

If you are a snake collector, don't grab for the tail of a rattler going down a hole; the head end may already have turned and be facing outward.

For precautions around the home and children's play yard, see chapter 12 on control.

Don't let the fear of rattlesnakes keep you out of the country, or from hikes or picnics. Rattlesnakes are more plentiful in this book than in the wild. But be careful and alert.

The Rattlesnake in the Blankets

In the early days of the western migration, when the travelers often slept on the ground, there must have been a lot of worry about rattlers crawling into the blankets, to judge from the frequency with which the hazard is mentioned in the books of the day. Whether the danger was as serious as pictured may be doubted, though there is plenty of evidence that there was some slight risk that a sleeper might awake in the morning with a rattler for a bedfellow. It is true that on the desert the rattlers are most active in the early evening; with the coming of the midnight chill in the spring and fall, or daylight in the summer, they seek refuge, and might avail themselves of so unexpectedly adequate a concealment as that provided by a pile of blankets with a man at the core. But it may be suspected that the trapper or cowboy at the evening campfire did not minimize the risk to the attentive tenderfoot, particularly one gathering material for his next book on the hardships and hazards of Western travel.

From my own correspondents, I have had the following comments:

□ I once was sleeping out in the open not far from Hot Springs, South Dakota, and just as the sun was peeping up I heard the familiar sound of a rattler in my bed with me. Well, I just want to say that I never got out of a bed quicker than I did that one. This only goes to show that they will bunk with you without being invited.

☐ A rattler in my bed wrapped around my ankle, and I didn't dare jump out of bed till it crawled off.

How Illegitimate Bites Happen

Just as is the case with legitimate bites, the reader can best be impressed with the things not to do—if he must handle snakes—by the citation here of some actual occurrences. Some of these accidents happened to amateurs; others involved persons who might have been presumed, through practice, to know how to handle rattlesnakes.

Handling Supposedly Dead Snakes. A hunter stoned a snake, and, thinking it dead, picked it up and was bitten.

Pulling Snakes from Holes or Seeking Them in Holes. A rattler was escaping under a rock. A collector reached down to seize it by the tail to pull it out and was bitten by another rattler, hitherto unseen under the rock.

Mistaking a Rattler for a Harmless Snake. A woman was bitten by a small rattler as she was explaining to her daughters, aged eight and nine, that it was a harmless garter snake.

Picking Up Snakes. A collector, using a short, straight stick in the usual way, pressed a rattler's head into the ground so as to pick it up back of the head. But the ground was soft and the snake's head partly hidden, so that when he picked it up, the snake, with an inch or so of neck free, was able to bite a finger.

Snakes Biting While Held in the Hand. A man picked up a large rattler by the neck without controlling the posterior part of the body. The snake threw a coil around his wrist and was able to work enough of its neck free to bite the man's hand. A man holding a rattler by the neck, in order to show some visitors its fangs, had his attention distracted and was bitten in his other hand.

Putting Snakes in Containers or Carrying Them. Snake collectors usually carry their catches in flour or similar sacks, since this is a much handier method than any other. Although a rattler will rarely bite through a sack, such cases as the following demonstrate the necessity for caution: a man poked a red diamond (*C. r. ruber*) with his finger, and was bitten through the sack. A boy was carrying a pigmy rattler in a sack that was fastened to his belt. Even this small snake succeeded in making an effective bite through the cloth of the sack.

One man was bitten while trying to put a prairie rattlesnake *(C. v. viridis)* into a bottle.

Milking Rattlesnakes. An operator was demonstrating the removal of venom from a snake to a small audience. He became so interested that he used one hand to gesticulate and got it within range of the subject.

Bites by "Fixed" Snakes. Rattlesnakes purchased by snake shows from regular dealers may have had their fangs removed by order of the purchaser. Accidents happen when the fangs grow in again, or where a careless job has been done by the seller. A carnival-pit operator was bitten by a large western diamond *(C. atrox)* that was supposed to be fangless. Whether replacement fangs had grown in, or the snake had been shipped untreated, it was impossible to discover.

Multiple Bites

Multiple bites are of two kinds: those involving two or more bites in the same accident, and those in which the victim receives several bites on successive occasions.

Typical incidents of the first kind are these: a woman gathering strawberries was bitten three times before she could get out of range. A child, bitten in the leg, stooped down to grab the leg and was bitten in the hand. A soldier, lying down, was bitten four times by a *C. horridus.* A boy was trying to force a small prairie rattler *(C. v. viridis)* into a quart jar and suffered two bites, one in the forefinger of each hand.

Bites on many successive occasions, suffered by those who handle snakes, are the rule rather than the exception. One handler claimed to have been bitten forty-nine times. Tex Sullivan, a well-known collector, had been bitten many times—twenty-two according to a letter written in 1950. Preachers of a snake-handling cult claimed to have been bitten from 250 to 400 times.

Unfortunately, these successive bites, as I have mentioned elsewhere, seem not to develop any immunity, at least not to an important degree. Sullivan's twenty-second bite was an exceedingly serious case.

HUMAN ATTITUDES TOWARD RATTLESNAKES
Snake-Handling Cults

The incorporation of snakes into religious ceremonies is exceedingly ancient and widespread throughout the world. Several American Indian tribes participated in such ceremonies, using rattlesnakes.

A Christian snake-handling cult in the Southeastern states gained considerable newspaper notoriety in the 1930s and 1940s, when it made the handling of venomous snakes a test of faith, based on the Scriptural admonition: "And these signs shall follow them that believe ... they shall take up serpents and ... it shall not hurt them" (Mark 16:17–18). The sect is said to have been founded in 1909. It eventually spread to small scattered rural groups in Virginia, West Virginia, Kentucky, North Carolina, Tennessee, Georgia, Louisiana, Alabama, and Florida.

At the services, amid scenes of fanatical fervor and mass excitement, rattlesnakes and other venomous snakes were passed from hand to hand to demonstrate the faith of the disciples and converts. The adherents admitted a fear of snakes; only under the hypnosis of these emotional revivals were they able to handle them. The snakes were not restrained in a manner calculated to prevent their biting; on the contrary, they were held at mid-body so that they were free to bite the holder or anyone within range. If a man were bitten it proved his lack of faith.

No doubt most of the followers were carried away by religious piety with a touch of exhibitionism, but this could not be said of the audiences that came to see a free snake show. Later, when several of the states, or possibly all of them, had passed laws forbidding the handling of snakes in religious ceremonies, the onlookers were augmented by those who hoped to see a riot when the state police arrived to interfere.

Unfortunately, these demonstrations of emotionalism by the mountaineers were not harmless affairs, and a number of innocent neophytes, who had not yet reached an age of independent judgment, were seriously hurt. Even babies were permitted to touch the rattlers. Thus the prohibitory laws became necessary and were enforced, despite claims that they infringed the freedom of religious worship. The extent to which services of this type are still being held surreptitiously I do not know. The sect is no doubt still in existence, but with less startling rituals. While they lasted, the newspaper reports were extensive. Altogether, it is said that over a hundred people were bitten, seven fatally, between 1934 and 1948. I have found newspaper confirmation of six fatalities. The record of bites may be an underestimate as some of the preachers claimed to have been bitten up to four hundred times.

Recent newspaper accounts indicate a revival of snake-handling rituals, either by the same or other cults. A case of snake bite during church services was reported in Greenville, South Carolina, in August, 1953. Two fatalities were recorded in the Southeastern states in the summer of 1954, one at Trenton, Georgia, the other at Fort Payne, Alabama. In connection with the latter case, a lay preacher was fined $50 for handling venomous snakes and endangering other members of the congregation. In the summer of 1955 there were at least two additional fatalities; one in July at

Altha, Florida, and the second at Fort Payne, Alabama, the scene of one of the deaths of the previous summer. These deaths were followed by memorial services in which the deceased were praised for their undeviating faith. In these funeral rites, attempts were made to handle the rattlesnakes that had caused the fatalities, despite the refusal of the authorities to permit these features of the obsequies. It is apparently a tenet of the cult that the deaths are justified and sanctified if the other followers carry on by a continuance of snake handling as a demonstration of faith.

Murder, Suicide, and War

In ancient days, long before the time of Cleopatra, snakes were occasionally the fatal instruments in suicides, murder, and in war. They were used in the torture and the execution of criminals and were employed to safeguard treasure.

Rattlesnakes have never had any popularity for these purposes. On very rare occasions the newspapers may report a suicide attempt, but I have heard of only one success. Rattlesnakes are not deadly enough. A bite may not mean an escape from life, but a very painful and prolonged escape from death. However, according to one writer, a man at Colfax, California, succeeded in gaining entrance to a building one night and jumped into a snake pit housed there. He deliberately annoyed the rattlers until he had been bitten eighty-five times, after which he went home and died.

One murder in Los Angeles is often referred to as "The Rattlesnake Murder," but actually the victim of this sordid crime failed to die of the bite of a rattler and was drowned.

Rattlesnakes have had important parts in the plots of many books and stories, the most notable being Oliver Wendell Holmes's *Elsie Venner*, of which the plot depends on the effect on an unborn child of a rattlesnake bite sustained by the mother.

Protective Devices

In ancient days much confidence was placed in certain repellents or barriers — materials and devices so hated and feared by snakes that they could be used not only to keep snakes out of homes and camps, but would protect the carrier from being bitten by a snake encountered in the field. Such were the innumerable botanical repellents recorded by Pliny. Or, one might be protected by eating a viper's liver, wearing a deer's tooth, or carrying burned vulture feathers. In modern folklore, an onion is the most popular repellent;and the hair-rope barrier around the camp is no doubt still used here and there. But all of these devices must be relegated to myths and folklore, for none has ever been shown to have the slightest value.

Protective clothing is in a quite different category, for its value is unquestioned and important. Often it entirely prevents the fangs of a venomous snake from reaching the flesh of the person at whom a strike is aimed; and even when there is penetration, the clothing may absorb part or all of the venom, or reduce the depth of penetration to an important degree.

There are ample statistics to show the importance of foot and leg coverings. In this country, where most people go about well shod, there are about forty percent more bites in the lower limbs than the upper; but in some countries bites in the legs exceed those in the arms by three or four to one.

The hunter or fisherman going into the field in rattlesnake country will probably wear boots, leggings, or a shoe-puttee combination, and it will pay him to give attention to their impenetrability and length as well. No hunter will be seriously inconvenienced by foot and leg protection below the knee that is virtually impervious to a rattler's fangs.

Much, of course, depends on the size of the snake. Thin and soft shoe leather will protect fairly well against young rattlers, but more is needed to stop the fangs of the adults of the larger species. To learn something of the possibility of fang penetration, I experimented with a western diamondback *(C. atrox)* four feet eight inches long and weighing three and a quarter pounds. Its fangs were about one-half inch long. The snake was caused to strike a man's oxford shoe that had been stuffed with cloth so as to afford a firm target. The leather was about 1 mm. thick. The snake struck readily enough and with considerable force, for it was backed against a wall, which aided in the forward drive.

In no case—there were several strikes—did the fangs penetrate through the leather; in fact, the penetration was not deep enough to interfere with the snake's withdrawal, nor did the fangs catch sufficiently to cause them to be pulled out of their sockets. There was enough depth to cause a snapping sound as the fangs were withdrawn, and one small solid tooth remained imbedded in the leather.

A somewhat smaller snake of the same species not only failed to penetrate the shoe, but marked it so little that the point of contact of the fangs could not be located with certainty. The venom injection by this snake was uneven. There was considerable left on the shoe at the first strike, then little or none for several, and then again some in each of several strikes.

I think it clear from these tests that leather of moderate thickness cannot be penetrated by a rattler of considerable size, and that heavy boots or a heavy shoe–puttee combination will afford adequate protection from rattlers to the part of the leg covered. Yet some experienced hunters prefer a boot with the upper part loose and flexible rather than stiff, as they think the latter may be more easily penetrated. This I doubt because thickness is the most important feature. All authorities agree that trousers should be worn outside of, rather than tucked inside of, boots, since the loose cloth will offer a greater interference with penetration. Also, the snake may close its mouth in an attempt to bite when it feels the cloth. In considering the protective value of boots or clothing, it should be remembered that the point of a fang is solid; the discharge orifice is above the point and may be blocked by leather, or the venom may be absorbed by cloth, even though the point of the fang may have penetrated the skin of the victim.

Here are the experiences of some of my correspondents:

☐ I have never had a snake fang penetrate an ordinary boot top, although I have had many try it. I have had these prairie rattlers hang up a heavy blow on the bottoms of my Levis until I trampled them to death, but have never been scratched. Had one slap me one time squarely on my hat brim as I was climbing a rimrock while marking timber. It gave me a bad scare.

☐ A rattler struck at a hunter walking in the brush, the fangs piercing — in the order named — a thickness of overall leg, a leather boot top, and two thicknesses of wool sock. The skin of the man's leg was barely scratched, but enough venom entered this slight wound to make the man feel somewhat ill and nauseated for about three days; however, he did not seek the aid of a physician.

Soldiers are usually well protected by adequate leg coverings. It is said that the medical statistics of the Civil War do not disclose a single death from snake bite on the Northern side.* At army camps in the United States during World War II (years 1942 to 1945 inclusive) there were 1,910 hospital admissions for bites or stings of venomous animals, including, of course, rattlesnake bites. This was at the rate of one man out of every ten thousand exposed. There were no fatalities. This entailed nearly twenty million man-years of exposure, many of which were spent in the rattlesnake-infested areas of the Southwest.

Special snake-proof boots are available on the market. They are expensive and slightly cumbersome, but may be advisable for hunters in some particular areas of the Southeastern states and Texas, where very large diamondbacks may be encountered. Trousers impervious to snake bite are being produced experimentally; they have a removable lining of wire cloth. They have the objectionable features of being heavy and stiff.

Heavy gauntlet-type gloves are of unquestioned value for anyone doing much climbing or working among rocks.

---

*This throws doubt on one of the better rattler yarns to the effect that after a certain Civil War battle — was it Shiloh? — 60 Union soldiers were found dead from rattlesnake bite, but no Confederates. This resulted from the well-known fact — or is it a fact? — that rattlers particularly dislike blue. But I was told by one professional collector that he always went afield in blue trousers; he had absolute confidence in their repellent effect.

The Human Fear of Snakes

It is generally agreed, from tests and observations, that children have no natural or instinctive fear of snakes. There has been much discussion of this point, but it goes without saying that no child should ever be allowed to handle a venomous snake.

Since most children cannot distinguish venomous from nonvenomous snakes, the inculcation of a moderate fear of snakes is probably justified as a safety measure. But to instill such a horror and revulsion at the sight of a snake that, when they become adults, people cannot bear to look at a snake in a cage is a mistake. This insensate fear is so widespread — it is incorrectly believed to be instinctive — that many newspapers and magazines will not print a picture of a snake, nor can one be shown in the movies without public protest.

The cultivated fear of snakes has had results quite the opposite of those desired. It causes people to become so paralyzed upon encountering a rattler in the field that they cannot take the most elementary safety precautions. If bitten by a snake they are in no condition to judge whether it was venomous or harmless. This is a matter of importance in deciding whether painful and even dangerous remedial measures are necessary, for the bite of a harmless snake requires only a touch of antiseptic, as would a scratch or a splinter. Even in areas of the United States — and they are extensive — where rattlers are the only venomous snakes, victims usually cannot even report with certainty whether or not the culprit had rattles on its tail. These are the people who destroy all harmless snakes — the natural competitors, and even the destroyers of rattlesnakes — thus aiding in the protection and the increase of the dangerous snakes they so greatly fear. So children should be taught to avoid snakes, not to be terrified by them, and eventually we may have a more understanding adult population.

That the shock caused by the bite of a harmless snake can be serious and even fatal is well authenticated. It is obvious that such an eventuality is more likely to happen to persons in whom the fear of snakes is exaggerated. Two such cases have been recorded in San Diego County. In one, a man was bitten by a harmless gopher snake and almost died of fright; in the other a hunter was stuck by the barb of a wire fence, thought he was bitten, and barely survived the shock.

On November 1, 1952, near San Diego, a Mexican agricultural worker illegally in the United States, because of his fear of rattlesnakes, refused to sleep with his companions in a railway culvert where they were hiding from the immigration officers. Instead, he preferred the tracks, where no snake might be hiding, and was killed by a train. In another accident near Blythe, California, a boy, sleeping on the pavement of a road to avoid snakes, was run over and killed.

Then there is the fantastic story of a man who put on his son's jacket by mistake. It was too small and he thought he had been bitten by a rattler and had swelled up. He became seriously ill, but recovered when he found himself able to don his own jacket. A prospector was struck at by a rattler that missed its target. The man walked away and fell dead of heart failure. I have mislaid the source of this story; it may well be folklore.

Some of the exaggerated fears of rattlesnakes arise from equally exaggerated stories of their prevalence, their viciousness, and the inevitable fatalities from their bites. In truth, they are rarely as common as people think; they are timorous creatures that will bite human beings only if hurt or frightened; and a bite is rarely fatal if properly treated. In 1803 it was said that if the travelers' tales of the danger from rattlesnakes were true, America would be uninhabitable.

The Effect of the Rattlesnake Threat on People's Actions

A question arises as to whether the hazard from rattlesnakes is considered by explorers, hunters, fishermen, ranchers, and other outdoor people, to be serious enough to affect their actions. We are dealing here, not with the actual risk, but with what people think the risk to be. If the historians and travelers are to be believed, the rattlesnake hazard has been taken seriously by many.

In an account of a journey in Canada in 1669–70, it is claimed that La Salle was so frightened by seeing three large rattlers that he was taken with a high fever. In 1716 it was maintained that in King Philip's War the soldiers were more afraid of rattlers than of the Indians they were fighting. On the survey of the Virginia–North Carolina boundary, it was reported that when it was discovered that the rattlers had come out of hibernation, further work was postponed until fall.

Describing a hunting trip into the Sacramento Valley of California in 1855 one writer said: "The dread of rattlesnakes destroyed in great measure the pleasure of our sport, for we lost many a good shot from looking on the ground—which men are apt to do occasionally when once satisfied of the existence of a venomous reptile, the bite of which is by all accounts mortal."

People are known to have abandoned the purchase of property when they learned that rattlers were occasionally found in the vicinity; and instances are not unknown when they have disposed of homes already owned, upon the discovery of a rattlesnake in the vicinity. Upon the advent of antivenin in the United States in 1927, a commercial reporting agency sent out a statement to its clients advocating that publicity be given to the remedy, as it might make areas popular, either for settlement or vacations, that had hitherto been shunned because of the actual or reputed presence of rattlers.

RATTLESNAKES AND DOMESTIC ANIMALS
Animal Fears of Snakes

A few tests have been made to determine the natural fear of snakes evidenced by monkeys and apes. The tests apparently indicate that they do have a natural fear.

In the early natural histories, it was stressed that all animals, including the European horse, were frightened intuitively by the sound of the rattlesnake's rattle. But other, more objective modern writers, state that much depends on the conditions under which the horse discovers the rattler. Horses are generally alarmed at sudden sounds or movements, and so are likely to be frightened if a rattler nearby suddenly rises into its striking coil and sounds its rattle.

Dogs show much the same variability in their reactions to rattlesnakes as do horses, some becoming confirmed rattlesnake killers.

A study made on an experimental range in central California showed that cattle were not afraid of rattlers, even after they had been bitten.

One writer observed that a kitten showed no fear of a diamondback. Another found that prey-sized mammals introduced into a rattler's cage showed no sign of fear, which duplicates our experience at the San Diego Zoo.

Stock Losses from Rattlesnake Bite

Next to the frequency and mortality of rattlesnake bites among human beings, the most important phase of the rattlesnake's life history to man is the economic loss suffered through damage to his livestock and domestic animals. Of course, there is no question concerning the desirability of killing rattlesnakes about human habitations and farms, but whether the stock raiser will be benefited by having the rattlers killed on the open range is a matter of doubt. The increased rodent population may well do more damage to the grazing value of the range than the rattlers that would have kept them in control might do to the stock.

Horses and cattle are usually struck in the head by rattlers while grazing. Colts and calves seem to have a fatal curiosity that causes them to sniff a snake, and their elders, also, are sometimes too curious. Occasionally an animal will lie down on a rattler and be bitten.

Symptoms of Rattlesnake Bite in Animals

Most rattlesnake-bite cases in animals, with the exception of those suffered by riding horses or by dogs, occur without the presence of a human witness. However, stockmen and farmers are so familiar with rattlesnake-bite symptoms in stock that their diagnosis of snake bite is usually correct.

Nearly all descriptions of rattler-bite symptoms stress swelling as the most noticeable symptom. Here are a few from correspondents:

☐ The usual result of a rattlesnake bite is that the head is swelled to nearly twice its normal size for a few days.

☐ When cows and horses are bitten in the head, there is a bad swelling and they may be partially blind for a while; also, they have spasms, become jerky, and have a running nose.

☐ Cows that are bitten sometimes peel off and lose their hair, and get terribly thin.

☐ One of my horses showed up one day with a badly swollen head and nose — so large that the horse resembled pictures of a hippopotamus. I concluded the horse had been bitten by a rattlesnake, perhaps a day or two before.

Rattlesnakes and Dogs, Coyotes, and Small Animals

Dog-rattler affrays are complicated by more than the usual uncertainties. There is, first, the great variability in dog sizes, and the almost equally important factor of hair length and density.

Dogs are most susceptible to rattlesnake bite when they are engaged in some activity that is absorbing their attention and senses, so that they are not fully alert to the danger of a hidden snake. Although the odor of a rattlesnake is imperceptible to a human being, it is no doubt evident to a dog. Thus it is that hunting dogs and sheep dogs, because they are bent on assigned duties for their masters, are the ones most often bitten. Others are bitten protecting children; and many that have become rattlesnake killers eventually lose their caution and make a careless move.

It seems appropriate to touch on coyotes after the dog discussion. In south Texas it was estimated that twenty percent of the coyotes showed rattlesnake scars.

Small animals and birds, whether domestic or wild, that are of sizes that the rattlesnake seeks for food, if struck squarely, usually die in less than ten minutes. But there is much variability in the time, depending on the size of the snake and the effectiveness of the bite, so that some may survive for as much as an hour, or escape entirely.

Note: The section "Recommended Treatment for Rattlesnake Bite" (pp. 214–217) is from *Snake Venom Poisoning*, by Findlay E. Russell (Philadelphia; J. B. Lippincott Co., 1979).

CHAPTER TWELVE

Control and Utilization

Rattlesnakes are of economic value by reason of their destruction of small mammals, such as rabbits, ground squirrels, and gophers, that would otherwise cause even greater inroads on agricultural fields and grazing ranges or spread diseases. However, the rattlers are a menace to both man and his stock because of the danger from their bites; and were it possible to replace every rattler with an equally useful, but harmless, bull snake, king snake, or racer, the balance of nature would be preserved, yet with the elimination of the risk of snake bite. Failing in this, the agricultural and other authorities in each area must determine whether the damage by the rattlesnakes outweighs their usefulness, and therefore warrants their destruction. Of course, there can be no question regarding their elimination near towns and cities; here the possible danger to persons, especially children, is such that every rattler should be destroyed. At the same time, educational campaigns should be instituted to discourage people from killing harmless snakes, for every one killed makes room for an additional rattler in the scheme of nature's economy.

Various methods of rattler control or elimination have been proposed, such as bounties, poisons, traps, the destruction of food supplies and refuges, and the encouragement or importation of competitive predators. Other means of suppression that are not the result of any conscious or organized effort are such conditions as fires, the casualties of traffic, and agricultural developments unfavorable to snakes. The relative effectiveness of these factors, organized or incidental, will now be considered.

Rattlesnakes may be hunted solely to achieve their destruction, or to effect their capture alive for such commercial or scientific value as they may have for snake shows and zoos. The methods of finding the quarry do not differ greatly for these divergent objectives, except that few people are willing to use the time and effort required to catch a rattler, as compared with that needed merely to kill it. The raiding of dens is almost the sole method of rattlesnake hunting sufficiently fruitful to attract those who wish only to destroy the snakes, but this is only one of the methods used when live snakes are sought.

METHODS OF CONTROL

Campaigns for Killing at Dens

One of the most effective methods of rattlesnake control is to organize campaigns for killing them when they are concentrated at their dens, either as they gather to hibernate in the fall or when about to leave in the spring. Here the rattlers may be taken alive, or clubbed or shot, in great numbers. This method is possible only in the colder sections of the country where the rattlers are gregarious and the annual concentration draws individuals from considerable distances. There is almost no danger to those whose legs are properly protected with puttees or heavy boots. The more timid can use shotguns or .22 rifles. A .22 repeater, using dust-shot shells, is to be recommended as it minimizes the danger to one's companions. The dust shot will disable even a large rattler, so that it cannot escape until killed with a stick or hoe. Repeated visits are necessary as the snakes come in on successive days, and some may escape into the holes and crevices.

Several of my correspondents have written that groups of men in their communities make a regular practice of raiding nearby rattlesnake dens in the fall or spring. One stated:

> On rattlesnake-hunting days, we would gather up a party of from ten to fifty or seventy-five people and go out to hunt. We took along our lunch and made a picnic of it. We got plenty of snakes and you would hear rattlers singing all day.

Fig. 41 Dead prairie rattlesnakes (*C. v. viridis*) destroyed in a raid on a den. (Reproduced by courtesy of A. M. Jackley)

These seasonal, community-wide rattlesnake-control campaigns are by no means a recent development; they have been popular and effective since colonial times. Such a project even became known as a "rattlesnake bee."

On a rattler hunt in Iowa in 1849, the people were divided into two contestant groups. Each participant put up two bushels of corn to be won by those getting the largest numbers of snakes. Two men got ninety rattlers in an hour and a half, and the total killed in one year was 3,750.

One writer tells of an extensive raid on a den in New York in 1810, and how he and his comrades tied a powder-horn bomb to a rattler's tail, then lit the fuse and let the snake crawl into the den, where an explosion followed, presumably with good results. The blasting of snake dens with dynamite has recently been tried in various places. It has the disadvantage of not permitting the effects to be ascertained; also, any harmless snakes in the den will be destroyed. A further disadvantage is that the shattering of the rocks may open up new and better refuge crevices. Rattler dens are occasionally blasted in connection with the building of roads, dams, and similar works — an unintentional control.

Traps

Since a snake's feeding intervals are widely spaced and irregular, baited traps cannot be used with the same efficacy as for mammals and birds. Pits have been tried with indifferent success, since it is difficult to cause a snake to fall into one, and harder still to keep it there unless the sides are both steep and smooth.

The most successful trapping of rattlers can be effected at dens, particularly when the snakes are issuing from hibernation in the spring. The simplest method is to block all exits but one, and pipe this into some such receptacle as an oil drum.

Fences

Fences may be of some value in rattlesnake control if the area to be protected is relatively small, as for example, a children's playground. However, only adequate maintenance will assure even a moderate degree of exclusion.

A rattlesnake-proof fence (I do not say snake-proof) should be of one-fourth-inch mesh from the ground line to a height of at least two feet; above that the mesh may be one-half inch. The fence should be carried to a total height of at least six feet. The posts should be on the inside, and every effort should be made to make the exterior as smooth as possible to eliminate climbing holds. There should be no overhanging bushes or trees, for although rattlers do not climb as readily as many snakes, they do climb to some extent. The clearance of brush from the fence should be at least two feet at the bottom and four or five feet at the top.

The fence should be carried at least one foot below ground level, although the mesh of this section may be increased somewhat, since the purpose is to keep mammals from burrowing under, thus eliminating holes through which the snakes might gain ingress. Examinations should be made, from time to time, to determine whether tunnels have been made; this is another reason for having the ground clear along the fence. Where deep-burrowing rodents are present, even two feet below ground line may be found inadequate, and holes must be plugged upon discovery. Special attention must be given to the gates of snake-proof fences, both to see that they are not left open, and that the clearance is not sufficient to permit a rattler to enter.

Gases and Poisons

The killing of rattlers at their dens by the introduction of a poisonous gas, such as hydrocyanic acid, has been tried with some success. But snakes are not so susceptible to poisoning as are mammals, and there is always a difficulty in finding and plugging all of the crevices so that the gas will not escape. Obviously, so dangerous a poison should be handled only by persons of experience. A professor at Northwestern University reported one instance in which a Great Basin rattler *(C. v. lutosus)* suffered no ill effects from thirty minutes' exposure to a phosgene concentration that would have killed a man in about ten seconds.

The slow metabolism of rattlesnakes (in common with other snakes), as compared with that of mammals, not only makes them less susceptible to poisoning by gas but permits them to withstand a reduced oxygen supply for some time. One writer put a rattler in a glass jar for a week without fresh air and reported that it suffered no ill effects. Another placed one in an atmosphere of carbon monoxide for half an hour, from which it emerged unharmed.

Encouragement of Competitive Predators and Enemies

One of the most practical methods of rattlesnake control is the encouragement of competitive predators. The rattlesnake population in any area depends largely on the available food supply, and the more rodents consumed by other animals — harmless snakes, for example — the fewer there will be to support the rattlesnakes. One would think that this obvious fact should serve as a protection to the harmless snakes, since every time someone kills a harmless snake, he is only making life easier and safer for some rattlesnake. It is astonishing how impossible it is to convince such people that it is quite easy to distinguish rattlers by merely looking at their tails; there seems to be a widespread opinion that they frequently leave their rattles home on the hatrack.

Fortunately, in most agricultural districts the farmers have become fully alive to the economic value of harmless snakes, especially gopher or bull snakes, both as rodent destroyers and rattler competitors, and remove them only when they become a threat to poultry or rabbits.

The encouragement or importation of rattlesnake enemies has been suggested as a means of control. Among wild mammals, deer and badgers would probably be effective; of the birds, the red-tailed hawk; and, of the snakes, the king snake would serve both as a rattlesnake enemy and as a competitive predator. But no bird (and, indeed, no mammal or reptile) in the United States feeds habitually on snakes, to the exclusion of other food. So if we are to secure any measure of rattlesnake control through the activities of other wild creatures, it must be through their competition for the rattlesnake's food, rather than through their direct predation on rattlesnakes.

Of the domestic animals, the hog has most often been credited with being effective in rattlesnake control. Although the efficiency of hogs as rattlesnake destroyers may have been exaggerated, it is probably true that they may be counted on to eliminate the rattlers within a limited area.

The question is frequently asked why the mongoose is not imported into the United States to destroy rattlesnakes. The mongoose does not live exclusively on venomous snakes. It is omnivorous and may become a serious pest. Even as a snake destroyer, it is partial to elapine snakes such as the cobra, and avoids viperine snakes.

Elimination of Food Supply and Cover

One of the most effective methods of eliminating rattlers in specific areas, such as around homes, is to curtail their food supply and cover.

Since most species of rattlesnakes feed largely on mammals, the elimination of the rats, mice, gophers, ground squirrels, and other rodents that attract rattlers, will naturally discourage them. Similarly, as the snakes are by nature secretive, the removal of brush, rocks, boards, trash, and other hiding places tends to keep them away. Mammal holes should be plugged wherever found in an area one wishes to protect.

Bounties

The eradication of venomous snakes by the offer of bounties for their destruction has been tried with some success, although it has certain objectionable features. It is even now being used to a limited extent in some states for the control of rattlesnakes.

The principal objections to the bounty system as a method of rattlesnake control are the following: snakes are brought in from beyond the borders of the bounty district; snakes may be bred and raised for bounties; females are held until they have given birth to young, thus increasing the bounty; claims are made for snakes found dead on the road; snakes are gathered in wild areas, where they do no harm, rather than around human habitations; harmless snakes are killed under the mistaken impression that they are venomous and eligible for bounties.

Fires

Forest, brush, and grass fires, as a means of destroying rattlers, have been suggested. It need hardly be said that the destruction of the watershed cover, to say nothing of harmless and useful wild life, would cause damage far outweighing the possible benefit of killing off the rattlers, some of which would surely escape into ground holes or rock crevices.

Traffic Casualties

In these days of fast, multiple-lane traffic over smooth-surfaced roads, on which snakes cannot secure adequate traction and therefore progress slowly and with difficulty, every major road is an almost impassable barrier, for scarcely a snake ever crosses in safety during hours of heavy travel. Thus the borders of the highways soon become virtually denuded of snakes, although the adjacent areas, through population pressure, continuously feed new material into them. Some years ago I estimated the traffic casualties in San Diego County, California, at over ten thousand snakes per year. This figure has now increased because of the greater density and speed of traffic, so that I should now place the casualties at about fifteen thousand per year. Of these, roughly eight percent, or twelve hundred per year, are rattlesnakes. A corresponding figure for the entire United States would be well over eight hundred thousand rattlesnakes per annum. Obviously, there are great local differences because of variations in both rattlesnake and traffic density. Eventually road casualties will decline — this is already true of major highways — because of past destruction.

METHODS OF CATCHING RATTLESNAKES

As I have said initially, by all odds the most productive method of collecting live rattlers is to locate them at their dens. But when collecting is done in areas having mild climates, where the rattlesnakes seek individual retreats, rather than communal dens, other means of hunting must be used. Even in these areas, spring will be found most fruitful, since it is mating time and also because the snakes are hungry after their winter fast, and therefore less cautious in roaming about. The late fall is also a good season for hunting, as the young of the year are to be found, and occasionally an entire brood may be discovered. To be successful, the collector must become familiar with his territory to discover the seasons when snakes are most active, in what surroundings, and at what time of day or night they are most likely to be found. The important thing to remember is that rattlesnakes are naturally secretive, and, unless actively in search of food or mates, are likely to be partially or entirely hidden. Thus, in the Southwest, the hunter looks for them under bushes, in clumps of cactus, in rock crevices, or under flat stones or overhanging banks. In this kind of collecting his equipment should include not only the means of catching and carrying his snakes, but also tools for investigating these hiding places without danger to his hands or feet.

In desert areas, where most snakes are nocturnal, an efficient method of collecting is to drive on the paved roads at night, picking up the snakes seen crossing in the glare of the headlights.

In some sandy areas in the Southwest, professional collectors regularly catch rattlers by following their tracks. When rattlers are hunted in woods, brush, or fields, the problem is to get them to declare their presence by rattling or moving, for otherwise their concealing coloration renders them hard to see. (Although many dogs are rattlesnake killers, some, with their keen sense of smell, can be trained to find rattlers without attacking them, and thus become efficient adjuncts in their capture.)

Rattlesnakes can sometimes be located by watching for a specialized rattler-warning chirp and posture used by ground squirrels. Domestic poultry, especially turkeys, have characteristic ways of reacting to the presence of snakes, which their owners soon learn.

Tools and Equipment

Where the snakes to be captured are well concentrated, as they are about the dens when entering or leaving their winter refuges, the best tools are first, a hook, or a narrow rake with four or five teeth, for the member of the party who is to keep the snakes from escaping. The handles should not be so long as to be clumsy—not over five feet at most. A "potato rake" conforms closely to these specifications.

The "picker-up" may use a hook or a pair of forceps or tongs; for either a forked stick, or a loop-snare, such as may be employed in the capture of a single snake, is much too slow for this operation. If the weather is moderately cool and the snakes not too excited, a hook works very well; for when it is placed under a rattler at mid-body, the snake will usually drape itself over the hook, and then can be raised into the container. Some collectors have used nets, made like a butterfly net, but of heavier material. These are not satisfactory in stony places.

When a hunter is operating where his quarry is much more scattered than around the dens, somewhat different equipment will be more suitable. For picking up the snakes, various devices are preferred by different collectors. Some use a forked stick, although this simple device is less often employed than published accounts would lead one to suppose. Most hunters find some form of snare both safe and useful.

Some collectors dispense with all collecting tools. They merely seize the snake behind the head with one hand and with the other grasp it near the tail, and then by stretching it out, render it helpless. If a snake is coiled ready to defend itself, it is worried until it uncoils in an endeavor to escape. Other collectors pick up a rattler by its tail and hold it at arm's length to prevent its slashing at its captor. The rattler may attempt to climb up its own body to reach the hand that holds it, which can be prevented if one jiggles the tail. This scheme is most dangerous with half-grown snakes. It is interesting to note that this method of picking up rattlers was reported to be used by the Indians long ago. It need hardly be said that this writer recommends none of these foolhardy practices, but prefers some form of snare-stick.

Whatever the means for picking up the snake, some container must be readily available. If the collector is afoot, a sack will best serve the purpose, otherwise boxes with hinged lids, or cans are to be recommended. A can should have a depth equal to the length of the largest snakes to be caught to discourage their climbing out. If other snakes have already been collected, it should be expected that they will try to escape or may even strike, when the container is opened to drop in the new captive. Therefore, one must be alert to see that the prior occupants are not in a threatening position when the box is opened. Mixing rattlers with harmless snakes should be avoided, not because they may injure each other, but because the more active snakes, particularly racers, may be so difficult to control that they will increase the normal hazards of bagging the rattlers. If excelsior is put in the bottom of the collecting box, the occupants may hide under it and be less likely to rear up in an endeavor to escape when the box is opened.* There should not be too much crowding or the younger snakes will be crushed.

Although rattlers will rarely bite through a muslin sack, it is unnecessary to take chances, and therefore sacks should not be allowed to dangle against one's body when they are being carried. Snakes can enlarge and squeeze through a surprisingly small hole so the coarse weave of burlap or jute bags is not to be trusted. Canvas is satisfactory if not too heavy and stiff.

One writer suggested wrapping the rattles of captured specimens, thus curtailing the sound produced so as to prevent confusion with other snakes one may meet. I confess I have sometimes been startled by the sound of a rattle nearby, only to find it was a sudden activity upon the part of a snake I was carrying beside me in a sack.

Collections of Preserved Snakes

Live rattlers are, of course, interesting exhibits in zoölogical gardens and snake shows, and some useful observations of their habits may be noted there. But studies are also made on preserved specimens—studies of morphology and anatomy, of rattles and fangs, stomach contents, or reproductive cycles and sex differences. For these basic studies preserved specimens are of fundamental importance, and their collection and preservation constitute an essential activity in herpetology.

Specimens for scientific collections are preserved, either in formalin (thirty-seven percent commercial formaldehyde diluted about ten to one), or in about seventy-five to eighty percent ethyl alcohol.

Every effort should be made to have the specimens harden in neat circular coils fitting around the periphery of the jar, for otherwise much space will be wasted in final storage.

Shipping Rattlesnakes

No attempt should be made to ship live rattlesnakes—or any other live snakes, for that matter—by mail, since this is contrary to postal regulations. Rattlesnakes, as well as other live snakes, may be shipped by express, if placed in containers prepared in accordance with specifications of the Railway Express Agency. Several rattlers may be sent in a single

---

*One of my more pleasurable hunting memories is of the time during prohibition days, when I was stopped on a lonely road just north of the Mexican line, by two men of the Border Patrol. They opened the rear doors and leaned in on either side to examine my gear, particularly a hinged box on the floor. Said one: "What might there be in the box?" When I replied: "Rattlesnakes," they grinned at each other as if to say: "This is the best one we've heard yet." So they lifted the lid, and a couple of big red diamondbacks that I had caught an hour or so before, rose up on their tails with enthusiasm for prospective freedom. The examiners suffered nothing more serious than a couple of bumped heads, for even in those days car doors were low. The only remark I heard was (excessively expurgated): "My goodness—they *were* rattlesnakes!"

container, if first put in muslin bags. Excelsior or shredded paper, placed loosely in the box, will aid in preventing injury. Neither food nor water is required en route.

The greatest hazards to a successful shipment are overcrowding and overheating. Under no circumstances should the boxes be so small that large snakes will be piled more than three or four inches deep, if evenly distributed over the bottom; and in the case of small snakes not over two inches. With rattlesnakes, there is not only the chance of their being crushed if too crowded in shipment, but occasionally their tails become entangled at the rattles. I have had this happen in the case of large shipments to such an extent that an hour or more was required to untangle them, and some were fatally injured.

It is well to place on every shipping container a note to keep it in a cool place. A box left on a station platform, exposed to the sun in summer, will result in many fatalities. Small specimens may be safeguarded by moist moss in the container.

When specimens within a box are to be segregated because they are from different localities, they may be tied up in separate muslin sacks, such as flour or meal bags, with an appropriate locality memorandum dropped in each sack. In fact, some shippers prefer to use sacks within a box, for this method reduces the likelihood of the snakes piling up in one corner and crushing each other; and even if the box be damaged the snakes cannot escape.

COMMERCIAL UTILIZATION OF RATTLESNAKES AND THEIR BY-PRODUCTS

Because of the emotional reactions of many people toward snakes, rattlesnakes particularly, publicity dealing with the commercial utilization of these reptiles or their by-products is usually tinged with sensationalism and exaggeration. A single instance of the sale of a few snakes or their venom is likely to be exploited in the newspapers as a regularly established business with a sound future. The press is by no means the sole offender, for those participating in such uncertain ventures are often exhibitionists by nature, and may, indeed, profit from the lurid publicity. Unfortunately, however, people who read these exaggerated reports are struck by the profits to be gained from creatures otherwise considered liabilities around their homesteads, and they are thereby encouraged to undertake snake farming as a business. So every time the wire services carry one of these stories to the effect that it pays better to milk rattlesnakes at $12,500 per quart of venom, than cows at twenty cents per quart of milk, reptile departments of museums and zoos are besieged for advice on this lucrative but hitherto neglected source of income.

Here are some extracts from letters from people at whose earnestness, I assure the reader, I have no desire to scoff. My complaint is directed at the misleading publicity that gives such unwarranted encouragement:

☐ I am writing in the interest of my son who will shortly be back from the war. He has always been interested in snakes, and I think he would like to raise them for their poison, because I see it is valuable. Please tell me how to start a rattlesnake farm and how much ground you need.

☐ There is a rockpile above my house that is full of rattlesnakes. I kill a dozen or more there every year. I see there is money in raising them for their skins and virus. Can you tell me how to raise them, what their feed is, and how to get the virus out? Can they climb over a fence?

☐ I read in the paper where a woman in Casa Grande is making lots of money raising rattlesnakes in a snake farm. How many rattlesnakes do you need to start a farm? Could I start with one pair? How do you tell the males from the females? Do you have to have an incubator for the eggs and how long before they are full-grown? How do you get the oil and the gall, and where can I sell them?

The answers to all these queries follow the same pattern: there are no successful rattlesnake farms in the United States, because such live snakes or by-products as may be marketed are secured from a continuous accession of fresh snakes newly caught in the wild. In general, snakes are not easy to care for in captivity; they require much attention, expensive food, are subject to epidemic diseases, and do not reproduce readily. As to markets for snakes and their by-products, these are variable and sporadic. Such buyers as there are naturally look to sources that can be counted on to furnish material at any time and in any quantity. So the business has gravitated to a few firms that enlarge and equalize their sales by dealing in other animals as well as in snakes.

The Live-Snake Market

In any snake show, rattlesnakes are a major attraction. So it is that every snake display, from the scientific exhibits in the reptile houses of the larger zoos to the snake pit in some traveling carnival, requires a foundation of rattlesnakes. These are generally secured from commercial dealers, some of whom have been engaged in the business for a considerable time, and have become well known in the various strata of the amusement trade. Most dealers are to be found in Texas, with a few in Florida, Arizona, and California; for the availability of stock as well as climatic conditions afford these states an advantage with respect to both local and exotic species. None of these dealers handles rattlesnakes exclusively; they also supply other kinds of animals for exhibitions or the pet trade, and deal generally in animals and their by-products.

The prices of rattlesnakes fluctuate with other commodity markets, and, in addition, vary with conditions in the amusement trade. In the early 1930s there was a distressed market in snakes, and rattlers could be picked up at twenty-five cents a pound or less. At this rate even a big five-foot rattler brought hardly more than a dollar, packed and ready for shipment.

Today (1955) western diamonds are generally quoted at $3 to $18, the smaller sizes selling at $2 per foot, and the larger, which always command a premium, at $3 per foot. All of these quotations are for fresh stock, packed and ready for shipment.

An amateur, coming upon a big rattler in the brush and looking upon it as a $10 commodity, is doomed to serious disappointment, whether he sells the snake to a dealer — if he can — or goes into business for himself. For there is a big spread — and rightly so — between the raw material under a mesquite bush and the packaged goods in the express office. The commercial dealer must develop both sources and markets. Since he must be able to ship at any time, he must carry a stock and sustain the losses thereby involved. He must advertise in the trade papers. Necessarily, his overhead is considerable. Unfortunately, sensational newspaper articles fail to distinguish between field and retail prices, thus encouraging those unfamiliar with the trade, or its uncertainties, to expect high returns. This is especially true of someone who, finding a large number of rattlers at a den, counts them as so many five-dollars bills, and wonders why this gold mine has been so long neglected.

Rattlers can be bought from dealers either "hot" or fixed," the latter term meaning with the fangs removed. Fixed snakes are generally purchased by carnivals or snake pits where the show, as staged, involves handling the rattlers. Fixed snakes are not wanted by zoos, since the mouth is likely to be injured, and fangless rattlers are short-lived. Some dealers charge extra for fixing — about fifty cents per snake — whereas others include this in the regular price.

The safety procured by fixing depends upon the method used. The commercial dealers usually jerk out or break off the functional fangs. This renders the rattlers fairly safe for a few days or weeks at best, that is, until the next replacement fangs grow into place, after which the removal must be repeated, or the snake is hot.

Simple fang removal is not at all difficult. The snake is caused to open its mouth, the fang sheaths are slipped up, and the fangs are broken off with snips or tweezers as close to their bases as possible. The points of the next reserve set usually protrude from the sheath, and these also may be easily pulled out. If one fears to handle a snake for fang removal, its functional fangs may sometimes be drawn by causing it to strike a piece of leather or cloth, which is then jerked while the fangs are still entangled in it. However, this is not a trustworthy scheme.

The removal of the magazine of reserve fangs is possible, but requires the skill of a delicate operation and is seldom attempted; besides, a snake so treated would usually survive only a short time.

Snake Shows

Snake exhibitions are of several kinds: the varied exhibits at the zoo; the life-like groups at the natural history museum; and the traveling pits of the circus sideshows, the carnivals, and the county-fair circuits. The latter are probably the largest customers of the live-snake dealers whose operations I have discussed. These shows, to attract a crowd, often incorporate various sensational features beyond a mere handling of the snakes, such as "swallowing" small snakes, drinking venom, or allowing rattlers to bite the pit keeper. Some of the snake-show operators have had interesting experiences with rattlesnakes and have a considerable knowledge of their habits, but it is difficult to learn anything of value from them. For the spirit of showmanship eventually gets the upper hand, and they will not, or cannot, differentiate what they have seen from what they think the listener would like to hear, or what would most amaze or shock him.

Various methods of deceit are practiced in these shows. Most rattlers in pits of the traveling type are either fixed, or have had the venom pressed out of the glands before each show. Rattlers that still retain their fangs are referred to as "hot," "green," or "unfixed." The man in the pit is a "geke," while a snake swallower is a "glommer." Glommers are said to be former glass blowers, whose throat pockets are so stretched that it is easy for them to retain a small snake there during the act. When a geke shows spots of blood on his wrist or arm, where a rattlesnake has bitten him, it is the rattler's short solid teeth that have drawn the blood, for the fangs are absent.

What with fang removal, lack of food, and generally rough treatment, rattlesnakes in pit shows are usually short-lived. The operator takes this wastage as a routine expense; he has no facilities wherewith to feed the snakes, nor would they be likely to accept food under the conditions in which they are kept. Standing orders calling for periodical shipments of fresh snakes are usually placed with the suppliers, and it is by this means that a presentable exhibit is maintained.

One special fake occasionally seen in snake shows is the horned rattler. This creature—not to be confused with the true horned rattlesnake, or sidewinder (C. cerastes), which has a short hornlike scale-covered projection above each eye—has a vertical horn in the center of the head. Actually the snake is a western diamond, and the horn a rooster's spur held in place by the skin of the snake's head that has been slit and allowed to grow around the base of the spur. Or it may be attached with collodion. At one time these reptilian unicorns sold for $10; I presume they are now higher.

A type of snake show whose popularity has lately caused it to multiply is the roadside pit, like the motor court, a product of new modes of travel. These exhibits vary greatly in size and excellence; many are well worth stopping to see, particularly those that specialize in the local fauna. Lectures are sometimes given. Usually the pits are adjuncts of such other activities as gas stations, cold-drink stands, and the like. Both the snakes and their by-products, such as skins, rattles, and vertebrae necklaces, may be offered for sale; in fact, some pits are merely the showcases of a general reptile business. Often the shelves around the cages are decorated with jars filled with fangs, dried venom, and rattles, the results of past extensive operations. The advertisements of these shows take the form of roadside billboards with such legends as "World's Largest Rattlesnake," "Reptile Exhibit 10 Miles," or, "Stop at Smith's for Rattlesnakes and Gas." Generally the intervals between signs become shorter as the exhibit is approached, and the signs themselves more startling. Shows of this kind are especially numerous along the southern border, from Texas to California, although by no means restricted to this section. But it is here that the signs are a most conspicuous part of the barren desert landscape.

In recent years, snake lectures have become sufficiently popular, particularly at high schools, to encourage some naturalists to undertake lecture tours as a means of livelihood. With a few live specimens, some sample by-products, and either stills or movies, a program both instructive and entertaining can be staged.

However unnecessary it would seem, I judge from inquiries I have received that I should warn parents not to permit their children to keep live rattlesnakes in their vivariums. Almost anything to be learned from observations of captive snakes can be learned as well from the harmless species. Rattlesnakes are dangerous creatures; familiarity breeds contempt; and, to judge from an accumulation of newspaper clippings, even the most experienced reptile-house keepers and snake-show demonstrators occasionally have cause to regret a single moment of carelessness.

Rattlesnakes sometimes provide the basis for a community attraction having the nature of a fiesta. Such an event can usually count on more free publicity than many a more entertaining celebration. For example, there are well-advertised gladiatorial combats between rattlesnakes and black snakes, rattlesnakes and king snakes, or rattlesnakes and Gila monsters. These generally end in an endeavor of all combatants to escape, after showing a decided lack of interest in each other. The humane societies are surreptitiously warned in advance, in the hope that they will publicize the affair by trying to stop it.

Another entry in the attempt to draw the wandering motorist is the rattlesnake derby. The contestants are released in the center of a fifty-foot ring and then roused into motion by an electric shock. Without repeated urging they will not move far from the starting point. The audience is attracted by the hope that the rattlers will streak for the other side of the ring and scare the daylights out of their fellow observers. But they never do.

In connection with a television show in which live rattlesnakes were used, it was observed that the liability insurance carrier required signs to be placed at all entrances to the studio, warning of the presence of the rattlers. Some life insurance policies require the payment of an extra premium by those engaged in handling venomous snakes, the penalty of a hazardous occupation.

Venom

The venom market is another phase of rattlesnake utilization, the possibilities of which have been overstated in newspaper and magazine articles. Relatively high prices are reported paid for all venom offered; as one popular writer put it, large snakes yield a dollar or more at each milking. During World War II, which did lead to a temporarily increased demand, it was generally rumored that the "government" had made a firm offer of $50 per gram for dried venom. I know of no such price paid by any branch of the Federal Government, nor, indeed, of any regular demand. But the rumor persists; and inquiries concerning the big profits to be made in this phase of snake farming continue to reach the herpetological departments of our scientific institutions.

Since it is difficult and expensive to feed rattlers in captivity, and their venom yield declines markedly with repeated milkings, rattlesnake farms, as such, would not be commercially feasible, even were there a steady demand for venom. Only the rarest kinds of snakes are worth milking more than once or twice; and what production of venom there may be depends on a continuous acquisition of fresh specimens, which, after milking, are used for other purposes. Also, although dried poison, as a curiosity, is easy to produce, to prepare venom suitable for scientific purposes, such as the production of antivenin, introduces complications. First, it is necessary to segregate the venoms of the several species milked, and the accurate identification of rattlesnakes, in areas inhabited by more than one species, requires some study and experience. Second, although the actual removal of venom from the snakes is simple, its purification is difficult and requires expensive apparatus.

Rattlesnake Oil and Fat

Rattlesnakes have long been used as the bases of various medicinal preparations. Some of these originated with the Indians; others were employed, and in fact still are, by the white population, especially in rural areas. These pharmaceutical uses do not produce any regular demand for snakes. Aside from an occasional purchase of rattlesnakes by the Chinese in America, the only medicinal use today, involving any market of commercial significance, is for the production of rattlesnake oil.

The use of snake oil is neither new nor confined to the Americas. It was, in fact, both ancient and widespread; it was a well-known remedy long before rattlesnakes had been brought to the attention of Europeans, viper oil being commonly recommended for the same afflictions for which rattlesnake oil was subsequently favored.

As the Indians had applied rattlesnake oil in pre-Columbian times and were using it in the days of the explorers, it was natural that the earliest descriptions of rattlesnakes should stress these beneficial properties, for the information of European physicians. Thus the colonists, with the precedent of the use of viper oil in their former homes, soon adopted rattlesnake oil as a sovereign remedy for various ills, particularly as an ointment or liniment to reduce stiffness or pain. No doubt the snake's fundamental litheness led to the assumption, upon the part of primitive practitioners, that the creature's oil must contain the synthesis of its graceful flexibility which might, therefore, be transferred, at least in some degree, to a human user. It is a principle of primitive medicine that every animal and plant has at least some property beneficial to mankind, and that it is the duty of the scientist and physician to discover it. Rattlesnake oil, which was thought to be particularly penetrating—it was believed that it would go right through a man's hand—seemed to satisfy this requirement of an otherwise antisocial reptile.

These are some of the ills for which rattlesnake oil or fat has been used: to absorb tumors or swellings; for relief of frozen limbs, lameness caused by falls, bruises, aches, and sprains, rattlesnake and mosquito bites, wounds, rheumatic and other pains; internally for hydrophobia, and externally for ringworm, sties, sore eyes; for toothache, by insertion of fat in the cavity; for goiter, deafness, croup, arthritis, gout, lumbago, and neuralgia.

Rattlesnake oil was one of the standard remedies sold by those primitive vaudeville acts, the traveling medicine shows, often operating under such designations as Blank, the internationally famous doctor, or Eaglefeather, the Indian medicine man, whose flaring torches lit the first theatrical performance ever seen by many a Western boy. Here we heard "the great soprano just come from Covent Garden," the guitar virtuoso from the Deep South, and witnessed the unexpected discomfiture of the comedian's straight man. Then followed the appearance of the great doctor himself at the tail of his wagon, and the rush of the shills in the audience to buy the almost—but never quite—last available bottle of snake oil. Who could deny his family this sovereign cure for every serious ill, when it cost only a dollar?

Today, rattlesnake oil is principally used as a pain reliever for rheumatism, and as a liniment for circus performers, acrobats, and the like. From the frequent inquiries I receive as to places where it may be obtained, I judge it is still a popular home remedy. Snake oil is, at best, only a lubricant having no greater efficacy than goose grease or any other animal fat.

There is a story that a customer entered a store in Atchison, Kansas, in the early 1860s and asked for half a pint of rattlesnake oil. After the satisfied buyer had left with his purchase, the druggist remarked that prescriptions for rattlesnake oil, bear oil, and lard oil were all filled from the same barrel, so all customers' requirements were easily satisfied.

Rattlesnake oil also has nonmedicinal uses. One writer says that snake oil makes an excellent gun oil that will not clog or run.

Rattlesnake oil is rendered like any other oil derived from animal fat. If the snakes are freshly caught and healthy, they have a rather plentiful supply of fat that lies within the body cavity along the sides in the form of white flakes or lobes. It is not difficult to separate the fat from the flesh. The hibernating season in the fall is the best time in which to catch the rattlers, because they are both congregated and fattest at that time.

The oil that I have seen is slightly cloudy. The odor was neither pungent nor unpleasant.

As to prices, the following have been mentioned: 1833, $4 per ounce; 1906, $4 per pound; 1947, $1 per ounce. Present quotations are highly variable. One recent newspaper advertisement quoted pure oil at $16 per ounce. A well-known dealer in snakes and accessories lists rattlesnake oil at $1 per ounce or fifty cents in twelve-ounce lots.

Rattlesnake Flesh as Food

As I point out elsewhere, the Indians ate rattlesnakes to some extent, either as a matter of necessity, when food was scarce, or in conformity with some tribal ceremony. Similarly, today, rattlers sometimes may save some hunter or trapper stranded in the wild without food; or serve as a sensational course in some back-yard barbecue. But they are not an important food source. Although the flesh is quite palatable, only a relatively small quantity is secured, even from a large snake.

Two of my own correspondents likened the meat to frog's legs. Some, with whom I talked, thought it more like chicken or canned tuna. Personally, I have found it more like rabbit than chicken. Much, of course, depends on the method of preparation.

Cooked rattlesnake has been given several euphonious names. In 1839 it was claimed to be on the menu of a frontier inn at Kaskaskia, Illinois, as "musical jack." One writer said that when eaten on the plains it was called "prairie eel." Another tells of a traveler in colonial days who was fed "cold eel," which he hugely enjoyed. But, upon being told on the following day what it was he had eaten, he became violently ill, lost his hair, and subsequently sued the landlord, recovering 20,000 pounds of tobacco as damages.

Rattlesnake may be cooked in a manner similar to rabbit or chicken. It should be soaked in brine overnight before cooking. It may be fried, baked, or served as a stew or soup. It is tasty fried with bread crumbs. One author gives the following recipe for rattlesnake steaks: required to serve six persons—five pounds of rattlesnake meat, flour, salt and pepper, Louisiana red hot, fat, and vinegar. Use only large, healthy rattlesnakes (three to five pounds live weight preferred).* Decapitate with an ax about six inches behind the head. Remove the skin and viscera, cut the remaining body section diagonally into one-inch thick steaks. Soak the steaks in vinegar for ten minutes, remove and sprinkle with hot sauce, salt and pepper, roll in flour. Fry in deep fat. Serve immediately.

Articles on eating rattlesnakes often caution against eating one that has bitten itself, which, so it is said, would render the meat highly dangerous. Since the poisonous quality of snake venom is destroyed by heat, this precaution may be ignored.

If about fifty percent of a rattlesnake is assumed to be waste, the following quantities of meat should be available from large snakes: length four feet, one pound; five feet, two-and-a-half pounds; six feet, four-and-a-half pounds; seven feet, seven-and-a-half pounds; eight feet, eleven-and-a-half pounds. However, no one should be so foolish as to kill such a rarity as a seven- or eight-foot snake for food; it would be better saved for a zoo or snake show.

George K. End established a rattlesnake canning business in Arcadia, Florida, in 1931. The product, advertised as "genuine diamondback rattlesnake with supreme sauce" was given much publicity because of its sensational implications, and enjoyed quite a vogue as a fad. Selling at $1.25 for a five-ounce can, it could hardly fall in the category of a grocery staple. Fifteen thousand cans were sold in 1940; twenty-five hundred diamondbacks were canned, which would give a yield of somewhat less than two pounds of meat per snake. Fifty people were said to be employed as hunters and canners. The by-products were skins, rattles, venom, and oil.

End was subsequently bitten by a rattler and died. The business was purchased by Ross Allen of Silver Springs, Florida, who has continued the sale of canned rattlesnake meat. In keeping with the prices of other food items, the price has been advanced and is now $1.50 per can. I was informed by Mr. Allen that in 1946, sales continued at the rate of fifteen thousand cans per year.

*Snakes of this large size are usually obtainable only in a few species (*adamanteus, atrox,* and *atricaudatus*) in the southern United States.

During World War II, when many youngsters were in training in Texas camps, the roadside stands catering to servicemen made a specialty of "rattleburgers." This invited the usual newspaper publicity, and fears were expressed that the rattlesnakes might be exterminated.

Skin Products

Rattlesnake skins, like those of many other snakes, because of their beauty of color and pattern, have long served for various ornamental purposes. Only their relative fragility and small size have prevented their wider use where ornamental leather is required, for no snake is more decorative than such rattlers as the southwestern speckled rattler *(C. m. pyrrhus)* or the northern blacktail *(C. m. molossus).*

The use of rattlesnake skins for ornamentation is an old one. I describe elsewhere how the Indians employed them, when not prevented by taboos. Skins have also been used as scabbards for swords and bayonets. Audubon mentions rattler-skin shoes. Fur trappers adorned themselves with skins in the pioneer days. Later, rattlesnake-skin belts and hatbands were standard equipment for the cowboys.

Anyone with a skin he particularly prizes as a trophy should place it in the hands of an experienced taxidermist. Such a skin should be mounted on a sheepskin or other suitable backing. Some preparators will tan snake skins for fifty cents per foot; others will furnish tanned skins at from fifty cents to $2.25 per foot, depending on width and rarity.

Leather products of rattlesnake skins are now available in great variety, the most popular being belts, hatbands, wallets, billfolds, purses, handbags, key containers, comb cases, watch fobs, cigarette cases, and tobacco pouches. Articles of clothing such as sport jackets, caps, and neckties are made, but these are too showy to be considered more than fads.* On the other hand, snake-skin-covered coat buttons, and shoes for women, are quite attractive and appropriate. Some of the more bizarre articles fashioned from rattler skins have been holsters, gauntlet tops, lamp shades, book covers, and table runners. Souvenir knife cases, playing-card cases, and bookmarks are occasionally seen.

Miscellaneous Products

Miscellaneous rattlesnake products offered for sale by dealers include such things as skulls, rattles, and fangs. One merchant makes a specialty of vertebrae strung into necklaces; these are said to be used as charms to facilitate the teething of infants. Preserved specimens are sold to educational institutions. Plaster casts are also prepared, as well as photographs and slides showing species differences and typical postures. Even tools for

---

*Peter Gruber, known as Rattlesnake King in the Pennsylvania oil fields in the early days wore a rattlesnake-skin suit that was said to have cost $650,000.

catching and handling rattlers are among the items listed by the larger dealers.

Rattlesnake rattles, always popular trophies when a snake has been killed, were extensively used for personal and home adornment in early days in the West. Sometimes the bead portiere of the better homes of the 1880s became a rattle portiere, a noisy affair in the slightest breeze. Pasted on boards to conform to suitable designs, rattles comprised a type of art so fantastic that it hasn't even yet returned to popularity, as have most of our bygone crudities. A notable series of these pictures was exhibited in a Texas saloon at the turn of the century, and represented, in several panels under glass, a deer, an Indian brave and squaw, an eagle, and a number of insignia and appropriate mottoes. The deer panel was five feet four inches wide, by four feet six inches high, and used the rattles of no less than 637 snakes. The eagle required 574. Altogether, more than fourteen thousand rattlesnakes died to decorate this early cocktail lounge; and a shipment of 18,460 additional rattle strings had been received from a ranch on the

Fig. 42 A panel representing a deer made up of the rattles of 637 rattlesnakes

Mexican border. These were subsequently used in additional designs, and the entire collection of pictures and insignia was eventually transferred to the Buckhorn Curio Store in San Antonio, where they still are (1955). More than thirty-two thousand rattle strings were used in the composition of the various panels.

Photographing Rattlesnakes

If possible, the field collector should always carry a camera to record the rattlers in their natural surroundings. On rare occasions he will be fortunate enough to find rattlers feeding, mating, or indulging in the male combat dance. Necessarily, however, these field pictures seldom include the detail essential to illustrate rattlesnake patterns. For reproduction in more formal papers, it is necessary to depend on laboratory shots. In these, it is inadvisable to simulate natural backgrounds, since they tend to obscure the patterns that are the primary objectives of this type of picture. I prefer snakes photographed on a glass base, thus eliminating confusing shadows.

L. C. Kobler has made a large number of excellent photographs at my request during the past twenty-five years. He has minimized, rather than exaggerated, the patience necessary to secure successful shots. Time after time, when all adjustments have been made for field and focus, the snake will again endeavor to escape and a new beginning must be made. Rattlesnakes, having some confidence in their ability to defend themselves, are more ready than most snakes to rest quietly in a watchful coil. Sometimes a snake may be soothed by placing a dark cloth over it. This gives it

Fig. 43 Eagle-cactus panel made of the rattles of 847 rattlesnakes. The pointed eagle feathers required unbroken rattle strings, including the buttons. (Figs. 42 and 43 published by courtesy of the Buckhorn Curio Store, San Antonio, owner of the panels, and the San Antonio Zoo, which provided the photographs)

a chance to settle down in what seems to be obscurity. Holding a snake firmly in a confined position, sometimes in the exact pose desired, for a minute or so will often cause the subject to remain quiet for an extended period. Few living objects are more difficult than snakes to keep entirely in focus, because of the sharp details comprising scales and pattern over every part of the body.

Particularly troublesome subjects may sometimes be quieted by an initial cooling in a refrigerator. We have also used ether with some success, although too much will lead to an unnatural pose.

CHAPTER THIRTEEN

Enemies of Rattlesnakes

The principal natural enemies of rattlesnakes are such mammals as deer and badgers, certain birds such as hawks and roadrunners, and, among the snakes, king snakes and racers. Various domestic animals also destroy rattlers. When the extent of control exercised by these animals is surveyed, it must be remembered that, with the possible exception of badgers, they never kill for fun, or with the altruistic idea of dispatching a creature dangerous to man. Some, such as deer and antelope, kill rattlers for their own protection or that of their young; whereas the carnivores nearly always kill for food. The greatest destroyer of rattlesnakes is man himself.

One difficulty, when the rattler-killing proclivities of some animals are to be determined, is to know whether the victims were actually killed by these enemies, or were found by them on the road already crushed by passing automobiles.

Within the past few years, studies of the stomach

contents, pellets (indigestible remnants of food disgorged by birds), and scats (excreta) of predators, have added much to the knowledge of their food habits. In no case have rattlesnakes been found to comprise more than a minor part of the diet of any animal — and recent research has shown that such potentially dangerous creatures as rattlesnakes can fall prey to seemingly weaker creatures, that many birds and mammals are quicker and more alert than rattlesnakes, and can seize and render them helpless without danger to themselves.

MAMMAL ENEMIES

Deer have long been reputed to be rattlesnake killers, and a number of my correspondents have had personal experiences fully verifying these reports:

I have seen buck deer kill rattlesnakes on three different occasions. They just back up and run and stomp them; then they turn and do it again. After a buck deer gets through, there isn't a piece of rattlesnake over one inch in length. A buck deer really makes sure that they're dead.

Domestic hoofed animals are also reported to kill rattlesnakes occasionally. Some correspondents call attention to the fact that rattlers are scarce or absent on sheep or goat ranges. This might be caused by actual destruction by these animals, particularly of juvenile rattlers; or it might be because the sheep reduce the feed available to rodents, thus, in turn, starving out the snakes. One correspondent says:

Sheep kill quite a few snakes. The snakes strike at the sheep and get hung up in their wool.

Occasionally a horse may kill a rattler. In the *San Diego Tribune-Sun,* March 24, 1951, appeared an account of a mare, which, while grazing in a corral, was seen to snort, rear, and stamp down with her front feet. She sniffed the spot and then resumed browsing. Her owner, who had been watching, went over and found a three-foot rattlesnake, mangled by the horse's hoofs.

It is presumed that cattle often kill rattlers by stepping on them inadvertently. However, one of the members of the San Diego Reptile Club reported that he had seen a cow jump on and kill a rattler with her forefeet; it seemed to be done with intention.

Hogs have long been reputed to be killers of snakes, rattlesnakes particularly, and I have received several firsthand accounts of such incidents. One is as follows:

On one occasion, while hunting squirrels, I saw a large sow with pigs kill a rattler by the simple expedient of stepping on it with her front feet and grasping the head of the snake with her mouth. The sow and pigs then finished the job by eating the rattler.

The peccary, Mexican wild hog, or javelina, is frequently cited in the literature as being the enemy of the rattlesnake. Few confirmatory field notes have been reported to me, although I have read one account of the killing of a rattlesnake by a band of peccaries. They jumped on the snake and cut it to pieces with their hoofs, and then ate the remains.

There seems no question that coyotes occasionally kill rattlesnakes, usually for food:

Coyotes kill rattlers, snapping and shaking them until the snake is literally shredded. This I have observed several times.

Some believe badgers to be the most important rattler killers in South Dakota. A correspondent mentions:

We had a pet badger that loved to kill rattlesnakes and drag them all over the place.

The opossum may be at least an occasional rattlesnake killer. One writer stated that when an opossum was put in a cage with an eastern diamondback and a brood of young, the opossum ate the brood of eighteen young rattlers the first night, and on the second killed and ate a part of the mother snake.

I have been surprised at the large numbers of sheep and farm dogs reported to be rattlesnake killers, some with a considerable regularity. And there seems to be a somewhat standardized method of attack. The dog dances around the snake, barking and making threatening advances. As soon as the snake abandons its defensive coil and stretches out to crawl toward some refuge, the dog rushes in, bites the snake, sometimes on the neck behind the head, but often at mid-body, and then, before the snake can retaliate, relinquishes his hold by tossing the snake in the air. This maneuver, rapidly repeated, soon finishes the snake.

The following account of a house cat that killed rattlers was sent me by a correspondent in San Antonio, Texas:

In August, 1924, an employee of the Apicultural Laboratory at San Antonio came to the office and asked if we would like to see "Momsie Cat," as she was called, kill a rattlesnake. The entire office force proceeded to a piece of flat ground not very far from the Laboratory building, where we saw the mother cat with three half-grown kittens in a battle with the rattlesnake, which was approximately four feet long. The employee said that his wife had heard the rattlesnake making a great deal of noise near their house and on looking out of the door saw the cat watching it. The cat waited till the rattlesnake was coiled and then made a spring toward it. When the rattler struck, the cat dodged the blow and as the snake stretched itself with the blow the cat would bite at the back of its neck. At the time we were called the rattlesnake lay stretched out on the road and was rattling as best he could and attempting to get into some brush. The old cat would walk up slowly until she was about a foot from the rattlesnake's head and then would suddenly jump on the snake's back and bite its neck and then run. She repeated this a number of times. Each attack left the snake in a much weaker condition and at last he coiled up and ceased to rattle. The old cat approached cautiously and catching him by the back of the neck started to drag him toward the house. When she became conscious she was being watched, she left the rattlesnake and came over to where we were watching. She seemingly was exhausted by the fight she had had with the snake. On examination, the back of the snake's head was found to be full of punctures from her teeth.

The reports of several writers indicate that bobcats or wild cats occasionally attack and kill rattlers for food.

Reports from correspondents also make it evident that rodents occasionally attack rattlers. Some of the animals mentioned are woodchucks, squirrels, and rats.

Those who keep rattlers in captivity soon learn that it can be fatal to the snakes to put live rats or mice in with them for food, unless the rodents are also well supplied with food for themselves, for otherwise at night they will attack and kill the lethargic snakes. The following account is quite typical:

We have a record in our files of a captive rattlesnake being killed by a mouse. The snake did not appear aggressive when the mouse was dropped into the cage as food. The next morning the snake was found dead with a wound in back of the head, and the mouse alive.

A variety of other mammals attack rattlesnakes, presumably to eat them. Among these are ground squirrels, raccoons, skunks, and the ringtailed cat. Even the lowly rabbit will defend its young:

One Sunday afternoon I heard a rabbit squeal, and looking in the direction of the sound I saw a mother rabbit jump onto a spot and give a terrible rake with her hind feet. Hurrying to the place, I saw a large rattlesnake swallowing a small rabbit. Blood was oozing out of marks on the snake's back caused by the toenails of the mother rabbit. I killed the snake and found three young rabbits inside.

BIRD ENEMIES

As might be expected, birds of prey that feed in part on snakes occasionally take rattlesnakes. Among them are eagles, although snakes form a smaller part of their diets than is the case with some of the hawks.

One writer tells in detail of an encounter between a bald eagle and a large prairie rattlesnake. The eagle attempted to beat the snake with its wings; also to injure it by dropping it from a moderate height. The eagle eventually won, and flew away carrying the rattler.

Hawks, especially red-tailed hawks, prey quite regularly on rattlers. Here is one observation:

> I saw a chicken hawk pick up a rattlesnake off the ground and fly straight up in the air with it. When the hawk got about three hundred feet up in the air, it dropped the snake and let it fall about one hundred feet and then caught it with its talons. This went on for some time, until the snake had most of the life taken out of it, and was ready to make a meal for the hawk.

Audubon asserted that he watched snakes hide under rocks to avoid being seen by vultures or falcons flying overhead. Like others of Audubon's observations on snakes, this is hardly to be credited.

The roadrunner is the most famous rattlesnake killer in the Southwest, but many of its exploits are quite mythical. There is no doubt that roadrunners dispose of young rattlers as they would feed on other young snakes, but adult rattlers would be too large for a roadrunner to cope with.

Wild turkeys, particularly when in flocks, will sometimes destroy a rattler:

> I observed four turkeys march around a coiled rattler and kill him, then pick the rattler to pieces. When the rattler struck at the turkey in front of him, one or two behind would dart in and peck him in the head.

Domestic turkeys react toward rattlers in somewhat the same manner as the wild ones, except that they rarely will attack a rattler. A writer who herded turkeys for a while, reported that they would always surround a rattler and set up a clamor, to which he would respond by killing the snake.

Chickens also occasionally will dispose of young rattlers. A zoologist reports:

> I was standing in my yard (my place is located in thick woods, five miles from town), when, suddenly, one of my chickens came running from the woods in the yard with what looked like a spotted garter snake hanging from its head. I took after it. The chicken ran in the woods and would not drop the snake until I had overtaken it. To my surprise, I saw it was a fine pigmy rattler, striking right and left, unharmed except for a few rattles missing. So after searching for years for one while on snake hunts, one of my chickens carried one into my yard for me.

On several occasions I have noted ravens eating young rattlers by the roadside, but these probably were traffic casualties, and I cannot be certain whether these birds would attack a live snake. Crows, bobolinks, mockingbirds, herons, killdeer, and shrikes are among the birds reported to have attacked rattlers.

Several early writers tell of vultures preying on rattlers, probably through confusion with hawks. An interesting early tale is to the effect that turkey buzzards kill rattlers by means of their intolerable stench.

REPTILE AND AMPHIBIAN ENEMIES

King Snakes

It is well known that certain snakes, particularly king snakes and racers in the southern United States, will eat rattlesnakes. Such occurrences have often been noted in the wild, and the remains of rattlers have been found in the stomachs of predatory snakes found run over on the road.

In order that we might observe the method whereby a king snake attacks and swallows a rattler, we fed young rattlers to king snakes on six occasions at the San Diego Zoo. The following notes describe how a king snake ate a rattler on one occasion. The king snake was a male, about four feet long. It had been in captivity for nine years and had been fed mice at fairly regular intervals. These it accepted readily. Once it had been fed a fledgling swallow. It had been fed no snakes.

The rattler, a young female southern Pacific rattler (*C. v. helleri*) with a complete string of three rattles, and therefore about eight months old, was about sixteen to eighteen inches long. It seemed to be heavy-bodied and well fed.

When the rattler was placed on the ground the king snake, expecting its usual mouse, seemed somewhat at a loss. It first came up to the rattler, which was lying quietly, and stopped with its snout about an inch from the rattler's mid-body. It remained thus for a minute or more. Then it moved ahead slightly until its head was resting on the rattler's body. At this moment C. B. Perkins, in charge of the snakes, returned from feeding the other king snake, whereupon the snake again came to the door and rose up for its prospective mouse. It then lowered as if looking about the cage for the mouse.

The rattler was then moved slightly with a stick, to attract the king's attention. The rattler elevated its body somewhat. Immediately the king snake seized the rattler at mid-body. This was 12:16 P.M. The rattler bit the king snake at once; and, as the king snake threw two coils around the rattler's body, posterior to its head-hold, bit it again toward the neck. The fangs could be seen to take effect. The king snake now threw a third coil around the rattler's body, and then began to work its mouth along the rattler's body toward its head. At 12:17 the rattler, with only a part of its neck and head still free, bit the king snake in the head. The king snake, working it upper and lower jaws alternately and laterally, almost let go of the rattler in these successive motions, but never quite did so.

At 12:20 the king straightened its neck in front of the rattler so that it was directly facing the prey and started to swallow the head. Then the king snake straightened the rattler's anterior part of the body by pulling back with its own head and neck, the rattler being still held by the coils at mid-body. The rattler's neck was seen to stetch appreciably by the tension, but it is doubtful whether it was injured. There was no evidence that the rattler's neck was broken, as is sometimes stated (probably inaccurately) to occur. At 12:22 the rattler's head had been swallowed; it was still alive but was not moving.

The king snake now started a protracted swallowing process. Its method of working the rattler down was to move its own head from side to side continually, but always keeping it horizontal. The upper jaw advanced continuously with alternating cross holds, each one being a little forward of the last.

The rattler was somewhat over half-way down at 12:25. It was quite motionless and upside down. Meanwhile the king snake was slowly crawling forward around the cage always as if to get a better purchase, but as there was a U-loop in the neck of the king, its head was always moving back from the rattler. At 12:31 about five inches of the rattler were still protruding from the king snake's mouth, and the rattler wriggled its tail slightly. The king snake was still operating its alternating cross holds and the teeth snapped audibly as they pulled loose.

The king snake rested at 12:32, and although it worked a little during the next three minutes, it seemed quite tired, and some rests lasted a half minute or more. At 12:38 the rattler's tail still protruded and was squirming. Suddenly there was an eversion of the rattler's scent glands and a discharge, after which they were again reverted. The king snake, with the rattler almost engulfed, did not seem to use its neck muscles to work the prey down the throat.

At 12:41 only the end of the tail was showing; it continued to squirm to the last. Now the neck muscles were brought into play and the rattle finally disappeared at 12:44. After this the neck muscles were seen to work much more actively so the rattler was forced down the gullet. When it seemed well down, the king snake made an angle or fold in its neck, and sent this backward as a wave, thus settling the rattler better in the stomach. At 12:46 this was completed and the king snake opened and closed its mouth as if to test it and get it settled properly in place after the strain. The tongue was protruded in a natural way several times. There were still some side motions of the body as if to adjust the meal. The king snake yawned again at 12:48, and then lay extended along one side of the cage. The swallowing had required thirty minutes. The king snake in the following days showed no after-effects from the bites it had received from the rattler.

In another case a female king snake ate a young southern Pacific rattler (*C. v. helleri*) about twelve inches long. In this case the king snake got a mouth hold in advance of its holding coils and had little difficulty, although tired enough by the operation to take frequent rests. After seizing

Fig. 44 A California king snake about to seize a young speckled rattler (*C. m. pyrrhus*)

the rattler it worked its jaws to the rattler's snout within four minutes, so quickly, indeed, that it frequently almost released the rattler in working its jaws along. The king snake had the customary U-shaped loop in its own neck and stretched the rattler violently by pulling against its own holding coils. The entire operation took thirty-six minutes. The king snake was then put in a sack and carried to my home. Sometime within the twenty minutes en route it regurgitated the rattler, which was entirely unhurt, notwithstanding the severe stretching it had received.

Both the literature and my correspondence contain reports of rattlers attacked and killed by king snakes of such size that they could not possibly have swallowed the rattlers. At the San Diego Zoo we have repeatedly offered rattlers of moderate to large size to hungry king snakes without success.

Fig. 45 The king snake seizes the young speckled rattler, simultaneously biting it and throwing coils around its body.

Some published accounts even speak of "crushing the rattlesnake to a pulp" or "breaking every bone in its body," such damage being done to rattlesnakes much larger than the king snake. I also think this quite incredible, and beyond the muscular power of a king snake. In only one of our six experiments did the rattler seem to be seriously injured by the king snake before it was swallowed. In this instance the king snake was twice as long and twice as heavy as the rattler, and had maintained uninterrupted pressure on the rattler's heart-lung section for nearly an hour and a half. And, of course, in another case the rattler emerged ininjured after a squeezing and swallowing that lasted over half an hour, plus several additional minutes spent in the king snake's stomach. In some of the stories heard of rattler–king-snake battles, the rattler, sometimes outweighing the king snake by at least ten to one, is crushed to shapelessness within a few minutes. In considering the evidence, one should remember that a king snake cannot remotely approach the squeezing pressure that a man can exert with one hand.

Other Snakes

Other snakes reported to eat rattlers include racers, black snakes, and bull snakes.

The belief that the large snakes known in the West variously as bull, blow, or gopher snakes are deadly enemies of all rattlers is widespread, particularly in the upper Missouri Valley. Yet these snakes not only den with rattlers regularly, but they are primarily feeders on mammals, with birds and birds' eggs next in order. Certainly bull snakes eat snakes and lizards much less often than do king snakes, and their eating rattlers must be of rare occurrence.

Rarely rattlesnakes are cannibalistic, usually through the unnatural conditions of captivity. I once received a shipment of small western diamond rattlers and found that one had swallowed another measuring slightly less. This was later regurgitated. Occasionally snakes may swallow each other through starting to eat the same prey simultaneously.

Miscellaneous Reptiles and Amphibians

Lizards may occasionally kill small rattlers. One author gives an account of a collared lizard found with a neck hold on a banded rock rattler, *C. l. klauberi*. When the author endeavored to catch both, the snake got loose and bit the lizard, which died in five minutes.

Swimming rattlers sometimes become the prey of alligators. A Louisiana correspondent reports that alligators shake the snake vigorously from side to side, after grasping it in its jaws, in order to kill it completely before attempting to swallow it.

An account (with a photograph) of a bull frog that swallowed a young diamondback *(C. adamanteus)* was published in 1952. The frog had been put in the cage as food for some large snakes. Thus it is indicated that small rattlers may in rare cases become the prey of bull frogs.

MISCELLANEOUS ENEMIES

Since, as discussed elsewhere, rattlers swim quite readily in streams, ponds, and lakes, young ones may become the prey of fishes. One fisherman caught a four-pound rainbow trout with a nine-inch rattler in it.

There are several reports of spiders that have killed and partially consumed small rattlers in Latin America. It is evident that large venomous tropical spiders have sufficient venom to kill a small rattler in a few minutes. Consuming the snake by sucking, as is the way of these spiders, takes many hours, if not days.

The *New York Times* of July 27, 1938, contained a dispatch from Ely, Nevada, stating that swarms of Mormon crickets killed and ate three rattlers.

PARASITES AND DISEASES

Although reptiles, including rattlesnakes, are subject to various diseases, and are afflicted with many internal and external parasites, researches on these parasites have been less intensive than on the parasites found afflicting other groups of vertebrates, since they affect man less directly. Of those that may be of importance to man, as far as the rattler parasites are concerned, there appear to be only two: the linguatulids or tongue worms, which have been occasionally transmitted to human beings, probably from rattlesnakes; and some of the Salmonelleae, a group of bacteria causing certain poultry diseases, of which snakes, including rattlesnakes, are intermediate hosts. Other parasites that afflict rattlesnakes include flatworms, tapeworms, round worms, ticks, and mites.

One of the most serious bacterial diseases of snakes in captivity, including rattlesnakes, is one variously known as mouth rot, mouth canker, or osteomyelitis. It is mentioned by almost every work on the care of captive reptiles. A pneumonia affecting rattlesnakes has been mentioned by several writers. Tuberculosis and tumors have also been encountered.

CHAPTER FOURTEEN

Indians and Rattlesnakes

Since one of the purposes of this book is to discuss the various contacts between rattlesnakes and humans, I shall outline in this chapter the attitudes that the Indians held toward these snakes, as far as may be judged by the accounts of explorers, the findings of ethnologists, or the customs and legends that some tribes have retained.

Rattlesnakes, although they were never so important economically in the pre-Columbian Indian communities as were many other animals—particularly such mammals and birds as furnished major components of food supply, clothing, or shelter—nevertheless played their part, especially in medicine, myth, and legend. As do all primitive peoples living close to nature, the Indians duly observed each animal and fitted it into their scheme of life and religion, and in the case of the rattler this position was accentuated by its venomous nature.

Generally speaking, the Indian attitude toward rattlesnakes varied from mere aversion or toleration, through appeasement and propitiation, to reverence and even worship, always tinged with the animism (assignment of supernatural powers to many creatures and objects) that colored the Indian view of nature. Besides its place in tribal myths, the rattler had a place in war, in medicine, in art and decoration, and even, to a minor extent, in food supply. In all of these contacts and reactions there were major tribal and territorial differences, so that examples rather than generalities must be cited. Usually the rattlesnakes were given a status apart from and above that of other snakes, for, in extensive areas of the western United States they were the only seriously venomous snakes, and throughout the country they were the most dangerous. Besides, the presence of the rattles simplified recognition, and also gave them a distinction above others of the serpent tribe.

A question naturally arises as to the accuracy of the published reports of the Indian–rattlesnake relationships — of the Indian's knowledge of rattlesnakes and his reactions to them. Without doubt the reports of the early explorers and colonists on the Indian attitudes toward rattlers were deficient and erroneous for the same reasons that their accounts of the rattlesnakes themselves were inaccurate: they were untrained in observation and given to emphasizing the lurid; they were credulous and passed on, as firsthand observations, stories based on hearsay and the deliberate exaggerations of the campfire. Their statements of Indian ideas and attitudes were often colored by their own preconceptions or misconceptions.

Many Indian myths and usages with respect to rattlesnakes have now found their way into white American folklore, beliefs, and customs. This is particularly true of methods of snake-bite treatment and ideas concerning rattlesnake habits. And as Indian beliefs have influenced white folklore, so also white contacts have affected the Indian beliefs. (Many of them have resemblances to folklore in other parts of the world.) There are probably few tribes that still retain unchanged their pre-Columbian attitudes toward rattlesnakes.

It should be clearly understood that this presentation of the ideas of the Indians respecting rattlesnakes is only an effort to assemble these ideas for review. Their inclusion is not to be taken as an acceptance of their accuracy, for many, if not most, are obviously fantastic, as was well known to the ethnologists from whose accounts they have been abstracted.

RELIGION, SUPERSTITION, AND FOLKLORE

Serpent worship has been, and still is, widespread among many peoples and races.

Usually among the North American Indians rattlesnakes were believed gifted with supernatural powers, usually for evil, but this was also true of many other animals. They were cast as characters in innumerable myths and legends, and were the subject of extensive taboos. But seldom were they accorded major positions in any tribal religion.

Rattlesnake Protection and Appeasement

In reviewing the preternatural relationships of Indians and rattlesnakes, I find there is one practice almost universal among the Indians, from coast to coast — a rather surprising taboo against killing these dangerous creatures, particularly surprising among an outdoor and nomadic people to whom rattlesnakes were a not inconsequential hazard. This attitude of avoidance and even protection seems to have stemmed from three somewhat distinct assignments of the rattlesnake as an avenger: (1) avenging a hurt done a brother rattler; (2) avenging an affront to the gods of weather — involving also a supplication for favorable weather, rain especially; and (3) avenging, as an agent of a higher god, infractions or derelictions in religious duties.

As early as 1709, it was reported that the Tuscarora Indians of North Carolina were afraid to kill rattlers lest the snakes' relatives return to wreak vengeance.

In 1764 this experience was reported among the Ojibwa: the writer found a rattler and wished to kill it, but the Indians protested. They surrounded the snake, calling it "grandfather," and blew smoke toward it, which he says the snake appeared to enjoy. They asked the snake to take care of their families in their absence, and to cause the British agent to supply their canoe with rum. Later, when they were caught in a severe storm on Lake Huron, they blamed their predicament on the writer's having threatened the rattlesnake. They sacrificed two dogs to it by throwing them overboard, and likewise offered some tobacco. They prayed to the rattler-god. In this case the snake seems definitely to have been deified.

In 1846 an interesting account was given of the ceremonial killing of a rattlesnake by the Menominee — a mixture of propitiation and profit. The finder of a snake made a proprietary signal. He then welcomed the snake as a messenger from the spirit land. Tobacco powder was sprinkled on the snake's head as an offering. The snake was then seized by the neck and tail, and every vertebra broken by a single quick jerk. The fangs were removed, and the snake's body was cut into small pieces, which were distributed among the rest of the Indians to be carried in their medicine bags as amulets against evil agencies. The skin of the rattler was attached to the hair of the finder and worn as a trophy.

One writer supplies an incident in which an Arkansas Cherokee refused to kill a rattler; in explanation he repeated a long legend, of which the general tenor was: "I let them alone and they let me alone."

The Comanche would kill a rattler only when it was sluggish and failed to rattle when approached. Then it was presumed to be on the warpath and lying in wait for an enemy, which justified its death.

The Tarahumare believed rattlesnakes to be the companions of sorcerers, who meet and talk with them. The Tarasco believed that rattlesnakes must never be touched, much less killed.

Among the Tolowa of Oregon, as a penalty for killing a rattler one had to drink and eat from separate vessels for five days, for the "rattlesnakes are people" and this was the same method of purification required of an Indian who had slain an enemy in battle.

The Nisenan of central California had a genuine terror of rattlers. If one of their people were bitten, he would be excluded from camp for several days, for otherwise the snake, having tasted blood, would follow him in. The Yokuts never destroyed a rattler. One Indian caught one on the plains and turned it loose in the mountains where it would be less likely to be found and killed by white men. Another, seeing a white man about to dispatch a rattler, frightened it into the rocks so that it might escape.

The second basis for the propitiatory or protective attitude of the Indians toward rattlesnakes stems from the close connection between rattlesnakes and weather in the Indian cosmology. It is implicit in the details and paraphernalia of the Hopi Snake dance, as well as in lesser-known rituals of other tribes. In the arid areas of the Southwest, which often suffer from insufficient rainfall, the ceremonials usually had the purpose of assuring an adequate water supply; in more favored sections having adequate rainfall, floods might be caused by angered rattlesnakes, and this could be prevented by propitiation. The resemblance of lightning to the snake's strike, and of thunder or the hiss of rain to its rattle, made inevitable the connection, by the symbolically minded Indians, of the rattlesnake with the rainstorm.

The Micmac had a legend that thunder is produced by seven flying rattlesnakes, crying to each other and waving their tails as they crash across the sky. A flash of lightning is produced when they dive for their prey.

The Cherokee feeling toward snakes was one of mingled fear and reverence, because they were thought to have an intimate connection with the rain and thunder gods. Every precaution was taken against killing or injuring one, especially a rattler. He who killed one would see others, and when he killed a second, more would come until he was driven crazy. To destroy a rattler was to destroy one of the most prized ornaments of the thunder god. As the rattlesnake was regarded as the chief of the snake tribe, he was to be feared and should be killed only under extreme necessity; one had to ask the pardon of the snake's ghost or another would come to avenge him. The rattles, teeth, flesh, and oil were greatly prized for occult and medicinal purposes, but the snakes could be killed for them only by certain priests after appropriate ceremonies. When a rattler was killed, the head had to be cut off and buried an arm's length deep in the ground, and the body hidden in a hollow log. If they were left out, rain would come in torrents and the streams would overflow their banks. The Flathead believed that if one whips a snake it will rain.

While the Navaho did not have a snake dance, they, too, linked snakes with lightning and rainfall, and under no circumstances would they kill one. An object that protected from lightning also protected from snakes; so long as a man carried his shell and turquoise charm, a snake might enter his hogan but would not bite him.

Finally, as a third basis for the Indian treatment of rattlers, an attitude of protection was adopted because the rattlesnakes were thought to be agents of higher deities. Among some tribes, particularly in southern California, the rattlesnake was an officer of punishment, carrying out the penalties imposed by other gods. The southern California Shoshoni believed, when a rattler bit a dog near one of their homes, it must have resulted from a neglect of religious observances. The story was told of a young Paiute who failed to take the proper religious course after a nightmare. Having dreamed that he was attacked by a rope, he should have taken a cold bath and communed with the spirits on the following morning. Failing this, on that day he was bitten by a rattler, and, despite treatment by the most skilled tribesmen, he succumbed on the third day.

Among the Luiseño and Diegueño of southern California, rattlesnake bite was also attributed to dereliction of religious duties, and its prevention could best be assured by the performance of ritualistic dances. If several people were bitten, the priests were blamed for not holding these ceremonies oftener.

With children, rattlers were sometimes used as bogies. Among the Pima if a child put his foot in a mortar, the mother said, "A rattler will bite you." In instructing the Diegueño children, the priests told them if they were good and got bitten they would succeed in getting home, but if they were bad they would never return.

An occasional belief in the rattlesnake as an instrument of punishment is heard of among tribes farther east. Among the Meskwaki (Fox), one who had more than four skunk-cabbage roots in his possession at one time would have his home invaded by rattlesnakes, which would bite his family.

Rattlesnake venom sometimes was employed in intratribal punishments. Among the Omaha a sharp stick, poisoned with dried venom, might be jabbed into an offending brave in the confusion of a crowd as a punishment for disturbing the peace or disputing a chief's authority.

Sometimes the man was first given a chance to reform by the poisoning of his horse, as a threat of worse to come.

Among the few tribes not fearing to destroy rattlesnakes were the Huchnom, who simply killed them as dangerous pests.

A seasonal postponement of certain ceremonies, or the avoidance of relating particular myths and legends during the rattlesnakes' season of activity, seems to have been common methods of forestalling the snakes' resentment. The Penobscot shared the almost universal American Indian belief that legends must not be related in summer lest the snakes overhear and bite the offender. The Catawba of the Piedmont of northwestern South Carolina would not talk about snakes after dark, lest it invite annoyance from them. If a snake heard such tales he would lie in wait in the path. If he heard travel plans he would also wait along the way; to deceive him you had to say you were leaving day after tomorrow when you really meant tomorrow.

Certain dances of the Cherokee, being deemed possibly offensive to the snakes, were held in the late fall after the rattlers had gone into hibernation. Among the Navaho and Ute, cat's cradle could not be played in summer since to do so would invite snake bite. The Kato believed that anyone singing rattlesnake songs in summer would be snake-bitten.

There were a number of customs governing terms to be used in speaking to or about rattlesnakes in an endeavor to appease them. Apologetic phrases and figurative names were often employed, the latter as a method of concealment or evasion. Some even approached the formality of rituals.

The Ojibwa called the rattlesnake "grandfather." The Cherokee honored the rattler with a name that signifies "the bright old inhabitant." The Cherokee avoided saying that so-and-so had been bitten by a snake; rather, it was said the victim had been scratched by briers, thus deceiving the listening rattler spirit.

The Chiricahua Apache had a dread of rattlers that went beyond a mere fear of snake bite. If they met one they addressed it as "mother's father," or sometimes "mother's mother"; they requested it to keep out of their way. They never mentioned rattlers except in invectives; and even when one said, "I wish a snake would bite you," the wisher might have gotten a sore mouth for saying it.

In California it was the custom of the western Shasta, if one of them killed a rattlesnake, to make an apologetic speech and cast the body in the direction of an enemy tribe. The Yuki used the apologetic term "grandmother" when they narrowly missed stepping on a rattlesnake. Their figurative name for the rattlesnake indicated that to them the source of the rattlesnake's danger lay in menstrual blood, which they considered the deadliest of all poisons.

The eastern Pomo, when coming upon a rattler, said, "Grandfather, I am not going to bother you; let me pass safely." One Diegueño shaman referred to a rattler as an "old man with a basket hat on his head." Such euphonious terms were thought to improve the rattlesnake–Indian relationship and understanding.

Necessarily, in this outline of rattlesnake protection and appeasement by the Indians, I have included only the citations that refer to rattlers. But it should not be thought that these practices were restricted to rattlers or, indeed, to snakes. On the contrary, propitiative and ritualistic treatments, protection, and terms of kinship, were often applied to other animals, such as the bear, deer, puma, and eagle, to name a few. The kinship term "grandfather," for example, was not restricted to rattlesnakes.

Some of the names used for rattlesnakes by the Indians were less propitiatory and more prosaic. The Cayuga referred to the massasauga as the "picture of the sun" or "dappled" because of the round spots in its pattern. The Menominee called the rattlesnake the "rattling-tail," while to the Lenni Lenape (Delaware) it was the "frightener." The Catawba referred to the timber rattler as "rattlesnake bristling," and the ground rattler was the "rattle tail contining one thousand." The Cherokee called the timber rattler "now you see snake." The Maricopa referred to the sidewinder as the "left-handed snake."

Legends and Tales

The Indians, lacking a literature, were much given to the recitation of tales and legends, many of which were passed down as oral traditions unchanged through the generations. In a considerable number of these, the rattlesnake appears as one of the characters, although much less often than several other animals, notably the coyote in the West.

Some of these tales are quite amusing, such as the Algonkin story of how a deity fixed a string of wampum to the tails of rattlesnakes, thus forming the rattles and making it impossible for them to move without sounding a warning; the Tewa on why rattlers will not bite in summer when the moon shines, and how the rattle can be heard two miles away; the Pawnee story of the little rattlesnake crying because his rattle wouldn't make a noise; the Wishram on how the rattlesnakes ate some kind of grass in summer to make the bite more venomous.

Miscellaneous Superstitions and Beliefs

The Indians had a number of rattlesnake myths not essentially religious or propitiatory in character. Among these was that of the rattler with a jewel in its head. One writer was told by the Cherokee that their most treasured and brilliant stone came from a snake's head. This was an enormous snake, protected by a number of followers. A brave Indian, encased in leather armor, succeeded in killing the snake and securing the stone.

Another writes of a great rattlesnake seen swimming in the Mississippi with a carbuncle in its head. It made a brilliant flash of light in the sunrise. The snake was shot, but although the body floated, the stone was lost. There is a myth of the great rattler of the Arbuckle Mountains, with a diamond in its head, and others studding its sides.

An interesting myth of the Chitimacha of southwestern Louisiana is this: in ancient times two Indians saved a pair of rattlers in a great flood. Ever after, the rattlesnakes remained the friends of man. Each home was watched over by a rattler in the owner's absence; when the owner returned, the rattler went about its business.

The Cherokee believed that seeing a snake at the start of a journey was an omen of death. Should you stretch your hands toward a rattlesnake, and it appeared cross and evil, it meant you had not long to live. But if it were calm, you might pick it up and set it down; then if it crawled west it signified death; if east, a long life.

The Omaha (southern Sioux) believed that some rattlesnakes could shoot or project venom for a distance of at least a hundred feet.

The Hopi and some Pueblo tribes thought the smell of a woman to be highly offensive to a rattlesnake; it angered him so that he would bite.

The Pima Indians credited the rattler with the ability to select the direction toward the best mesquite beans. The author wonders at the source of this idea. (It might be guessed that the densest mesquites would have the most abundant beans, and would be chosen by the snakes as the best refuges from the heat.) The Pima considered it unlucky to come upon two rattlers, one after the other, while searching for something.

The Maricopa believed jackrabbit blood contains rattlesnake venom, for the cottontail, a relative of the jack, was bitten by the rattlesnake in the tribal creation myth.

A number of unrelated rattlesnake superstitions and beliefs were prevalent among the California tribes. The following are Nisenan beliefs: while bull snakes are hatched from eggs, rattlers are born by way of the mother's mouth—a belief obviously related to the swallowing-the-young myth. Rattlesnakes and squirrels may interchange bodies. An Indian killed one in the process of transformation; it had a squirrel's head and a rattlesnake's body.* The wood dove is the rattler's niece; if you mock her a rattler will bite you. If one person in a family is bitten, another will be. When one recovers from rattlesnake bite, he must give away everything he had with him when bitten or be doomed to bad luck forever. The

---

*A rattler, while swallowing a squirrel tail first, would have just such an appearance.

Nisenan told of a man who boasted he would eat a rattler's heart. He did so and ever afterward was quick-tempered like a rattlesnake.

The Yurok believed that an eclipse is caused by a rattlesnake swallowing the sun. After biting a person, a rattlesnake recuperates by steaming itself on hot rocks.

The Maidu thought rattlesnakes were harmless while drinking, as they laid aside their poison until finished. This fantastic theory is well known in white folklore.

The Owens Valley Paiute believed that if one was bitten by a blind rattler (while shedding) he also would become blind. One who has been bitten would become sick at the same time next year (this is also an ancient white myth).

The Northfork Mono held that if one kept rattles in his home or sang a snake song, rattlesnakes would be attracted and move in.

The Navaho, like the Apache, believed in transmigration, and that the spirit of an especially mean Indian would find lodgment in a rattlesnake. There was an Indian tradition that the number of rattles borne by a snake recorded the number of people it had killed. Certain Paraguay Indians believed that young snakes grow from the bodies of dead adults, so the bodies should always be taken far away.

Although changed by time and white influences, the Indian superstitions respecting rattlesnakes are still retained by many tribes and individuals.

Dreams

Snake dreams, and especially rattlesnake dreams, were usually deemed ill omens. The Cherokee believed that sickness might result from dreaming about a snake. The Oto held such a dream to mean that other tribes were about to attack, and that the camp should therefore be moved at once. The Pomo thought it good luck to dream of a rattler or bull snake, but bad luck, on the contrary, if the snake left you in the dream.

The Tübatulabal thought rattler dreams were sent by witches. Anyone who dreamed of a rattlesnake was required to go out into a canyon on the following morning to make a speech to the creature, whereby the bad dream might be rendered ineffective. The Paiute had a closely related belief, as illustrated by this story: a man on a hunting trip dreamed of being attacked by a rope. Next morning he failed to make the proper ritualistic gestures of appeasement. That day he was bitten by a rattlesnake and died, despite the best ceremonial treatments. Among the Mohave, to see a rattler in one's dreams forewarned of aches and pains in old age.

The Navaho thought it not serious to dream of snakes unless one was bitten in the dream.

The Maya had a queer dream sequence: if a man saw snakes in a dream he would quarrel with his wife; if he dreamed of an unclad woman he would see a rattlesnake next day.

PROTECTIVE MEASURES

Repellents, Charms, and Taboos

So intimately is the avoidance of rattlesnake bite related to its treatment, in Indian procedures, that it is difficult to separate preventive from curative measures. The Indians naturally extended their remedies to uses as repellents, for to them the idea was obvious that a substance which counteracted rattlesnake venom should also be objectionable to the snakes themselves. Hence, to keep rattlers away from camps or to avoid them when traveling, the tribes that depended on plant cures carried with them some of the same plants used in snake-bite treatment.

The herb called the "rattlesnake masterpiece," which was used in snake-bite treatment, was considered to be so effective that, if it were rubbed on the hands, any snake encountered would be unable to move. The Menominee had a root which they pounded up and put on a stick; the snake would bite at this and be killed by the poison of the root.

The Iroquois constantly carried dried roots that they chewed and spit on their hands to repel snakes, as well as to apply to the wound in case of a bite.

The Maricopa had a plant known as "rattlesnake afraid"—an aromatic shrub resembling rhubarb; this was chewed and spat on horses' hoofs to protect them on journeys. The Kaibab tied pieces of the root of lovage, wrapped in buckskin, to their moccasins to prevent snake bite.

The Indians of southern California thought the odor of the plant they called "chucupate" so repugnant to snakes that they habitually carried a piece of its root for protection. The Kato used dried seaweed and octopus for protection against rattlesnakes. The Atsugewi used angelica root; also they rubbed turtle liver on their legs, and the women attached turtle heads to their skirts.

In Central America, the Muskito chewed a few guaco leaves daily as a preventive against snake bite. If it were spit on a snake's head the creature would be stupefied by this infusion.

Repellents were sometimes used about habitations to keep the rattlers away. The Chippewa (Ojibwa) sprinkled around their camps a decoction of calamus root mixed with wild sarsaparilla; they also used an infusion of rattlesnake fern sprinkled about the wigwam. The Navaho splattered the floor of a hogan with a plant of the parsley family, as the snakes are supposed to dislike its odor.

Several tribes used tobacco, about both their persons and living quarters, on the theory that it is deadly or unpleasant to rattlesnakes. The Menominee thought the smell of tobacco made rattlesnakes sick, thus rendering it easy to catch them. The southeastern branch of the Yavapai, when moving into caves, burned strong tobacco to keep the snakes from issuing from the crevices. The Tarahumare attributed great repelling power to tobacco, and often smoked before going to sleep, to keep the rattlers away at night.

Various animal and mineral products were also used for protection, although more as charms than repellents. Among the Menominee a piece of flesh from the neck of a turkey buzzard was dried and applied to the body as a powder. When an Indian was so protected, a rattler would not come near — much less bite him. According to Cotton Mather in 1716, the Indians anointed themselves with fat of the kite, a bird that feeds on rattlers, and then handled rattlers as safely as they might have handled eels. Some tribes wore feathers on their moccasins, so that the rattlers, fearing roadrunners, would flee.

The Cora of northwestern Mexico wrapped a rattler fang in an oak leaf and wore it as a love charm, and rattlesnake venom was used as an aphrodisiac.

The Indians of the Rogue River Valley, in Oregon, knowing the antipathy of rattlers for hogs, used to ask the settlers for pieces of pigskin to be placed around their ankles when gathering berries. (These might be of some service, not because of the hog-rattler aversion but, rather, the mechanical protection afforded by the leather.)

Charms may enter the Indian–rattlesnake relationship in two forms: first, as protective amulets against the snakes; and, second, as charms formed of rattlesnake parts or effigies worn to ward off other evils. I shall now pass to the second category.

When approaching the enemy, the Menominee warrior would take a snake skin from his war bundle to give him the serpent power of stealthy approach. The Nez Percé Indians invoked a charm against evil by placing a rattler's head on hot coals in the earth and covering it with fresh liver and gall from wild beasts. During the steaming the liver was thought to absorb venom from the head, which was then carried as a talisman in a buckskin bag.

The Karok wore rattles on the head as a charm against illness, since the rattlesnake never gets sick. Among the Zuñi the rattle was worn as a protection against human enemies; the fat or oil of the snake was rubbed on the face on the theory that the enemy would fear and flee from this as he would from the snake itself. A medicine of snake droppings was given to Acoma war captains, both to increase their strength and to give them power to foretell events.

Rattlesnakes and rattle charms were definitely associated with eye troubles and blindness. For example, a man whose wife was pregnant found and kept a rattle—something he should not have done. The son became blind two or three months after birth, and died when still young. In another instance, carrying a bright rock that looked like a rattle caused rheumatism and blindness.

The protective values of various charms were in some instances based on actual or fancied resemblances to rattlesnakes. Thus the Wintu of California attributed good luck to stones having a coiled shape suggesting a rattler. Among the Yana certain small round stones having bright-colored bands were interpreted as being rattlesnakes, and were much sought as charms. They were supposed to cure disease, and to bring luck in gambling and hunting.

Aside from the widespread ban on killing rattlesnakes that I have discussed elsewhere, several tribes had interesting taboos. The Modoc believed it dangerous to whistle around elderberry bushes, lest the whistler be bitten by a rattlesnake. Among the Maidu the marriage of first cousins—considered incest—would surely be punished by rattlesnake bite.

Although rattlesnakes were said to have been killed by the Kato whenever met, they had to be clubbed rather than shot with an arrow, or lameness would befall the offender. The Yuki would not use the topknot of the California quail in their feather regalia because the topknot is used as a club to kill rattlers by the quail. (Actually the topknot is feather-soft and delicate.) Among the Pomo, it was taboo to kill rattlesnakes except in the spring when they were mating, and then both must be killed or the survivor would avenge its mate, as well as in resentment through having been found in a compromising situation.

Immunization

Various methods of immunization as a protection against snake bite have been attributed to the Indians. The Indians near Tampico, Mexico, inoculated themselves with rattlesnake venom as a precaution against future bites. Some Mexican Indians understood immunization by use of increasing doses of venom. The Indians may have used this protection in colonial days. Certain Colombian tribes inoculated themeslves with snake ashes.

SOCIETIES AND CEREMONIES

Dances and Exhibitions

Rattlesnakes were used by the Indians in a number of dances, either actually, or through effigy or symbolism. The Pomo of the north had a dance, including an episode in which a large defanged rattler was waved in the faces of the watching women to frighten them. The wizards of the Yokuts tribe held a rattlesnake dance each year. Four of them capered about with live rattlers in their hands, chasing and threatening each other. It was presumed by the observer who reported this ritual that they had either defanged the snakes or had denied them water for several days, which he apparently believed would render them harmless. The credulous tribesmen thought the priests invulnerable and pressed forward in an endeavor to secure, from the dancers, immunity from snake bite during the coming year.

An anthropologist has described this rattlesnake dance of the Yokuts in considerable detail. It was essentially a demonstration of magic power by the rattlesnake priests or shamans. On the first evening the priests danced while each carried a snake in a bag, and tests were made to determine which of the audience would be the victims of snake bite during the ensuing year. Next day these victims were treated in advance by suction of the prospective bites, and by the process of securing the emission therefrom of the tiny rattler that causes the pain — usually represented by a rat's tail or a small dead snake transferred from the mouth to the hand of the officiating priest. This was followed by an episode in which the real rattlers were danced with; they were so teased and annoyed that they were frequently seen to bite their tormentors, hanging by their fangs to fingers or hands. The reporter makes no suggestion as to the source of their immunity from serious results.

In a snake dance held by a subtribe of the Yokuts, money was put up by the audience, maybe $50 or $60, to see a shaman bitten. He would throw his snake on the ground to anger it, then permit it to bite him and dangle from his arm by its fangs, after which his fellow doctors would cure the bite with a sucking ceremony.

The Pomo dance seems to have been devised by members of a secret society to impress outsiders. It involved an extensive ritual centering on a single rattler, after the latter had been fed beads and feathers, which it was expected to eat with enjoyment. The ritual concluded with the supposed swallowing and regurgitation of the snake by a priest trained and designated for the purpose. Following the ceremony the snake was turned loose, for it was taboo to injure it. The Nomlaki, a Wintun subgroup, sometimes held an emergency dance to cure rattlesnake bite.

Many rattlesnake medicine men or shamans demonstrated their ability to handle rattlers at times other than formal rituals or when effecting cures. No doubt this was to keep themselves before the public. One writer tells of a Northeastern Yavapai shaman who would pick up a rattler and put it in his shirt; he would then take it out and terrify people. In picking up a snake he would first draw a circle with dust across the snake's back. Once a shaman was bitten as he picked a snake up. He quickly tore it in half.

The Hopi Snake Dance

The Hopi Snake dance is the best known of all Indian rituals. The widespread publicity it has received results from a single spectacular episode that occupies but a half hour during an elaborate nine-day ceremony, only a few other parts of which are ever seen by the public. In this fantastic episode, which so greatly fascinates the spectators, some of the participating priests execute a form of dance around a ring while each holds a live snake in his mouth, gripping the snake at the neck with his teeth and lips. Some of the snakes are rattlers. The dance is held annually at two or three of the Hopi pueblos in central Navajo County, Arizona. Each year people come in large numbers and from great distances to see it, and the argument as to why the participants are not fatally bitten is unending.

The dance is carried on through the cooperation of two fraternities, or secret societies, known as the Snakes and the Antelopes.

Four days of the nine are given over to a ceremonial hunting of the snakes. The snakes may be either rattlers or of nonvenomous species. No observer appears to have accompanied the Indians on these hunting trips except for short periods, for they are much averse to being watched while catching the snakes. The snakes are sought in the area surrounding the village, one day at each of the four cardinal points, invariably in the order north, west, south, and east. During these hunts the novitiates, some of whom may be only boys, are initiated in the capture and handling of snakes.

The snakes that have been captured are placed in sacred clay jars, and are stored in the *kiva* (an underground vault used for a lodge room and various religious rites) of the Snake priests. Here, when not actively engaged in hunting, and particularly after the fourth ceremonial day of the hunt, the Snake priests live and engage in the making of prayer sticks, in observing various sacred rituals including ceremonial smokes, and in the preparation of their costumes.

Among the sacred paraphernalia there is one item of particular interest to the herpetologist; this is the snake-whip or snake-wand, a wooden shaft about eight inches long, to which is attached a pair of eagle feathers. From the first hunts to the final dance, these serve a practical purpose in soothing the snakes, or in herding them when it is desired to have them go in a certain direction, or to cause them to straighten out when they have coiled for defense. Eagle feathers are used because eagles are the masters of (prey on) snakes.

At noon on the ninth day occurs the secret rite of washing the snakes, in anticipation of their part in the dance. This takes place in the kiva of the Snake society. The snakes are taken in handfuls by the chief Snake priest and dipped in an effusion contained in an earthen bowl, the liquid having previously been the subject of a suitable ceremony. After the washing, the snakes are dried by allowing them to crawl on sand; they are permitted partial liberty in the kiva for as much as two hours, following which they are placed in cloth sacks awaiting the ceremony.

All reporters who have witnessed the washing state that the snakes are handled gently but fearlessly, except at Walpi, where they are hurled quite violently on a sand mosaic. There is no report of anyone having been bitten. During their brief freedom they are guarded by boy priests.

The Snake dance itself occurs at sundown on the ninth day, the time fixed by precedent. Prior to the dance, the snakes, carried in several cloth bags, are placed in the *kisi* (a temporary bower of branches shaped like a tepee) by the Snake priests.

On the day of the dance, the audience, which consists of the local Hopi, visitors from adjacent pueblos, Navaho, and, with improvements in roads, an increasing number of whites, selects vantage points from which they expect to view the ceremony. Many, especially the more timid, perch upon the housetops.

The number of participants in the final ceremony varies in the several pueblos, and at different times in the same pueblo. There may be as few as five Antelopes and eight Snakes, or as many as twenty-five Antelopes and thirty Snake priests. The number of reptiles used varies from about fifteen, where the priests are few, to fifty or more where the largest dances are currently held. At the turn of the century there were reports of as many as a hundred snakes being used at one pueblo. From a quarter to half of the snakes are generally reported to be rattlers; the others are harmless bull snakes, racers, or, occasionally, some rarer species.

The dance begins with the entrance of the Antelope priests, dressed in elaborate and symbolic costumes. They stamp and march accompanied by the jingle of their trappings and rattles made of buckskin.

After a short pause the Snake priests emerge from their kiva. They break up into trios, each containing one man who is usually referred to as the "carrier," a second called the "hugger," and a third known as the "gatherer." As the first carrier passes before the *kisi* he stoops and is handed a snake by one of those within. This snake he puts into his mouth, holding it with teeth and lips from six to twelve inches behind the head. The hugger now puts his left hand on the carrier's right shoulder, or about his neck, and together, the carrier continuing to hold the snake, they slowly dance, with a shuffling step, around the area, with the carrier on the inner side of the circle. Each pair is followed by a gatherer. After approximately one and a half times around, the carrier puts the snake on the ground and, in passing the *kisi,* receives another.

Meanwhile, other trios have followed the first, and there is a circle of dancing priests, who are receiving, carrying, and putting down snakes in more or less confusion. The hugger, while dancing around at the right

Fig. 46 The Snake dance at Oraibi, Arizona, in 1902. A carrier with a snake in his mouth in the central foreground; Antelope priests before the *kisi* at the right.

hand of the carrier, from time to time brushes the snake's head or the carrier's face with the eagle feathers of a snake-wand. This is presumed by some to be for the purpose of engaging the snake's attention, to keep it from biting the carrier. The hugger acts as guide, as well as protector, for the carrier's eyes are generally closed.

The gatherer belonging to each trio has been following his two fellow priests. When a snake has been put on the ground by his carrier he picks it up, usually six to eight inches behind the head, sometimes at once, but more often after sprinkling it with sacred corn meal. Or, if it coils, as if for defense, he brushes it with his feathered snake-wand or snake-whip, and, as soon as it has straightened out to escape, he seizes it. After he has accumulated several snakes in this way, some are handed to the Antelope priests, who hold them until the termination of the dance.

When all the snakes have been danced with and are now held by the gatherers or the Antelopes, one of the priests draws a circle on the ground with corn meal. Immediately all the snakes are piled into this circle in a seething mass; thence many try to escape but are carried or herded back within the ring. The women scatter sacred white meal upon the snakes. This is the part of the dance causing the most excitement among the audience, particularly those nearest the circle, since some of the snakes may reach them before being caught.

Now the Snake priests as a group rush to the squirming pile and seize the snakes by the handfuls until all have been picked up. Then they run in the four cardinal directions off the mesa and down the steep trails onto the plain below, where the snakes are liberated at some distance from the bases of the cliffs, usually at specified shrines, and thus the messengers to the gods are sent upon their way.

The Hopi Snake dance is not snake worship. It is a prayer for rain and the fulfillment of adequate crops; and the snakes are used as messengers to the underworld gods of rain, as required by an elaborate and ancient snake legend. Lurid accounts have occasionally appeared in the press of priests embarrassed by premature storms, or disconsolate over the failure of rain to follow immediately upon the conclusion of the ceremony. This is hardly an accurate judgment of what the dancers hope to achieve. The Hopi prays that an adequate number of thunder showers shall reach his own fields during the growing season, but he is not particularly concerned if one of these storms fails to follow immediately upon its conclusion.

The sight of the dancing priests, who hold live, venomous snakes dangling from their mouths, is what brings the curious, from many hundreds of miles away, to see this short ceremony, lasting but a half hour or less. Inevitably an argument ensues as to why there are no serious, if not fatal, consequences. The following are some of the theories that have been advanced to explain this absence of casualties:

A. Conditions Affecting the Audience

1. The audience is suffering from some form of group hypnotism.
2. The audience is not qualified to distinguish venomous from non-venomous snakes; no venomous snakes are used in the dance.

B. Conditions Affecting the Snake Priests

1. The priests have taken an internal protective medicine prior to the dance.
2. They possess a knowledge of antidotes, internal, external, or both, that quickly render rattlesnake bite innocuous and even painless.
3. Sucking, cauterizing, and tourniquets are resorted to in case of a bite.
4. The priests are so purified by a ceremonial emetic as to be immune.
5. They are smeared with a preparation so disagreeable to the snakes (as, for instance, in odor) that the latter will not bite.
6. They are covered with an invulnerable preparation, such as a thick paint.
7. They are so healthy from outdoor life that rattlesnake bite does not affect them.

8. They have an immunity resulting from a long fast prior to the dance.
9. They build up an immunity in each participant by increasing doses of venom, as is done with horses in the preparation of antivenin.
10. They have a mysterious hypnotic power over the snakes.
11. They are fearless of snakes, which, therefore, are without power to bite them.
12. They are protected by the religious exaltation of the ritual.
13. They are actually bitten, with serious results, of which outsiders are kept in ignorance.

C. Conditions Affecting the Rattlesnakes

1. The snakes' fangs, venom glands, or both, have been removed.
2. Their mouths have been sewed closed.
3. They have expended their venom on harmless snakes or on other objects in the kiva.
4. They have been milked of their venom in the kiva.
5. They are tame snakes, used repeatedly in successive years.
6. They have been lately tamed by handling.
7. They are doped or hypnotized.
8. They are startled into submission.
9. They are blinded by the sacred meal, or paralyzed by the tobacco fumes from the ceremonial smokes in the kiva.
10. August is the blind season for rattlers; they cannot see to strike.
11. They are invariably held in such a way that they cannot bite.
12. The eagle-feather snake-wands prevent their biting.
13. They cannot strike because they are not permitted to coil.
14. Their facial sensory pits have been plugged, which prevents their striking.
15. Rattlers are relatively innocuous anyway.

Many of these theories are hardly worthy of serious attention. With respect to the first group, though the majority are nonvenomous snakes, enough persons thoroughly familiar with rattlesnakes have viewed the dance so there can be no question as to their use in the ritual. These rattlesnakes are of two closely related kinds—the prairie rattler (*Crotalus viridis viridis*) and the Arizona prairie rattlesnake (*Crotalus viridis nuntius*). The prairie rattler, particularly if a large one, could produce a very serious bite—one that might well be fatal, especially to the youngsters who participate in the ceremony as neophytes.

So much for the potential danger; now how is it met? We may dismiss at once the so-called antidotes and other cures and protective conditions listed in Group B above. The Indians claim no antidote in the sense the term is used by us, for they consider their medicine a protective charm rather than a physical venom-neutralizer. We may also dismiss all those theories based upon the supposed inability of a snake to bite without coiling. A rattler held by the neck, as the priests hold these snakes in their mouths, could certainly bite the holder's cheek; nor would it be prevented from so doing by any protective paint or odor.

Rattlesnakes are much tamer and more lethargic than is commonly supposed, particularly after they have been in captivity for a short time; so that in any case bites from them in the dance would be rather infrequent. But that there have been cases of dancers being bitten appears to be without question. A friend of mine saw a dancer bitten in one of the ceremonies of 1939. Other observers have also reported bites.

The Indians have always denied that the snakes have had their fangs removed or are treated in any other way to render them innocuous. A thoroughly competent herpetologist examined a rattler selected at random in the kiva and found the fangs and venom glands intact. After the dance, two rattlers were chosen and sent east, where they were examined by a great authority on American venomous snakes, who reported the fangs present and the venom glands full. My son and I thought we saw the fang sheaths when the snakes opened their mouths; this would have been possible only if the fangs were intact. Various other observations have been made that seemed to weigh against any theory of defanging. However, a snake priest told an observer in 1922 that the snakes were defanged; and this was backed up in 1926 with five sworn depositions to the same effect, taken by an official of the Bureau of Indian Affairs. These also maintain that the snakes are milked occasionally prior to the ceremony, so that when the ritual takes place even a scratch by the stump of a fang would be quite innocuous.

This procedure, if its truth be verified, would account for the fact that the Indians have rarely permitted outsiders to accompany them on their snake hunts, and in the two instances when they were followed, the intruders were not allowed to see rattlers caught. Also, even in the days when friendly ethnologists were given access to the kivas, they were not permitted to witness certain of the snake-handling ceremonies.

Finally there comes the unimpeachable evidence of one writer, since reinforced by an additional incident. At Shongopovi in 1932, upon the conclusion of the dance, he followed one of the snake priests down from the mesa to the floor of the desert and succeeded in finding one of the rattlers shortly after its release. This was done by stealth and at considerable personal risk. Upon examination, the rattler was found to have both functional fangs and reserve fangs cut out by some sharp instrument, such as a knife. The snake was therefore quite harmless.

Then, in August, 1954, a member of the New York Zoological Society received a live rattlesnake that was said to have been captured by a spectator after the dance at Walpi in August of 1951. This snake was fangless, and Xrays showed that the replacement fang reservoir had been removed.

Thus we now know definitely that in at least two of the Hopi pueblos the rattlesnakes (or some of them) have their fangs removed. Whether this practice has always been followed, and whether at all of the pueblos, remains to be determined. I believe that thorough and repeated venom removal, immediately following the catching of the snakes and during the ceremonies in the kiva, to be the probable source of the Indians' immunity from serious accidents. I am more than ever convinced that, if fang removal is a comparatively recent development, say within thirty years or so, it was antedated by a system of milking, or, in some other way, of forcing the snakes to exhaust their venom.

SNAKE-BITE TREATMENT

The Indians devised many kinds of treatment for rattlesnake bite. Some of these were semireligious in character, involving rituals known only to special rattlesnake shamans or tribal medicine men; others were remedial measures dependent on the use of substances available to anyone. However, even when the curative materials were readily at hand, it was believed in most tribes that their effectiveness was dependent on special rites accompanying their garnering and application, as was characteristic of most Indian medical practices.

How did the Indians actually fare when snake-bitten? It is difficult to tell from the meager records. They naturally did not stress the failures of their methods; only occasionally is there mention of a death. Probably they fared about as well as white men without treatment. Their failures were usually attributed to deviations from prescribed rituals or the intervention of adverse spirits.

As it was customary in many Indian tribes to treat dream bites, it may be assumed that the bites of nonvenomous snakes were also treated. That these treatments were invariably successful probably led to the confirmation of many of their curative procedures, a situation by no means restricted to Indians.

Plant Remedies

By far the commonest Indian remedy for snake bite, especially among the Eastern tribes, was some plant, often a root, which generally was applied to the wound as a poultice and was taken internally as a sort of tea—that is, as an infusion or decoction. So universal was this type of remedy that it came to have wide acceptance—even among the whites, as well—although the nature of the plant differed from area to area, and from tribe to tribe.

Ethnobotanists have listed more than seventy-five different plants used for this purpose. Could we know the facts at this late date, we should probably find that almost every tribe had a different plant cure for rattlesnake bite, and some had several; yet all were thought by their users to be unfailing cures. And this variation was not the result of availability or nonavailability, for many of the plants had wider ranges than their reputations. The fact is that these snake-bite cures should be placed in the category of magical medicine; had any one been really effective it would inevitably have supplanted the others. Most of these plants have been tested in the laboratories, yet I know of no instance in which any marked curative value has been shown to be possessed by any of them.

Maybe the Panamanian Indians, said to walk backward when bitten and to eat the leaves of the first plant they touch, have as good a plant antidote as any. Without doubt this does emphasize the magical quality the Indians attribute to their plant remedies, which the white man forgets when evaluating them.

Remedies Derived from the Snake

There is a basic philosophy in primitive medicine, related to the eye-for-an-eye theory, which assumes suitable remedies always to be present in the instrument causing the injury. So, as might be expected, there were a number of Indian remedies for rattlesnake bite derived from the offending snake itself.

The Brazilian natives mashed the head of the snake, mixed it with the saliva of a fasting man, and applied it to the bite as a plaster. One writer recommends rattlesnake heart swallowed fresh; also the bruised liver applied to the bite. Later he amended the prescription: the heart was to be dried and pulverized, and drunk in wine or beer. These cures are doubtfully of Indian origin; they have a strong resemblance to some of the Old World formulas. Rattler fat rubbed into the wound was said to be beneficial, and was used by the Delaware and the Chippewa. Certain New England tribes used powdered snakeskin for a cure.

A method employed by the Opata of Sinaloa was described in 1763: catching the snake's head between two sticks, they stretched its body out by pulling the tail; the snake-bite victim then bit along the snake's body, whereupon the snake swelled up and died, and the man recovered. If the man broke every bone in the snake's body it was a sure cure. A Cheyenne Indian treated himself by eating the culprit.

The Tarahumare held the offending snake by the neck and bit it until the blood came. If the victim could prove he had teeth as strong as the snake's he would not die. The Havasupai had a modified procedure, along the same line: one who is bitten by a snake had to catch it, and, grasping each jaw, endeavor to tear it lengthwise into two strips. If he split it to the tail he would recover, otherwise he would die.

Among the Tolowa of northwestern California one remedy was to cut off the head of the guilty snake and press the stump of the neck to the wound, for raw rattler flesh is presumed to suck out venom. The natives of Rio, Brazil, preserved fangs in cachaca, an alcoholic beverage, which was drunk in case of snake bite.

Among the Chukchansi, of the San Joaquin Valley of California, it was believed that a person bitten by a rattler should kill the snake, otherwise it would continue to send poison into him.

Suction and Surgical Treatments

Suction, in the treatment of snake bite, is so natural a reaction to the first stinging pain, that it is no surprise to find it practiced in many tribes.

The Creeks and other Southern tribes were said to have used a suction treatment. The Paraguayan and Venezuelan natives used suction and a ligature. The San Carlos Apache, when they sucked a snake-bite wound, spit toward the four cardinal points and prayed that the patient would not be hurt. When the Creeks sucked a bite they kept tobacco in their mouths, evidently to protect the operator. The Delaware Indians applied suction, using a hollow piece of deer antler to suck through.

Such an extreme procedure as amputation was probably seldom or never attempted by the Indians before the days of white influence, since they had great confidence in their ritualistic cures. However, one writer explains the derivation of a geographical name (Honeoye = Finger Lying) from an Iroquois tale to the effect that an Indian, bitten in the finger while picking strawberries, cut off the injured member with his tomahawk.

Miscellaneous Cures

Lately I heard that the Kiliwa of Baja California use the following mixture: garlic, kerosene, salt, and the gall of the rattlesnake. Certainly much of this is of non-Indian origin.

The Cora caught a pig, cut off its snout, and applied the raw surface to the site of the snake bite. This resembles the split-chicken treatment. They then gave some of the animal's blood, diluted with warm water, as a drink. Certain South American Indians used the gall of the coati mundi for snake bite.

In the Northwest a number of the tribes treated rattler bite with mud poultices, a method possibly learned from observation of bitten dogs, which so often submerge themselves in mud. The Delaware were said to have cured rattlesnake bites by washing the wound instantly with water. But, in British Guiana, neither the injured man, nor any of his relatives, could go near or even drink water during his convalescence, or the bite would be fatal.

Some Indian repellents were also cures, such as a rattler painted on one's body and the dust of a crocodile's tooth taken in water.

A few tribes attempted no treatment, although this is not in accordance with the customary usages of primitive or magical medicine; the Nisenan had no regular medicine except along the Miwok border, although anyone bitten by a snake was put in a special house for sixteen days. While there, many snakes visited and frightened him.

RATTLESNAKE SHAMANS AND CEREMONIAL CURES

Rattlesnake shamans—medicine-priests especially trained or designated to treat rattlesnake bite—were to be found in many tribes, particularly in California and the Far West in general. Shamanistic power was acquired in a variety of ways—in some tribes by inheritance, in others through dreams or visions, or by the shaman's having been snakebitten.

If any element of their treatment for rattlesnake bite could be considered standard, it was the removal of a foreign object by suction, not necessarily, and even not usually, applied at the site of the injury. Nor was this suction method restricted to snake bite or to other afflictions involving the actual introduction of foreign matter into the body; it was used in the treatment of many other diseases. In some tribes the sucking technique was not used, singing or other rituals being preferred. It should be understood that even the sucking procedure was rarely employed in what we would term an efficient manner designed to remove venom from the wound, for the entire shamanistic concept, regardless of method, is founded on magical effects and supernatural relationships. Some Southern Paiute shamans, for example, applied suction to the top of the patient's head. The shaman, by elementary sleight of hand, would then produce from his mouth the foreign object, usually thought to be a tooth or tiny snake, that he claimed to have sucked out, and thus the cure was demonstrated. Almost always there were accessory rituals such as incantations, songs, and dances. In case the patient succumbed, the failure was usually blamed on some untoward event or mystic enemy, although the family of the deceased might attribute it to the evil intentions of the shaman himself.

Ceremonial Details

Some of the more interesting or peculiar treatments administered by the rattlesnake shamans, as described to investigating ethnologists by their Indian informants, were the following:

Among the Shasta the rattlesnake shamans were women. The method of treatment has been described thus: she first sucked out the venom. Then she told a friend of the patient she could see her rattlesnake guardian or familiar, but that he was angry and would not look at her.

The friend must then speak to the rattler, calling him by name and asking him to relent and be kind. The shaman then reported a change in the rattler's attitude by saying: "Now he lights his pipe." There was a further exchange of comments between the shaman and her mystic rattlesnake, involving payment of the fee by the friend of the patient. The shaman then dressed in her paraphernalia and danced, the rattler dancing in unison in his home far away. This completed the cure.

Among the Bear River Indians of northwestern California, the teeth of the devilfish were used to open the wound and allow the poison to escape. Then a rattlesnake rattle was burned and the ashes rubbed into the bleeding flesh of the patient. Chicken-hawk feathers were used in washing the bite to increase the efficacy of the devilfish treatment.

The Yaudanchi Yokuts believed that the offending snake injected gopher teeth, since the snake feeds on gophers, and it was these teeth that the shaman pretended to remove. The Chukchansi sent the patient away from the village with a virgin youth of the same sex. Or, the shaman could effect a cure by looking at the sun and whistling through a tube.

Several ethnologists have described the elaborate practices followed by the Yuki rattlesnake shamans. They considered the sun to be the deity of the shamans who cure rattler bite. The person bitten looked at the sun, as did the doctor as well. If it appeared milky to the latter, the patient would recover, so the doctor proceeded with the cure. First he abused the sun, and then painted symbolic rattlers on a flat stone. The wound was warmed with hot ashes, and the stone laid on for a time, after which suction was applied. This should draw out the tiny living rattler that caused the trouble in the victim's body. The little snake died just before it entered the shaman's mouth, so only a dead snake was produced to verify the cure.

Among the Maidu, if a person who had been bitten did not die at once, he went into seclusion for a month with an Indian who had been bitten and recovered. The patient must observe various taboos and be fed by the helper. Rattlesnakes and lizards came to visit him in his seclusion and these visitors aided in his cure.

When a Patwin was bitten, he immediately prayed to certain deities; this pleased both the deities and the snake, and had the effect of weakening the venom. A shaman who was called in then sucked the wound and brought out blood that appeared like bundles of small rattlers. If the venom was strong it caused the priest's mouth to swell.

The Diegueño ritual involved invocations and the waving of branches of sage; but most important, the shaman applied suction to the patient's back and stomach — without regard to the location of the bite — and this was presumed to extract the venom. One Diegueño used the following method: he sang four times, making four circuits of the patient.

All the relatives of the victim had to bathe daily for four days and fumigate themselves with chamiso brush. After this some cases recovered, but others died.

The Navaho cure was characterized by one of their standard ceremonials, in this case the "Beauty Way." This included sand paintings and songs. The treatment was sometimes staged before a bite had been inflicted—but one which was known to be inevitable—and in other cases long afterward; but, regardless of these time differences, it was carried out with the utmost concentration of purpose.

Sometimes the Southeastern Yavapai shaman called upon a snake to aid in making a cure. A rattlesnake might be taken to the shaman's camp, where it was told that the shaman didn't wish to make trouble; he knew that it had struck only because it was nearly stepped upon. The snake was then liberated.

A Papago woman, in an autobiography, tells how she was treated for snake bite by her son, who was a shaman. She was bitten on the wrist while picking beans. Her son chewed greasewood and put the paste on the bite. He then made a mark with a buzzard feather all around the wrist and ordered the poison not to go beyond this mark. (Such magical tourniquets are often encountered in the Southwestern treatments.) He then split the feather, tied the two pieces together, and bound this around the wrist as an additional ligature. The poison stopped here and she was well in two days.

At Santo Domingo Pueblo the Keres Snake society was an aggregation of those skilled in snake-bite treatment. The house of the society's leader was its headquarters. When an Indian had been bitten, he was placed in this house with the snake that bit him, together with additional rattlers caught for the purpose. This was to assure his becoming acquainted with snakes so they would not bite him again. After four days of this association, a meal painting was made on the floor. Near this the patient was placed, with the snakes lying in front. After a suitable ceremony, the snakes were all taken out and liberated, and the priests dined on food furnished by the patient's family. If the rite was successful and the patient recovered, he, in turn, became a rattlesnake shaman, for the healing ceremony was his initiation.

The following procedure is recorded among the Tiwa at Isleta, New Mexico: application for treatment was made direct to the snake father, or doctor, who must find the offending snake. The victim then spit into the snake's mouth, which caused the snake to cure the man, for, after the patient spit, the snake swelled until it burst and died, and the man recovered. No woman should approach the snake-bitten man, lest she instantly swell up.

The method of the Galaxy fraternity of the Zuni Indians, famous for curing rattlesnake bite, was thus: the victim must remain alone in a room, for if he saw a woman nourishing a child he would surely die. Three roots, chewed by a medicine man, were applied to the wound; the patient also chewed roots. If clouds gathered after one was bitten, he was more likely to die, for then the snakes were more vigorous and the limb would swell to the heart. But when the sun was hot the snakes were lazy, and in a few days the man would recover.

A real rattlesnake shaman, when they had one among the Surprise Valley, California, Paiute, did not have to cut or suck the bite. He just told it not to swell, for the rattler was his friend. He dreamed about rattlers and carried one around his neck. Sometimes he put Star brand of chewing tobacco on the bite, a modern touch. But now the last rattlesnake shaman is dead, and the present general-practitioner shamans cut around the wound, suck it, and apply the chewed roots of the corn lily as a poultice.

The sacred formula of the Cherokee for the treatment of rattlesnake bite followed this procedure: first (by song) say it is only a common frog that passed by and bit you; this is meant to deceive the rattler spirits that may be listening. Then rub tobacco juice on the bite, manipulating toward the left around the bite four times. Then repeat the ritual, blowing four times in a circle around the bite. A snake lying down always coils to the left, that is, counterclockwise [this is not a fact], and this motion around the bite is just the same as uncoiling the snake, which should produce a cure. If an Indian dreams of being bitten, the same treatment must be followed, otherwise, maybe years afterward, the same serious consequences will ensue.

There were various ways of maintaining the shamanistic power. The Arapaho medicine man, whenever he came upon a snake of any kind, stripped off the skin and ate the snake raw. But a Chemehuevi shaman lost his power when he treated his favorite dog, although it was thought the power might return to one of his sons.

Tribal Protection from Snake Bite

Some shamanistic rituals were for the protection of the tribe, to avoid snake bite rather than for the treatment of cases after occurrence. Among the Achomawi and the Upper-Sacramento Wintu, the shamans would talk to the snakes as they emerged from their holes, asking them not to molest their people. The women shamans of the Shasta would go through the villages in the winter and sing songs over the children's moccasins to protect them from rattlers. In one Miwok subtribe, special sticks were used wherewith to slap snakes; the rattlers feared these sticks and kept away from their vicinity at night.

The Yokuts shamans would go to a snake den in the spring and stamp and whistle. Soon the snakes would come out, led by a lizard. They were picked up and placed in baskets filled with eagle down. In the evening a dance was held, each participating shaman carrying a rattler in a bag.

From time to time the bags were placed on the heads of the spectators to find, by some appropriate sign, who would be bitten that year. On the following day, the threatened victims were cured in advance. A part of the ceremony also made it certain that in the future all rattlesnakes would give a warning rattle before striking. Among the Chukaimina sub-tribe of the Yokuts, the shamans also called the snakes out of their dens by means of a whistle. The snakes were then asked which ones desired to see the people. Such snakes as wished to participate in the ceremony answered by protruding their tongues and rattling.

Handling Rattlesnakes

From the close mystical association of rattlesnake shamans with the snakes, it might well be assumed that there would be a corresponding physical association, and such was indeed the case. In many tribes the shamans were reported to handle rattlers, often in public ceremonies, for this was an essential method of maintaining their standing and power in the community. Sometimes shamanism involved a reputed immunity to snake bite.

Among the Wishram a man with the rattlesnake spirit would not be bitten by one. Not only could he pick rattlers up safely, but he could kill a rattlesnake, mix the skin with his tobacco, and smoke it without ill effect. Among the Foothill Nisenan, the mere sight of a rattlesnake sha-man would kill a snake. The shamans of the Central Pomo could control rattlesnakes with song. The informants claimed that some rattlers were swallowed and thrown up again during the progress of a dance. The East-ern Pomo shaman could collect rattlesnakes by singing to them, making them come to hand. A Paviotso rattlesnake shaman was accustomed to carry live rattlesnakes around his waist next to the skin.

Among the Central American Muskito and Sumu, the best snake charmers were said to eat snake heads after removing the fangs; the lesser practitioners merely cooked their food with fat extracted from the snake's head. But, in spite of their supposed immunity, fatalities among shamans were occasionally reported.

Shamanistic Magic

Snake shamans were gifted with some magical powers not immedi-ately connected with their primary curative functions. It was said of the Paiute that in their dreams they could turn inanimate objects into snakes. A Kalapuya shaman was seen to blow on a piece of cedar bark and it crawled away in the form of a buzzing rattlesnake.

The Galice rattlesnake shaman could use a snake for soothsaying or to find lost articles. A rattler, released by him in a crowd, would make its way toward either the thief or the missing object. A Yuma shaman, called to treat a girl some distance away, exerted his power en route, so that when he arrived it took only an hour to complete the cure.

Shamanistic Fees

The matter of shamanistic fees was a delicate one. There seems no doubt that in some tribes they were exorbitant, for the patients and even their families were virtually stripped of their possessions to pay for their treatments. Disputes over fees often led to accusations of poisoning and the initiation of serious intratribal feuds.

Among the Klamath the fee for a cure was sometimes a horse. Among the Clear Lake Pomo, payments amounted to the equivalent of forty or fifty dollars in clamshell money. One-half the charge was rebated if the patient died. The Yana shaman was paid in hides, beads, and similar media of exchange. It was the custom of the Wintu to pile the goods the shaman was to receive before him at the start of his operations. If the cure failed, the goods were returned to the dead man's relatives. The Yavapai shamans required payment only from people of property. Among the Hopi, to join the Snake clan was considered the equivalent of paying the Snake chief for curing a case of snake bite.

Shamanistic Taboos

Shamanistic treatments by rattlesnake specialists were often surrounded by taboos similar to those affecting other Indian treatments. For example, among the Clear Lake Pomo when a tribesman was bitten, no one smoked tobacco until he had recovered. Also, no one who had recently eaten meat was allowed to be present at the spring propitiatory ceremony. The Cahuilla required anyone suffering from snake bite to avoid proximity to a pregnant woman; this taboo was current in other tribes as well.

Among the Chemehuevi, a snake-bite victim must not eat any produce from his own garden until his neighbors' crops had been harvested; if he broke this rule his garden would dry up and his fruits be flat and tasteless. Similarly, when a Washo was bitten by a snake and looked at the pine nuts before they were ripe, they dried up and were rendered worthless. Among the Paiute at Las Vegas, Nevada, the snake-bite treatment had to be given outside of a dwelling. Passersby must not approach unless they had bathed and were clean, otherwise the patient would be adversely affected. A restriction of some Central American Indians was that, during treatment, the patient must avoid foods of which snakes were supposed to be fond. It is obvious that these taboos were useful to the shaman, since he could explain a failure to effect a cure as having resulted from the infraction of some taboo by the snake-bite victim or his family.

Evil Shamans and Poisoners

As the shamans manifested their great powers through communion with, or control over, such dangerous creatures as rattlesnakes, it was only natural that their failures should be attributed to deliberate design. One who claims such power is sure to be accused of its misuse; and so it was that the rattlesnake shamans were not only blamed when their remedial measures failed, but were accused of causing all manner of mishaps and diseases. By no means all of these accusations were unjust, for some shamans did, upon occasion, use their power as an instrument for blackmail, failing which they endeavored to injure their victims or enemies by spells and incantations. These victims, in turn, retaliated on the shaman, by attempting to wreak vengeance for the injuries he presumably had caused.

Some of the evil methods attributed to rattlesnake shamans among the California tribes were as follows: among the Bear River people it was said that the rattlesnake poison with which shaman enemies were sickened or killed came from snakes caught in July, August, or September, for these are the months in which the rattlers can see; in the others they are blind. To make the witchcraft poison, the snake was hung by the head in the sun, or sometimes over a slow fire, and the grease was caught in a clam or abalone shell. To these drippings were added powdered rattler skins that had been shed in the spring, together with weasel toenails. A drop of this powerful mixture was placed in an elderwood pipe and, as the operator smoked it, he spoke the name of the intended victim and blew outward on the pipe. The poison would travel to the victim, where it would settle over him; in fact, it could see its way because it was taken from rattlers during their seeing months. One drop reaching the unfortunate objective would be fatal in a year, two drops in two years, and so on. But one large drop might kill the following day. The first notice the victim had of this fatal predicament was a buzzing like that of a rattlesnake in his ears.

Among the Nomlaki diseases often came from a magic poisoner, whose weapons were dried rattlesnake fangs and rattler venom rolled in ear wax. A person could be poisoned should the poisoner touch him or his excrement.

In a poisoning cult of the Pomo Indians, when a man was to be poisoned, a shaman was hired for the purpose. This was usually in revenge for a previous poisoning, one often supposed or imaginary, such as a natural death blamed on an enemy. With elaborate and secret rites the priest gathered suitable poisons, such as poisonous plants, rattlesnake venom, and the venoms of bees, spiders, red ants, and scorpions. They were placed in a mortar, and the blood from four rattlesnakes was allowed to drip on the mixture. Finally, when the poison was ready, it was applied to the victim in a variety of ways, such as poisoning a doll in effigy, or shooting a poisoned arrow over his house. The poisoner was well paid, but part of the fee was withheld until the victim died. Rarely, if ever, was the poison applied in any but a metaphysical sense, for the whole program was one of witchcraft, a sort of hexing.

In a method reportedly used by the Pomo, an enemy might be blinded. The eyes of a rattlesnake were removed while it was blind, preparatory to shedding, and an abalone shell was coated with the substance thus obtained. Then the shell was used to flash light in the eyes of the victim, accompanied by a prayer that he would become blind. Among the Atsugewi, a doctor with rattlesnake power would shoot at his enemy a small snake, that would bite the heart of the victim.

Among the Tübatulabal, witches were advised in dreams, by rattlesnakes and by other animals, how to make people sick, or to finish them off completely. A Chemehuevi shaman, when angered, could cause a snake bite by talking to himself and dreaming that a snake had bitten his victim. Such a bite was quite incurable.

Among the Kalapuya, shamans possessed of the rattlesnake-spirit power—one of the strongest—could shoot it at any enemy with fatal results. It was a tribal theory that any rattler that struck without rattling had been sent by a shaman. If it rattled, as a normal rattlesnake should, then it was just a plain snake.

A Shoshoni evil charm was prepared by placing rattler heads on hot coals and cooking them with other ingredients. When this dire mixture was preserved in a buckskin bag, it was possible for the possessor to cause the death of an enemy by looking intently at him and murmuring destructive incantations. A Zuñi witch was once accused of making a compound, including rattlesnake hearts, which was directed at children, often with fatal results. Much time was spent by other shamans in an endeavor to suck out these evil—but quite mystical—missiles.

Among the Omaha, a man who harbored a grudge against another might drop into the enemy's tent the figure of a rattler cut from rawhide. Shortly thereafter the victim would be bitten by a real snake.

Within this theme of curers gone wrong, we see what a fruitful field there must have been for murder and intratribal feuds bred of ignorance and superstition—of honest but primitive doctors destroyed because of the natural failures of their crude methods; of slander, blackmail, and personal revenge.

MISCELLANEOUS INDIAN-RATTLESNAKE RELATIONSHIPS

Rattlesnakes as a Cause and Cure of Disease

Not only did the Indians have various tribal cures for rattlesnake bite, but rattlesnakes were thought both to cause and to cure certain other afflictions. Often these effects were presumed to be caused by a spiritual rather than a physical intervention of the snakes.

Take first the sinister effects. There was a Cherokee belief that an eye disease, involving susceptibility to light, is caused by seeing a rattlesnake. The Eastern Cherokee thought that if one dreamed of a snake it would cause illness. Among the Creeks the rattler was thought to be the source of various diseases, including swelling of the face and limbs.

In the Midwest, the Omaha blamed stomach disorders on the rattlesnake and treated them accordingly. The snake's venom was thought to be projected through the air as an influence, and lodged in the patient's stomach as a liquid. This might come about if the victim incurred the annoyance of a rattlesnake doctor, or passed too near a real snake. The rattlesnake doctor usually effected a cure by sucking the poison from the patient's stomach, the skin remaining unbroken in the process.

The Tarahumare believed that a rattlesnake eats the soul of a child, which consequently sickens. In the lore of the Chiricahua Apache, if one handles a snake it will make the skin peel. Even handling a shed skin might result in sores on the inside of the lips, on the hands, and in fact all over the body. A person may be injured, not only by being bitten, but by being frightened by a snake, or by lying down where one has been.

Several tribes had beliefs respecting the unfortunate results of letting a pregnant mother look at a rattlesnake. Among the Thompson Indians of British Columbia the father of an expected child must not kill a snake of any kind or the child will resemble a dead person or ghost. The Sanpoil and Nespelem of northeastern Washington believed that if either parent killed a rattler during the period of pregnancy, the child would cry and writhe like a snake. The Menominee thought that if a pregnant mother looked at a rattlesnake the child would be marked. Among the Maricopa, neither the woman nor her husband should look at or touch a rattlesnake during pregnancy, lest the child be deformed and unable to walk, or even to stand, through a seeming lack of bones. The Hopi believed that neither parent should look at a serpent effigy, lest the child would not be born or would swell up.

The Havasupai would kill snakes, but in the case of the rattlesnake, care must be taken not to touch the blood or let it splatter the clothing, for if "inhaled" it would cause illness.

In California, the Central Pomo were of the opinion that if a rattler struck at one's shadow, dizziness would result.

In contrast to detrimental effects such as these, in other tribes rattlesnakes or their products were thought to be efficacious in the treatment of a number of afflictions. It is somewhat difficult, however, to differentiate such of these medical usages as may have originated with the Indians, from those that the colonists brought with them or devised. No doubt each group derived some practices from the other.

The story is told how a bedridden Carolina planter was cured of a mysterious and lingering ailment by an Indian using a defanged rattler. The treatment was only resorted to after the patient's family had given up hope. The rattler, a huge one, was placed around the man's waist where it squeezed "as if he had been drawn in by a Belt." Gradually the snake weakened; the next morning it was dead, whereupon the Indian said the man's "distemper" was dead also. And this proved to be the case.

Rattlesnakes were generally used by the Cherokee to protect their families in the case of epidemics. A roasted rattler was hung up in the house. Each morning the father bit off a small piece, chewed it, mixed it with water, and then blew the mixture as a spray over the other members of the family. An Indian doctor in New Hampshire prescribed a rattler's heart, warm and whole, for fever and ague.

Many of the cures were more specific in the affliction for which they were applied. One that seems to have been prevalent among the northerly tribes was the eating of powdered rattle to expedite parturition in difficult cases of childbirth. A Dakota explanation given for the efficacy of this drug was that the child heard the rattle, and supposing the snake was coming made haste to get out of the way. The Northwestern Indians used a powder made from the tail (rattle?) of the Pacific rattlesnake to produce abortion, as well as to expedite natural labor.

The use of rattlesnake fat or oil as a liniment or ointment (evidently with the hope of acquiring the lithe flexibility of the serpent) is very old. The Miami believed the oil drawn from the fat was so penetrating that, if poured on the hand, it would go right through to the other side. The fat was used by the Tarahumare to anoint a new-born child, so that light might enter its heart. The women of the Nahuatl tribe south of Mexico City stimulated hair growth by the use of rattlesnake oil. The hair would grow as long as the snake, but care must be taken lest, in a rainstorm, the hair twist like a snake and strangle the woman or her husband.

Various tribes used rattlesnake parts in connection with headache cures. Many tied rattles or skins to their heads, hair, or hatbands. The Sanpoil and Nespelem carried rattles under their hats.

The medicine men of the Verde Agency, Arizona, used a more spectacular treatment on the Yavapai seeking their help. They would hold a live rattler by the neck and tail, and apply it to the seat of pain. The snake would be caused to rattle and then, with a chant or the word "wisht," it would be swung away, evidently with the idea of carrying the pain with it. For headache the snake was put around the patient's neck, and then the same swinging procedure followed.

There is an interesting idea-sequence of the Shasta: if a person chokes on a fishbone he will recover if he calls on the rattlesnake, for the latter swallows its food without chewing. Another peculiar association is that of the Pueblo, who believe that a child's sore navel can be cured if a man who has been bitten by a rattlesnake blows upon the child or waves ashes around its head.

Rattlesnake flesh as a cure for consumption has been noted since colonial days. The Seneca gave the consumptive patient rattlesnake fat as food; and the same treatment was found among the Pomo of California, with the further requirement that he abstain from meat or fish for twenty to twenty-five *years,* for if he touches any fat except that of the rattler he will die. The Papago of southern Arizona also used rattler flesh in the treatment of tuberculosis. The snake was killed and the meat dried and used as a powder, a small quantity being put in the food while the patient was not looking. West Virginia Indians believed that tuberculosis might be cured if a live rattler were worn around the waist next to the skin. The fangs were pulled before this lively remedy was applied, which might require several weeks for completion of the cure.

Dried and powdered rattlesnake flesh was mentioned as early as 1657 by the Jesuit priests as being used by the Indians to cure fever. More recently it has been prescribed for leprosy in eastern Sonora.

Several Indian treatments for the teeth involved snakes, particularly rattlesnakes, the idea deriving presumably from contemplation of the strength and perfection of their fangs. An Indian belief in Vermont was that if a child bites through a rattlesnake from head to tail it will insure sound teeth. The Keres held the rattles of a rattler against the teeth to cure headache or toothache. Among the Zuñi, it was customary for a person, who at some time had been bitten by a rattlesnake, to rub the gums of a child suffering from retarded dentition. The Maya used rattlesnake parts to ease tooth-pulling.

The smooth sharpness and the strength of the rattlesnake fangs appealed to the Indians, who used them for various surgical purposes. This was noted as early as 1615 by Hernández, who reported their use by the Mexican Indian doctors in perforating the back of the neck as a cure for headache.

The Potawatomi thought the fangs were a charm against rheumatism; the mode of application was to scratch the affected part until it bled. The Cherokee employed the fangs to scarify a patient prior to the application of various remedies. They were also used in toning up ball teams before a game. The players were deeply scratched to note their nervous endurance and ability to withstand pain. This treatment was also thought to inculcate in the players the fierceness and swiftness of the rattler's stroke, making them more terrifying to their opponents; but it also had the unfortunate effect of making them cross to their wives.

War Uses: Arrow Poisoning

Arrows poisoned with snake venom were widely used in war, both in the Old and New Worlds. Pliny mentioned such a use among the Scythians in ancient times.

Since, in many areas of the United States, rattlesnakes are the only venomous snakes, it was natural that their poison should be employed for this purpose by the American Indians, and such use was a pre-Columbian development.

The prevalent pattern of procedure was to procure the liver of some animal, usually a deer, and to cause one or more rattlers to bite it, thus injecting their venom into the liver, after which it was allowed to putrefy, and the arrow points were then smeared with the product. By this means a larger quantity of poisonous preparation was available than there would have been from the venom alone, and the liver was presumed to increase its effectiveness. It was said that one liver would poison a thousand arrowheads.

Upon this fundamental theme several tribes introduced variations. The Seri Indians of Sonora and Tiburón Island confined a number of rattlesnakes, scorpions, centipedes, and tarantulas in a pit with a cow's liver. These creatures were then stirred up and infuriated until they had bitten each other and the liver. When the whole mass was in "a high state of corruption" the arrow points were dipped therein and allowed to dry.

The liver of a deceased Indian was sometimes used by the Mohave, and the mixture was enriched with the blood of a woman, this being considered extremely poisonous. The Southeastern Yavapai merely stuffed the liver with a rattler's head, as well as with spiders and tarantulas. The Chickahominy also added the snake's head to the liver-venom mixture, together with poison-ivy leaves for good measure.

Seldom do the accounts explain the method of getting the rattlers to bite the liver. Some tribes kept captive rattlers for the purpose, as reported by the early New Mexican missionaries. The Maidu and the Dakota took the deer (or antelope) livers out to snake dens in the spring, held them out to the snakes and allowed them to be struck repeatedly. Some presumed that the animal whose liver was used must be killed by the bite of a rattler.

A number of tribes, although using rattlesnake venom, did not employ the liver combination. The Umatilla pulled out the fangs and venom glands and dipped the arrow points in the venom, which dried on them like varnish. The Yana preferred a mixture of rattlesnake venom and deer milk. The Surprise Valley Paiute were said to allow the rattlers to bite the arrow tips directly. One somewhat doubts the accuracy of the informant in this case; it would be too difficult to get any quantity of venom distributed on the wood or stone by this means.

The Wishram were reported to secure the venom glands of a rattler and smear the arrowheads therewith directly. The Achomawi mixed a dried rattler head with the blood and flesh of a dog. The Klamath mixed rattler venom with the roots of a poisonous water hemlock, dipped the arrowheads in the moist mixture, and then dried them over a ceremonial fire. The Yuma took a rattler head, dried and powdered it, and then mixed the powder with clotted blood and red ants.

Several tribes used rattlesnake body parts or products on the theory that they were poisonous. The Karok dipped their arrows in rattlesnake brains. The Achomawi and Klamath were said to use dried and pulverized rattlesnake gall. The Ruby Valley Shoshoni used rattlesnake heart. The Hopi were reported to irritate a rattlesnake until it bit itself, and then dip the arrow points in its blood.

Some tribes used poisoned arrows in hunting, others only in war, but most employed them for both purposes. Few tribes had any fear that the poisoned arrows would be deleterious to the meat thus secured, and they were supposed to cause the game to succumb more quickly. The Catawba not only used venom for arrowheads, but also poisoned the points of sharp sticks or splints that they stuck in the grass as booby traps to injure pursuing enemies. They heated the venom to increase its toxicity.

The question naturally arises as to the effectiveness of these arrow poisons based on rattlesnake venom. I think we may say that the rattlesnake venom, as such, did not appreciably increase the peril from an arrow wound. Arrows were thoroughly dangerous weapons; the threat of infection was serious and may have been increased by the nature of some of the arrow-point preparations. But the poisonous properties of snake venom are destroyed by putrefaction; and as most of the Indian methods of preparation involved a deliberate putrefying of the product, it is probable that any true venom effects were eliminated in the process.

Psychologically the venom-poisoned arrow was undoubtedly a valuable weapon; for it was greatly feared by U.S. soldiers who had heard wild rumors as to its potency, and it emboldened the Indians. Some of the stories told of the effectiveness of the rattler venom preparations were quite terrifying. As early as 1763 it was reported that the Jova arrow poison would kill not only the stricken man but anyone who tried to suck the wound; and the Seri preparation would cause fresh blood to boil. The Western Yavapai claimed their preparation would kill within an hour, the Mohave within half an hour.

As a matter of fact, the Indians probably did not attribute any virulence they may have claimed for their poisoned arrows to the venom in the physical sense of a chemical compound having certain deleterious results, as we would view it; rather, they placed their confidence in its magical effect.

Art and Ornamentation

As might be expected of creatures so prominent in myth and legend, rattlesnakes had a place in the art, decoration, and symbolism of the American Indian tribes. Probably the earliest printed reference in English to the rattle of the rattlesnake — and therefore indirectly to the snake itself — mentions its use as a personal adornment by the Indians of Virginia. It was made by Captin John Smith in 1612. He said: "Some of their heads weare the wing of a bird or some large feather, with a Rattel. Those Rattels are somewhat like the chape [metal tip] of a Rapier but lesse, which they take from the taile of a snake."

The wearing of rattlesnake skins or rattles was general in certain tribes, but was taboo in others. Sometimes they were worn purely as personal adornment; in others they were thought valuable as fetishes or charms, of importance in the cure or avoidance of disease. In some tribes, the wearing of skins and rattles constituted the badges of office of rattlesnake shamans. To secure such ornaments was fraught with difficulty in tribes that had a taboo against killing these snakes.

Sometimes rattlesnake figures were used as painted decorations on the Indians' bodies. This practice was mentioned as early as 1761. In 1843 a Mandan dance was described in which the two participants had their backs painted with transverse black stripes to indicate that the dancers represented rattlesnakes.

Skins were sometimes applied to the decoration of hunting equipment. Crow Indians encased their bows with rattlesnake skins.

In the graphic arts, rattlesnake designs were widely exhibited. Some were realistic, others were conventionalized, or had mythical appendages such as horns, plumes, and wings. They dated back to the mound builders and to pueblos long abandoned. Rattlesnake carvings on stone discs and tablets, mostly from the mounds of the Southeast, have been described by several authors.

Some early travelers have testified to the presence in religious edifices of carved idols representing rattlesnakes. The Navaho made a small rattler image for use in the treatment of an infant, ill because of his mother's having seen — during pregnancy — the death of a snake.

Various utensils were decorated with animal figures, including rattlesnakes. An ancient stone pipe along the shank of which four rattlesnakes lie stretched was found near Santa Fe, New Mexico. Food bowls, pots, bottles, and vases, some of relatively recent date, were similarly incised with lively rattlesnake designs. A hairpin of bone or ivory, carved at the upper end to represent the tail of a rattler, was found by an explorer on Key Marco off the coast of Florida.

Rattlesnake designs were used in the baskets of various California Indian tribes, such as the Yokuts, Saboba, Paiute, Shoshoni, and Mono. An anthropologist, in discussing the Cahuilla baskets, points out that figures of men and animals, including snakes and lizards, were quite realistically woven in baskets currently made for sale, whereas in baskets for their own use the Indians preferred to retain their older and simpler symbolical designs.

Rattler designs seem never to have been so popular in blankets and bead work as in baskets. One might have expected them in the extensive Navaho output but, according to the Franciscan Fathers, such designs were taboo on blankets, as they were in pottery as well. The Huichol produced beautiful patterns for their blankets when inspired by putting their hands, first on a live rattlesnake, and then on their foreheads before

Fig. 47 Modern Cahilla Indian basket with rattlesnake design, showing both coiled and outstretched poses

they began to weave. The Seminole employed symbolic designs in their beadwork, two of which represented the diamondback and pigmy rattlers. The Klamath had a string figure or cat's cradle that was known as "A Rattlesnake and a Boy."

The Indians, both ancient and recent, frequently used flat rock-surfaces as canvases for artistic expression. On them they often drew animal figures, some of which were rattlesnakes.

Such interior paintings as the Indians made occasionally included rattlers among the other animals depicted. Many of these were lost because of their impermanent character. One writer mentions a painting of a horned rattler on a buffalo-skin lodge of the Dakota Indians.

Snakes were frequent elements of the ground paintings composed of colored sands made by many Southwestern tribes. The rattlesnake may occasionally be recognized, although often embellished with plumes or wings.

Although the reptilian figures so prevalent on Mexican temple walls and in the codices are stylized and embellished to such an extent that they often retain little resemblance to a rattlesnake — or, indeed, to any snake — the rattles may still be distinguished.

To the Indians, rattles were important instruments for the production of music. Many kinds were constructed of deer hoofs, turtle shells, gourds, and other hollow objects. Although the rattle of the rattlesnake was a noise producer ready to hand, it never assumed any great importance, partly through taboo, but more because the sound that may be produced with it artificially is too faint to be effective. To some extent it was used in California, possibly more for magical reasons than as an efficient noise maker. For convenience in shaking, the rattles were usually attached to sticks.

Indian Field Knowledge of Rattlers

The Indians imbued rattlesnakes with so many fanciful attributes — through myth, legend, and animism generally — that it is difficult to ascertain their relationships with the snakes on a purely materialistic plane, if, indeed, such a distinction can be made. As has been pointed out, many tribes had taboos against killing rattlers, usually through fear of retaliation. A few had no such reluctance to destroy these dangerous creatures, particularly dangerous to hunters roaming the woods without adequate protection on their legs.

Several of the early travel books or natural histories have stated that the Indians would not traverse the woods during wet weather, because, through dampening, the warning rattle would be almost inaudible. It is to be doubted whether this inhibition was widespread. They were too expert woodsmen to place dependence on a rattler's always sounding off before striking; like all good hunters they must have known their best protection lay in keeping their eyes on the ground and watching their steps. One writer was told by the Indians of upper New York that rattlesnakes would not rattle while lying in wait for prey; and that they would not bite readily in midsummer. Another says that, among the Nisenan of California, young hunters were advised by their elders to go around logs rather than over them, to avoid the chance of putting the foot on, or within striking distance of, a concealed rattler, which is sound advice.

Several early writers credit the Indians with a belief in the rattler's ability to capture its prey by fascination—indeed, they are said to have first brought this widespread theory to the attention of the whites.

The Pinto Indians (near Veracruz, Mexico) considered young rattlers much more dangerous than the old ones, which they deemed almost harmless.

I find little published on how the Indians caught rattlers alive, upon the few occasions when they wanted them. It is said that some Mexican Indians caught them by the tail, holding them vertically as they squirmed and tried to reach their captors. [This can be done with safety if a snake be not so large that it can reach the holder's body.] Some Indians defanged a snake by causing it to strike a red cloth, which they then jerked while the fangs were caught in it. The Meskwaki (Fox) chewed the tops of mountain mint, and placed the cud on the end of a stick that was held under the snake's mouth. The rattler, now insensible to danger, could be easily caught, or so they claimed.

The Pomo Indians of California thought rattlesnakes sounded their rattles as a sign of good humor. The Wailaki washed themselves with a decoction of the fawn lily, which stopped rattlesnakes from having bad dreams, that is to say, reduced their irritability. When a rattlesnake gave warning he was considered to be a good snake and was left unmolested; if he failed to rattle he was killed. The Diegueño and Luiseño sometimes plugged rattlesnake holes to protect themselves. The Indians near Harrisburg, Pennsylvania, in about 1800, protected themselves from rattlers by sleeping on pallets raised above ground and surrounded by fires. The Bear River Indians of California used elkhide moccasins in their summer hunting as a protection against rattlers. Some Indians may have burned brush to kill rattlesnakes; this was reported by the Karok Indians of California.

The Indians seldom distinguished between rattlesnake species, except where there were extreme differences in size, color, or pattern, as was the case between the western diamond and the sidewinder.

In general, whether because of mystical attributes, or the actual danger from snake bite, it is probable that most Indians gave rattlesnakes a wide berth.

Rattlesnakes as Food

It is to be doubted whether rattlesnakes ever comprised more than a very minor part of the diet of any Indians, since the flesh available on all but the largest snakes is quite small in quantity. Probably the colonial writers thought the idea somewhat sensational, which was their reason for mentioning it. But where food was scarce, an opportune catch of a rattler might save a life, just as it sometimes did during the westward migration of the whites. Some of the early writers said the Indians were careful to eat no rattlers that had bitten themselves.

In some instances, rattlers were eaten because of some particular virtue thought to be contained in the flesh. One writer claimed that the Indians of New England, to refresh themselves while traveling, would seize a live rattler by the neck and tail, tear off the skin with their teeth, and eat the flesh.

An Indian belief, prevalent in Vermont was that eating rattlesnakes would promote longevity.

CHAPTER FIFTEEN

Post-Columbian Knowledge of Rattlesnakes

DONT TREAD ON ME

Europeans, coming to the American continent in successive waves as explorers, traders, and immigrants, gradually became familiar with the wild life of the New World. Among the novelties they found was the rattlesnake. Economically it was not important, even less so than in later years when agricultural development made rodent control more urgent. Probably the outstanding impact of these snakes upon the colonists was that they constituted an additional hazard, one, however, that could not be considered serious, compared with the major dangers from Indians, starvation, and disease. No doubt rattlers did impose some restraint upon the timid, abetted at times by the sensationalism of those who wrote accounts of their experiences in the new land. Rattlesnakes were often a subject of concern to people going into the woods. It is therefore of interest to examine some of the knowledge regarding them during the early colonial period.

One might have expected that when the earliest reports of the rattlesnake reached Europe there would be a considerable interest in so strange a creature—a snake with a peculiar caudal appendage with which it could make a violent hissing sound. However, it must be remembered that in those days natural history books contained apparently authentic descriptions of mermaids, seven-headed dragons, and other really noteworthy creatures. So the rattlesnake had severe competition as a novelty, and its advent into the European consciousness was slow and unimpressive, despite its being one of the distinctive oddities of New World animal life.* The early accounts of rattlers are of more than chronological interest. We may find in them the genesis of many of the queer beliefs so long current, some of which have not yet been rejected.

Countless early communications, never yet published, are still scattered through the archives and libraries of the Old World; among these there must be many a noteworthy comment on the fauna of the New World. It would be interesting to know in what letter or report, and by which explorer, the rattlesnake was first mentioned.

The earliest printed mention of the rattlesnake that has come to my attention is contained in *La Chronica del Peru*, by Pedro de Cieça (Cieza) de Leon, published in Anvers (Antwerp) in 1554. He says: "There are other snakes, not so large as this one, which make a noise when they walk [move] like the sound of bells. If these snakes bite a man they kill him." There are two evidences that the author referred to rattlesnakes: first, the use of the term "cascabel" (or "caxcabel" in another printing of the same date), a term already applied to the rattle of the rattlesnake, to judge from unpublished reports of the same period; and, second, because De Cieça traveled to Peru along the coasts of Venezuela and Colombia, where rattlesnakes are prevalent.

The next printed record is to be found in Pero de Magalhães de Gandavo's *Historia da Provincia Sancta Cruz*, an account of Brazil published in Lisbon in 1576. The remarks on the rattler are as follows: "There are others [snakes] of another species, not so large as the former, but more poisonous. They carry at the end of their tail something like a rattle ["cascavel" in the original] and wherever they go they keep on sounding it; and whoever hears this takes care to protect himself from them."

There is a similarity evident in the Cieça and Magalhães statements, although the latter leaves no doubt of the snake intended, as the "cascavel" is carried at the end of the tail. In both accounts the idea is conveyed that the rattle sounds continuously whenever the snake moves, a

---

*Charles A. Lesueur, the French-American naturalist, while at the New Harmony, Indiana, cooperative community in 1825, painted a drop curtain that pictured Niagara Falls and a rattlesnake as the two natural features most characteristically American.

common error of the early descriptions, and a further proof that Cieça alluded to the rattlesnake.

Following Magalhães, the next published work in Spanish or Portuguese that contained mention of the rattlesnake, this time with a full description, was Francisco Hernández *Quatro Libros de la Naturaleza, y Animales que Están Recevidos en el Uso de Medicina en la Nueva España. . . .* [City of] Mexico, 1615.

Some of the details given by Hernández in this first extensive description of the rattlesnake are as follows: it is called "la señora de las serpientes." It grows to a length of four feet or more. The color description is obviously that of the Central American rattler (*C. d. durissus*), but the author goes on to say that there are other species of various colors. One found near Colima, Mexico, has a retractile throat fan; evidently there has been a confusion here with some lizard. The bite is fatal if powerful remedies are not immediately available. When annoyed the snake throws itself into a defensive coil, from which it can face an enemy in any direction. The best treatment for the bite is to bury the afflicted limb in the earth until the pain ceases. This snake moves with great speed, so that the natives call it "ocozoatl," the name of a wind. It has a rattle for each year of its age. It has black eyes and a pair of curved fangs like a dog's teeth. The bite causes fissures in the body of the victim, followed by death within twenty-four hours.

The Indians catch and hold rattlesnakes up by the tail from which they dangle helplessly in spite of their squirming and rattling. People who keep them in captivity say they can live a year without food or water. The head, if cut off, will survive ten days or more. They are said to grow to a very large size in the province of Panico [Pánuco, Mexico], where people tame them for amusement. The venom comes out through the fangs, which are hollow. Some say they have live young but this is false [actually it is true]. When annoyed they rattle furiously, making a great noise, but they attack no one unless molested.

The Mexican (i.e., Indian) doctors use the fangs as lancets, puncturing the necks of their patients—at the back—to relieve headaches. The fat is used as an ointment to alleviate pain and reduce swellings. The Indians say the flesh is better than that of poultry. If wrapped in straw or cloth, rattlers become tame. The head of a rattlesnake, if bound to one's throat, will cure sore throat or fever. As a cure for the bite, a poultice of human excrement or of certain plants is to be recommended.

Although Hernández' account is typical of the early natural histories in its lack of order in arrangement, it contains much sound information as well as the first printed statements of several still-current myths.

Hernández not only published the first full-length account of the rattlesnake, but also the first picture, although this did not appear until

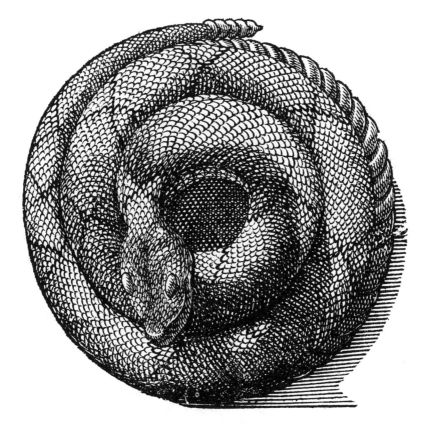

Fig. 48 The earliest illustration of a rattlesnake in a book. (From Francisco Hernández, *Rerum Medicarum Novae Hispaniae Thesaurus seu Plantarum Animalium Mineralium Mexicanorum Historia*, 1628)

the Latin edition of 1628. The engraving shows a rattler in its resting coil and is rather well executed.

The next full-length description of the rattlesnake to appear was that of Johann Nieremberg in 1635. Little need be said of this since it seems to have been appropriated in its entirety from Hernández, although no credit was given to the latter. The picture is in some details an improvement over Hernández', particularly as the tail is curved away from the body so that the rattle may be more clearly seen. However, the rattle has been rotated so that the wide side, instead of the narrow edge, is incorrectly shown as the top.

We come now to the rattlesnake description, made in 1648, of Guilielmus Piso, based upon researches in Brazil. He says the rattler makes a sound like a bell and is very swift. It attacks with either the teeth

Fig. 49 The second rattlesnake picture. (From Johann E. Nieremberg's *Historia Naturae*, 1635)

Fig. 50 An early crawling rattlesnake. (From Guilielmus Piso's *De Medicina Brasiliensi*, 1648)

or the rattle, the latter being the more quickly fatal. It adds a rattle a year to the string. Various antidotes are suggested, including incision, suction, and cauterization. There is a figure of a rattlesnake in a crawling position, with the usual vertical undulations of the old prints. The rattle, although quite small, is well executed.

The first published mention of the rattlesnake in English, as far as I know, was made by Captain John Smith in 1612 in his *A Map of Virginia, with a Description of the Countrey, the Commodities, People, Government and Religion*, where he says the decorations worn by the Indians include the rattle from the tail of a snake.

The first use of the word "rattlesnake" was also made in connection with an Indian custom. Edward Winslow in 1624, in *Good News from New-England*, wrote: "This messenger . . . leaving for him a bundle of new arrowes, lapped in a rattle Snakes skin;" and, further, "he signified to the Governour that to send the rattle Snakes skin in that manner imported enmitie . . ."

Of the rattlesnake as a live creature, the first English report is that of the Rev. Francis Higgeson (or Higginson) in *New-England's Plantation; or a Short and True Description of the Commodities and Discommodities of That Country*. This pamphlet of twenty-six pages was published in London in 1630. Says Dr. Higgeson:

> Yea there are some Serpents called Rattle Snakes that have Rattles in their Tayles, that will not flye from a man as others will, but will flye upon him and sting him so mortally that hee will dye within a quarter of an houre after, except the partie stinged have about him some of the root of an Herbe called Snakeweed to bite on, and then he shall receive no harme: but yet seldom falles it out that any hurt is done by these. About three years since, an Indian was stung to death by one of them, but wee heard of none since that time.

William Wood in his *New England's Prospect*, 1634, describes the timber rattler in some detail. The teeth are as sharp as needles, and, although the neck is no thicker than a man's thumb, the snake can swallow a squirrel. The bite causes death in an hour unless snakeweed is available; this is taken both internally and used as a poultice. He says it is reported that if the bitten person lives, the snake will die. Wood denies that the snake can kill with its breath, evidently a belief current at the time. He states that if a rattler swims a river it will die upon reaching the other shore.

Another early British reference is that of Thomas Morton in *New English Canaan*, 1637, who reports: "There is one creeping beast or longe creeple (as the name is in Devonshire) that hath a rattle at his tayle, that does discover his age; for so many yeares as hee hath lived, so many joynts are in that rattle, which soundeth (when it is in motion) like pease in a bladder, & this beast is called a rattlesnake."

Thomas Lechford (1642) says that when a person is bitten by a rattlesnake he turns the same color as the snake, "blew, white, and greene spotted."

In the late seventeenth century, articles on rattlesnakes began to appear in the scientific journals, particularly in the *Philosophical Transac-*

tions of the Royal Society. Several of these papers were physiological in nature. Edward Tyson (1683) was the first to publish a complete and, for the time, a remarkably accurate account of the anatomy of the rattlesnake. He mentions the facial pits, and notes that European vipers lack them. A paper in 1728 described the biting mechanism, including the venom gland and its operating muscles.

Captain Hall* (1727) was probably the first to conduct extensive experiments with rattlesnake venom on live animals. Using dogs, he proved that successive bites tend to exhaust the venom supply.

---

* I wish someone would supply the good captain's given name or initials; even the Royal Society lacks this information.

CHAPTER SIXTEEN

Myths, Folklore, and Tall Stories

Myths and folklore about snakes are of more than passing interest since they influence, to a considerable degree, the attitudes of people toward snakes. These attitudes, translated into action, often have important effects on economic and control measures, and on the snake-bite problem.

Some myths and legends arise out of exaggerations; others come from inaccurate field observations or interpretations. They become established through repetition. I have no hope of eradicating completely any of the myths repecting rattlesnakes so deeply imbedded in American folklore. I shall, however, examine some of their origins, and show how contrary some of these traditions are to various facts of nature. Many of these myths originated abroad centuries before any European had ever seen or heard of a rattlesnake. They were subsequetly transplanted to America and applied to the rattlesnake, a far more spectacular creature thatn the European vipers to which they had previously been ascribed.

Some of the myths and legends began with the Indians and were assigned by them to rattlesnakes before the advent of the white man. Few of these are so distinctive that they can be recognized as being non-European, or non-Old World in origin.

Some of the legends and myths were deliberately invented by hunters, guides, and cowboys to spoof the tenderfoot at the campfire. They subsequently spread like an uncontrolled campfire and came to be firmly believed by large elements of the population. One writer remarked that every ranch or camp conversation always got round to rattlesnakes.

Nearly all myths and legends contain some grain of truth; many merely stretch the truth. In such cases, in this work on the habits and life histories of rattlesnakes, it has not been easy to decide whether a particular item should be allocated to the appropriate factual chapter or to this one on folklore. Inevitably some duplication has been entailed.

No discussion of herpetological folklore would be complete without a word about Pliny (Gaius Plinius Secundus, 23–79 A.D.) His encyclopedia of natural history is an almost unparalleled example of industry and credulity. It is the funnel through which we can watch the ancient folklore pouring down into the medieval and modern worlds. So intact do some of the stories now applied to rattlers remain, that we should almost conclude that Pliny must have slipped over for a visit to America before Columbus.

Some folklore is harmless and entertaining, some is dangerous. A hair-rope insurance against the almost negligible risk of a rattler in the blankets is relatively innocuous, and if anyone is reassured and sleeps better with a horsehair reata surrounding his desert bunk, no harm is done. But a folklore remedy for rattlesnake bite is a dangerous folly, if it delays or complicates a really necessary treatment. Also, some rattlesnake horrendous incidents and tall tales related round the campfire are too often the prelude to practical jokes with live or dead rattlers, or with harmless snakes. Serious and lasting results can follow the frightening of inexperienced people in such horseplay, a practice that should be strongly discouraged.

MYTHS OF VENOM AND SNAKE BITE

Temporary Removal of Venom Glands

The myth that when venomous snakes drink, they temporarily remove their venom, venom glands, or fangs, is very old. It has been attributed to a book of allegories from the second century A.D.

In the rattlesnake legend, the snake leaves its venom glands on a rock, as it goes to a stream or pond to drink. When it returns, if someone has stolen them in its absence, it dies in convulsions or kills itself by striking its head against the rock. There are several variants. Mexican sheepherders said that rattlers remove their fangs when courting so as not to injure each other. The Sioux Inians believed that rattlesnakes remove their venom every night and put it back again next morning.

It is the theory that the snake removes its venom glands before drinking because otherwise it would, itself, be poisoned.

Effect of Death on Venom Virulence

Sometimes applied to rattlers is the old myth that, when a snake dies, any victim suffering from its bite immediately recovers. But in Florida the superstition is reversed; the venom becomes more virulent if the snake is killed.

Effects of Biting on the Biter

The serious result to a rattlesnake if it bites a man who has poison in his blood is a myth of some antiquity. As alcohol is an antidote for snake venom (another myth), it follows that if a man is bitten while drunk, the snake will die. Legend has it that Ethan Allen was bitten repeatedly while drunk. The rattler became cross-eyed and inebriated. When Allen awoke he complained of mosquitoes.

In the summer of 1951, a case in Texas attained nationwide publicity. An employee at an air base was struck by a rattler. The man was uninjured but the snake died. It was explained that the man was a metal plater whose blood was charged with sodium cyanide from his having continuously handled this dangerous chemical. He was immune but the snake wasn't.

There is a Southeastern folk belief that a rattler must always seek snakeweed after biting a man, as the latter's blood is fatally poisonous to the snake.

One Western tall tale concerns the rattler that strikes a man in the hip pocket where he carries a plug of chewing tobacco, naturally with fatal results to the snake.

Venom Transmitted through Mother's Milk

The effect of snake bite transmitted through mother's milk to nursing offspring has been the subject of a number of stories, some of the more recent of which deal with rattlesnakes. One of the most fanciful is that of a mother who was bitten in the fourth or fifth month of pregnancy. She recovered and bore a normal child, but when she nursed the child it turned the color of a rattlesnake and died. Several small animals were fed the milk with fatal results.

I know of no medical case report that mentions any adverse effect on a nursing child caused by a rattlesnake bite sustained by the mother.

There is a theory current in the Southeast that pregnant women are immune to snake bite. On the other hand, it has been stated that rattlesnake bite is always fatal to pregnant women, and that even the Indian remedies, usually so effective, are of no avail.

Preferences in Biting

Rattlesnakes have some peculiar, but often praiseworthy, preferences in biting. They won't bite children, at least until they have reached the age of seven. They won't strike a person from the rear, nor will they bite in water. Also, I was advised by one of my correspondents that rattlesnakes will bite only men, not women. This chivalrous attitude they must have inherited from their distant relatives the European vipers, who have long made this distinction. But our uncivilized rattlers seem not yet to have adopted the code of refusing to bite naked people, a myth descended from the ancients.

Transfer of Venom to Another Animal

The story that the venom of one creature, if ingested by another, makes the venom of the second doubly powerful is very ancient. It was most frequently reported as applying to vipers and scorpions. Occasionally it is heard today attributed to rattlesnakes, as mentioned by one of my correspondents: "Here in Medford [Oregon] they say that when a rattler is killed you must be sure to cut off the head and bury it, for if yellow jackets [wasps] eat the snake they will carry the venom to the next victim they sting."

Strange Venom Effects

Fantastic and unexpected effects of rattlesnake venom on man, animals, plants, and inanimate objects have often been reported. One of the oldest is the axhead story, first told, I believe, by Cotton Mather in 1714. In this, a rattlesnake strikes the metal bit of an ax. The metal changes color and, when the ax is used again, the discolored edge breaks off.

In another Mather story, a man kills a rattlesnake with a green branch of a tree; and is himself killed by the poison that reaches him through the branch. One story has it that rattlers bite the berries of the Taxus trees; birds eat the poisoned berries and thus fall ready prey to the snakes. There is a Brazilian belief that rattlers and other snakes sharpen their fangs on the trunks of young trees.

From this point on, the legends are of the frankly tall-story type, although some were not always so considered. In the oldest of these yarns, a hoe handle is struck by a rattlesnake and within an hour has swelled to such an extent that it pops out of the eye of the hoe. In some of the stories inflation really prevails; in one the farmer gets enough lumber out of his swelled hoe handle to build twenty five-room houses, and in another enough shingles to roof all his buildings.

In a yarn recorded in 1942, a toothpick in a man's mouth is struck, with the result that there is enough lumber to build a twelve-room home. Unfortunately, when the paint is applied to the new mansion, the curative property of the turpentine reduces the swelling to the original toothpick. In the tragic tale of Peg-Leg Ike, bitten in his wooden leg, despite the frantic efforts of his friends armed with axes to chop away the swelling timber, he was choked to death by its growth. The sorrowing survivors got enough kindling to last all winter.

Then there was the wagon tongue that was bitten. Fortunately it was possible to chop it away from the wagon in time to save the latter. A cypress stick swelled to produce enough railroad ties to build a mile of track; the disaster that occurred when rain washed the venom out of the ties may well be imagined. In one story, a bitten sapling swelled to the size of a California sequoia. And there was the mattress, which, when bitten, swelled into a featherbed.

A boy in an auto race around the Salton Sea was so unfortunate as to puncture all four tires in a cholla-cactus patch, but then ran into a bunch of rattlers that bit his tires, which swelled so that he didn't even have to slow down. Needless to say the youngster beat all the professionals. Sometimes the rattlers that bite tires are themselves blown up.

Related to the discoloration around a rattlesnake bite is the idea that, if a rattler strikes a piece of raw meat, the meat will almost instantly turn green.

HOW RATTLESNAKES CATCH AND EAT THEIR PREY
Charming Prey

Another old and controversial myth relates to the supposed power of a rattlesnake to charm its prey. This idea is not restricted to rattlers and was, in fact, attributed to vipers and other snakes before rattlers had been brought to the attention of Europeans. Nor must it be thought that this is a belief which time has dispelled; on the contrary, credence is still quite general today, as evidenced by some of the remarks of my correspondents.

Briefly, the belief is this: rattlesnakes have the power of so charming, fascinating, or hypnotizing prey that it makes no effort to escape, and even moves helplessly toward its doom. In its most acute form, the myth recounts how a rattler at the foot of a tree has only to fix a baleful eye on its prospective victims aloft, whereupon birds come fluttering down, or squirrels descend helplessly, and deliberately walk into the snake's open mouth.

One writer concluded that fascination must be true because rattlers are too slow and lazy to get prey by any other means. Another told of a laborer who placed a live rattler in a barrel. Later he leaned over the edge to look in. The eyes of the snake were fixed on him like balls of fire. He became sick at the stomach, but fortunately was pulled away by a companion before being completely overcome.

Occasionally, carnivals or snake shows, to stimulate the apprehensive curiosity of the customers, state that attendants will be on hand to lead out anyone who may be overcome by carelessly exchanging stares with a rattlesnake.

At zoos, there have been no observations on captive rattlesnakes with prey that verify any theory of fascination. Birds and small mammals placed in cages with rattlesnakes, as soon as they have become used to their new surroundings, pay no attention whatever to their reptilian cage mates. They run about, nestle against, or perch on the snakes; and, as far as rats and mice are concerned, if not supplied with food, they will frequently eat the heads, tails, or rattles of the snakes. It cannot be supposed that behavior patterns in the wild could be so different that snakes, which regularly fascinate or charm their prey there, would never show the slightest ability to do so in captivity.

Furthermore, rattlers require no such mysterious power to get their food. They are quite capable of obtaining it by more conventional methods.

Licking Prey

It is frequently stated that rattlesnakes—in common with other snakes—always lick their prey and cover it with saliva before swallowing it. As a matter of fact, such a procedure would be quite impossible with the slender tongue of a snake. The belief comes from having observed regurgitated prey that has become coated with digestive juices. No snake, rattler or otherwise, while eating under observation in captivity, has ever been seen to coat its prey with saliva. But it is well known that they do feel or sense their prey by repeatedly touching it with their tongues, probably for olfactory reasons.

There is a folklore belief that rattlesnakes cannot regurgitate. This is quite untrue; captured rattlers frequently throw up prey they have recently eaten.

Mythical Foods

Although it is known that rattlesnakes—like all snakes, except possibly one in South Africa—eat only animal food, there is a rather widespread tradition that they eat various kinds of berries, especially chokecherries and huckleberries, as well as cactus fruits. This may have originated from the discovery, in rattlers' stomachs, of some of the items of food which had originally been eaten by the snakes' prey. One correspondent killed two large rattlers and found watermelon seeds in their bodies, from which he concluded they must have been feeding on a broken melon. It is to be assumed that the seeds were the food remains of some mammals that had been swallowed by the snakes. One visitor to the San Diego Zoo was heard to remark that rattlesnakes are fond of honey.

Another belief is that snakes can milk cows; this is an extremely ancient and widespread myth, and is told of many different kinds of snakes, occasionally including rattlers. The idea is quite fantastic and is mentioned here only because of the often heard myth that rattlesnakes are fond of milk and will drink from a saucer. Conceivably, a thirsty rattler might do so to secure a substitute for water, although experiments have found that they refused milk even when denied water.

Continuing in the vein of the fantastic, there is the tale of the tame rattler in Central America that was so fond of milk it would climb up on its master's chair at breakfast to get its daily share. Kipling's *Jungle Book* did much to spread the idea that snakes are fond of milk. Conan Doyle's famous short story "The Adventure of the Speckled Band," in the Sherlock Holmes series, is another offender.

There is a story of a rattler that went about an Indian camp picking up fragments of provisions and licking platters. And not long ago there was a press release to the effect that a rattlesnake had found, on a window sill, a macaroni salad left there to cool, and had "devoured every bit of it. The snake, unable to navigate swiftly with so heavy a load, was easily killed."

Another well-known yarn concerns the snake that crawls into a hen house, swallows a china egg, and then finds the hole by which it entered too small for escape. This is occasionally told of rattlesnakes. It is true that snakes are sometimes prevented from escaping via their means of ingress, because of the bulge produced by the prey they have swallowed. A San Diegan gopher snake has been photographed trying in vain to crawl through the bars of a bird cage after eating the canary.

FOLKLORE REMEDIES DERIVED FROM RATTLESNAKES

Folklore remedies that involve rattlesnakes, either in whole or in part, have been prescribed for a variety of diseases.

Some Indian usages in treating toothache or improving the teeth still practiced by whites involve such expedients as biting a live rattler along the back; using a rattler vertebra as a toothpick or hanging it around a child's neck; or putting rattler fat in a tooth cavity. In 1950 I heard of the use of powdered dried rattlesnake, sprinkled on food, as a treatment for tuberculosis in northern Baja California, Mexico.

The following are rattler-based cures for other afflictions: the heart and liver warm and raw for palsy; a backbone necklace for chills and fever; a rattler drowned in wine for leprosy; rattler gall mixed with clay for fevers and smallpox. In New England, rattler gall was recommended for biliousness, also to promote longevity.

Dried rattlesnake in corn whiskey for rheumatism and a necklace of rattler bones for epilepsy are favored in the Kentucky mountains. Among the Pennsylvania Germans, to swallow the heart of a rattler will cure epilepsy. A bracelet or anklet of bones will forestall swelling from fatigue in Louisiana. One may cure goiter by wrapping a rattler around the neck and then allowing it to crawl away; this is a Nebraska folk cure. Behead a rattler and put it in a jar filled with rice whiskey; then after a year, drink the whiskey for rheumatism; this is an Idaho cure said to be derived from the Chinese.* Baked and powdered rattler flesh is a cure for venereal disease in Yucatán and eastern Guatemala.

Black folklore has a rattlesnake remedy in reverse, a method of hexing: if powdered dried rattlesnake is put on the enemy's food or in his coffee, he will be filled with little snakes in four months. Or, one may put dust under an enemy's hatband so that the dust will run down and blind him when he sweats. The dust should consist of powdered rattlesnake, or may be made from a rattler skin mixed with earth from the head of a grave.

There are many ancient myths of the curative power of snake skins. Some are founded on the basic idea that snakes restore their youth when they change their skins, and thus are virtually immortal. Some of these myths have been attributed to rattlesnakes.

The early colonial accounts, and some of the later ones as well, often failed to specify just what diseases the skin was supposed to cure, although fever, smallpox, rheumatism, and blood cleaning were mentioned. Usually the skin was pulverized and taken internally. Later a more mystical approach was evident, in that it was necessary only to carry or wear a skin to secure the curative effect. The use of a rattler skin as a hatband has been reported to cure headache (Kansas folklore), and rattler-skin belts were used in Kansas to prevent backache (it was essential that the rattler had not bitten itself). In New Hampshire the skin was wrapped around an ankle to prevent cramps while swimming. The Kentucky

---

*In my school days in San Diego, it was rumored that our Chinese residents would pay large sums for live rattlesnakes to be used for some mysterious medicinal purpose. Rheumatism was at least one of the afflictions for which the flesh was used, but the method of preparation was secret. I presume the rattlesnake, in America, replaced some indigenous viper that had been used at home in China.

mountaineers say that a belt will prevent rheumatism. Worn around the throat, a rattler skin will cure sore throat; or it may be wrapped around an injured limb to alleviate aches and sprains. Mexicans in Texas in 1854 were reported to use rattlesnake skins on the cantle of a saddle to prevent saddle sores. In New York and Pennsylvania a poultice of rattlesnake skin was used in the treatment of felons and carbuncles. Finally, witches will be kept at bay if one carries or wears a piece of rattler skin (Kentucky mountains, western Illinois blacks).

PROTECTIVE METHODS AND DEVICES

There are a number of myths and tales concerning how people may protect themselves against rattlesnakes, especially by the use of barriers and repellents. These have little relation to the really effective steps that may be taken described earlier.

The Protective Hair Rope

The protective hair rope, once a widespread myth of the Southwest, has now lost somewhat its vogue, presumably because fewer punchers sleep out on the ground today, and ordinary hemp ropes have largely supplanted the valued horsehair lariats of a bygone era. Briefly, the story was that a hair rope, if placed so as to encircle a bed or camp, would keep snakes out because the hairs would scratch their bellies, causing them to turn away.

Of course, the weakness of the scheme lies in the fact, which is easy to demonstrate by trial, that a rattler's ventral covering is so tough he would hardly feel the hairs on a reata; at any rate, they would not incommode a creature that does not hesitate to crawl through or over cactus. Moreover, researchers have reinforced denials of the effectiveness of hair ropes with photographs of rattlers in the act of crossing them.

One folklore report says that the rope must be striped with black and white to look like a king snake. An experimenter in 1919 tried a hair rope on a sidewinder and found that it would only cross the rope backwards, which is not surprising as this is the normal way in which a sidewinder crawls.

A Southwestern reptile myth gives this extension to the rope story: a rattlesnake has been known to squirt his venom at a victim taking refuge beyond a rope. In one instance, a dozen rattlers were found coiled next to a hair rope in the morning, as if waiting for its removal so that they might invade the camp. In a Texas tall story, some cowboys put a hair rope around their camp one night and in the morning there were 129 defunct rattlers that had tickled themselves to death trying to cross it.

The Infallible Shot

Another myth, widespread throughout the West, is that it is unnecessary to aim carefully when shooting a rattler, since the snake will automatically line up his head with the barrel of the gun and the first bullet will inevitably find its mark. An even less possible variant maintains that a rattler will strike at and hit the oncoming bullet.

I think the derivation of this myth lies in the fact that one ordinarily shoots a rattler at a much closer range than almost any other object. The aim therefore is generally good, although I can testify that it is quite easy to miss a rattler even at close range, especially with a derisive audience on the sidelines.

Repellents and Amulets

The belief that many substances will repel snakes, keeping them out of homes and camps, and even preventing their biting people who may be traveling through the woods, is an ancient one.

One of the most popular botanical repellents was the white ash. Another was the onion. Burning old shoes was recommended in Maryland folklore, in that of Arkansas, and by the Pennsylvania Germans. The latter also burned snakes for the same purpose. A string of snake bones worn around the ankle will prevent snakes from biting, according to Louisiana folklore. As snakes will not crawl near glass—so it is believed—glass jars have been used in the Ozark country.

In Texas the Mexicans throw a sombrero over a hole that a snake has been seen to enter, with the belief that this will prevent its coming out. I have heard it stated that in Texas some cowpunchers greased their boots with king-snake oil to keep the rattlers from striking. It was the custom in the pioneer Northwest (upper eastern Mississippi Valley) simply to put the curse of Adam on a snake, which would crawl away and die.

It goes without saying that none of the innumerable botanical repellents recommended from the days of Nicander and Pliny down to the present have been effectual beyond possibly affording the user a feeling of relief from fear.

MYTHICAL CREATURES
Giant Rattlesnakes

Rattlesnake lengths are usually exaggerated; in addition, some measurements are based on dried skins, which can be stretched up to half again as long as their original possessors. Thus we hear of eight- and nine-foot rattlers in Florida and Texas, where a seven-foot snake would be quite exceptional. But although a nine-foot snake is at least a possibility, some of the stories heard about large rattlers enter the realm of the fantastic and are to be considered myths. I should say that rattlesnakes ought to be placed in the mythical category at ten feet, thus allowing for some exaggeration.

Rattlesnakes of a truly legendary length were described at an early date. In 1714, rattlers seventeen feet long are mentioned. In 1750, they were said to grow to eighteen feet. There was one described in 1753 as twenty-two feet. The big ones, it was said, could eat a whole roe deer. One man claimed to have killed one twelve feet in length that measured fifteen inches around "at the shoulders." Another tells of a fellow hunter who was chased by a rattler about a rod (sixteen and one-half feet) long. One rattler, in 1858, was mistaken for a dead tree. When killed it was found to measure seventeen feet. One writer had heard of a rattler weighing 117 pounds (such a snake would measure about seventeen feet) with eighty-seven rattles.

In 1955 it was shown, by offering prizes for live Texas rattlers exceeding six feet, how rare such creatures are. At ten dollars per inch for each inch over six feet, the largest rattler brought its finder ninety dollars.

An interesting communication lately received is this from Baja California, Mexico:

For a number of years, we have been hearing stories about a very large rattlesnake said to live in some caves a few miles from our ranch. We have always listened to them with a smile and forgotten them. Last spring, a man who has worked for us a long time and whom we know to be truthful and not overexcitable, came and said he had seen the big rattlesnake. He said it was immense. He got terribly frightened and ran away. Now, three days ago, his son was gathering wild honey (the bees live in caves) when he saw the big snake. He had camped and was making a fire when he noticed his horse was excited. He looked and there, about twenty steps away, came the snake toward him. He ran to the horse and mounted and sat watching. The campfire was between him and the snake. The snake came on but when it neared the fire it raised its head four or five feet off the ground and hissed with such force it sounded like a bull. Then, it turned, went around the fire and down a draw, but kept its head up, looking back at him and hissing. The boy says it was no less than twenty-five feet long,* and he thinks it might have been thirty feet. Its rattles were as wide as his three fingers and about a foot long. The boy was terribly scared, horrified, and ran home at once. Next morning he came here to tell us. We have always told him that, if they saw the big snake at any time, we wanted to see it. So my two sons went with him. They hunted all day but could not find any trace of the snake.

There has recently been a report in circulation of a rattler twenty-one feet long that was killed near Poteau, Oklahoma. It was said to have

---

*A rattlesnake of this length, if of the same body proportions as an ordinary rattler, would weigh 275 pounds; a thirty-foot rattler would weigh over 400.

bitten a woman in the foot. I like particularly the Ozark story of the snake with a head the size of a water bucket and rattles as big as coffee cups.

From the *San Diego Sun*, July 30, 1881:

> One day I had my twelve daughters in a wagon, and the horse became frightened and ran to the very edge of a cliff. When I came up to see what stopped him I saw a rattlesnake had wrapped around his leg, and then had thrown a number of coils around a tree and set his teeth into the bark. I measured him and he was twelve feet long. Then I unwound him and he was only five feet long. You know how elastic a snake is.

A good folk tale is of the man who killed a fifteen-foot rattler, hoping to prove to the skeptics what he had seen, only to find that he had killed two normal snakes, one going into a patch of grass, the other coming out.

Reports of phenomenally large rattlesnakes are sometimes based on misidentified skins. I once saw in a sporting goods store the skin of a boa constrictor, with a card claiming it to be one of the largest rattlesnakes ever killed. The magazine *Texas Game and Fish*, in October 1954, contained a photograph of a rattlesnake skin that measured ten feet eleven inches without the head or rattles. It is not surprising that the rattles were lost, for the skin, to judge from the pattern, was that of a python.

Not long ago one of my correspondents in Alabama wrote:

> I have in my possession the skin of a rattlesnake that was killed by a train in Washington County, Alabama. The skin now measures eleven feet, but originally the snake was fourteen feet six inches long, as measured by the train crew. The rest of it was smashed by the wheels of the train and could not be saved. This snake had forty-three rattles, and it was estimated that about ten other rattles were destroyed by the train. The snake was thirteen inches across the middle of the body.

My correspondent kindly forwarded the skin to me for identification; it was that of an Indian python. No rattles accompanied the skin.

A suggestive feature of many of these stories of rattlesnakes of extraordinary length is that they have, equally extraordinary, numerous rattles, usually from twenty-five up. As a matter of fact, very large rattlers, say between six and seven feet in length, seldom have long strings, since they tend to break the rattles off when pulling them out from under their own heavy bodies. A seven-foot rattler will weigh upward of fifteen pounds.

Hairy Rattlesnakes

One occasionally hears tales of hairy rattlers, some taken seriously by those who tell of them, others as interesting items of folklore. I am told that the natives in Brazil believe that very old rattlesnakes grow feathers on their bodies. Presumably these stories start from observation of rattlesnakes that were shedding their skins in patches or were studded with cactus spines. I have seen them in both conditions, and the resemblance to hair was quite apparent.

Rattlesnake–Bull-Snake Hybrids

A myth of wide distribution, particularly in the Missouri Basin states, is to the effect that rattlesnakes and bull snakes have crossed, producing a particularly dangerous hybrid offspring having all the venom of a rattler, but with the speed and energy of a bull snake, and, of course, without the identifying rattles. Actually, there is not the slightest ground for this idea. It probably stems from the fact that the harmless bull snake of that region has a blotched pattern somewhat resembling that of the prairie rattlesnake. Furthermore, this snake, when frightened or threatened, will sometimes become the picture of viciousness, striking and hissing, and at the same time vibrating his tail against grass or dry leaves so that it makes a good imitation of its dangerous associate.

While there is not the slightest evidence of bull-snake–rattlesnake crosses, it is now known that different species of rattlesnakes may occasionally interbreed.

There is a myth or tall story current in upper New York State to the effect that black snakes are the male parents of rattlesnakes. One writer mentions another fictitious hybrid, the water rattler.

QUEER ACTIONS AND ATTRIBUTES
Rattlesnake Odor

Many of the early accounts of rattlesnakes dwelt at length on their offensive odor, particularly when congregated at their dens. It was most often likened to the smell of cucumbers, and was said to be perceptible at a great distance. This idea is still widely prevalent, but, as a matter of fact, rattlesnakes are virtually odorless, even when there are a number together. We have had, at the San Diego Zoo, more than three hundred in a single cage of no great size, and even with this concentration almost no odor was evident, as long as the cage was kept clean. Rattlesnakes, like other snakes, have special scent glands in their tails, which they use occasionally, possibly for sex attraction, and certainly as a defense mechanism, as discussed elsewhere. But the odor from these glands is neither so strong nor so unpleasant as that of many other genera of snakes, yet the latter have no such reputations as the rattlesnakes for noxious odor. I have repeatedly held my nose against a live rattlesnake and found the odor barely perceptible.

Some early writers say the odor is particularly noticeable after a rain; it is worse when the rattlers are coming out of hibernation; it is caused by surplus venom coming out through pores in the snake's skin.

Modern writers also dwell on the strong rattlesnake scent. One located a snake in a tree by its odor. He says even a man can smell a rattler twenty feet or so away, although the odor is not unpleasant. He once smelled a rattler in a woodpecker's nest in which he was about to put his hand. The heavy, nasty smell of an angry snake was mentioned in 1942. One rattler crawled between two sleeping men. Its odor was so strong as to awaken one of them.

Although the odor of rattlesnakes has usually been likened to that of cucumbers, some have noted a resemblance to the odors of a watermelon patch, a skunk, a goat, and garlic. A Wyoming pioneer stated that the cucumber smell was so strong that the presence of a rattler was detected in a load of hay.

Two small islands in Lake Erie were said to be so infested with rattlesnakes that the air was polluted. There is a fanciful story of a very large rattler with so strong an odor it caused a boy to faint.

Whatever validity there may be to all the stories of rattlesnake odors must have come from an occasional use of the scent glands by an annoyed or injured snake.

Rattle Myths

Almost the only snake myths and legends that apply solely to rattlesnakes are those pertaining to the rattle, for this is the only unique attribute of the rattlesnake. I have discussed earlier most of the superstitions and misunderstandings concerning the purpose and use of the rattle, and its formation and growth. But here are some of a different character.

There are a number of regional items of folklore regarding the use of rattles to cure or ameliorate disease. Carrying a string of rattles will prevent or cure rheumatism (Kansas, Texas, Illinois, Nebraska); will prevent smallpox (Mexicans in Texas; Nebraska), will keep a person from having fits (mountains of Kentucky); and is a general disease preventive, if worn around the neck (western Illinois). Rattles are worn in the Catskills, New York, to prevent sunstroke.

Another widespread belief is that rattles will soothe and pacify teething children. They may be worn as a necklace, placed in a bag and hung on the neck, or the child may be allowed to chew on them (Colombia, New Orleans blacks, Pennsylvania Germans, Texas, Nebraska). That magical medicine is involved is indicated by some of the specifications for the use of the rattles. Thus, in one of the prescriptions in Nebraska, there must be three rattles, and the cord around the child's neck must be red. Rattles were also used to prevent fits and convulsions in the child (New York folklore).

Although rattles are usually considered beneficial, there is one myth that they throw off poisonous dust that may cause blindness. This version came from Kansas: if you cut the rattles off a snake, the juice will fly in your eyes and blind you. This must have come from someone's experience with the rattler's scent glands, although I doubt that the spray is harmful in any way.

To return to the beneficial properties of rattles—they may serve as amulets or good-luck charms (western Illinois, Nebraska). If you receive rattles from someone, you will come to no harm while that person is near (Kentucky mountains). If you catch a rattler and rub the rattles on your eyes, you will always see a rattler before it sees you (Southern blacks). Also, Satan himself may be summoned to aid you in black arts if you get a button (rattle ?) off a graveyard rattler, sew it up with a piece of silver in a red flannel bag, and wear it over your heart (Southern blacks). To display a rattle from a dead snake will keep other rattlesnakes away (Southwestern folklore).

Rattles are often put in violins—called fiddles where the rattles are so used—to improve the tone, keep out dampness, or give the owner good luck (western Kentucky, western Illinois, Nebraska). But there is one belief that the rattles will blind the fiddler. In Brazil, a rattle in a guitar will improve not only the instrument, but the singer's voice as well. There is an item of California folklore claiming that a rattle tied to a banjo head will preserve the skin. The rattle-fiddle relationship is said to be derived from the fact that a swimming rattler elevates its rattles to keep them dry. But that it does so is also a myth.

Occasionally, we hear tall tales of the effect of music on rattlers, such as the recent newspaper yarn of the berry picker who used a mouth harmonica to render the rattlers harmless while he plied his trade. All of these are quite mythical, as rattlers, like other snakes, are deaf to airborne sounds.

Also, there is the tall tale of the Indian who trained a band of rattlesnakes to join him in song. With their rattles, they were able to carry four parts—soprano, alto, tenor, and bass. There are also a number of myths about snakes that have voices.

SOME TRADITIONAL STORIES
Rattler Encounters with Other Creatures

One of the most widespread of all snake legends is the one that describes a battle between some animal and a venomous snake; each time the snake bites its adversary, the latter runs off to a nearby bush, eats a few leaves, and returns, revivified, to the fight. Then a malignant onlooker appropriates the bush, and the next time the snake's opponent is bitten, it seeks the plant in vain and dies. But its sacrifice bears fruit, for by this means the life-saving plant is disclosed to mankind.

This story has been told of many different kinds of snakes and of attacking animals, probably most frequently of the cobra and mongoose. It also provides the explanation of how the efficacy of many plant cures was discovered.

A fairly fresh tall story has the rattler fleeing a king snake so fast the friction was setting the grass afire, but the trailing king snake was sweating so much in keeping up that it put the fire out. Their trail was a long, black streak through the grass. Another is that a rattler and king snake started to swallow each other simultaneously, and naturally both disappeared.

A Nebraska yarn tells of the bull snake that feints the rattler into striking. Missing the bull snake, it strikes itself and dies.

Rattlesnake Pilot

Though the legend of the rattlesnake pilot is subject to considerable variation, the general idea is that another kind of snake known as a rattlesnake pilot warns rattlesnakes of the approach of danger and guides them to safety.

The earliest reference to the pilot snake seems to be contained in a letter written by Cotton Mather dated June 4, 1723. The pilot is described as a snake that commands and governs the rattlesnake and, if the latter fails to obey, strikes it dead with a bone in its tail. Another early account called the copperhead the rattlesnake pilot, because it comes out of hibernation a week earlier than the rattlers and always precedes one in crawling about.

Various other kinds of snakes are credited with being rattlesnake pilots. Nowhere is there entire agreement as to which snake occupies the position. Rat snakes, king snakes, milk snakes, and the pigmy rattler have all been called the pilot.

There are differing versions as to how closely the rattlesnake follows the pilot. In some areas of the South the pilot is said to keep just ahead of its charge, so that if a pilot is seen one should stop in his tracks and look carefully about, to avoid stepping on the rattler. But in other stories, the rattler trails at some distance. As a reward for this guidance, the rattlesnake, when successful in securing prey, shares its catch with the pilot. The legend fails to state how this could be possible with creatures that swallow their prey whole. There is a Louisiana oil-field story that when rattlesnakes issue from hibernation in the spring, they are sleepy, have poor vision, and consequently are unable to strike accurately. It is the duty of the pilot to aid a rattler in determining when and in what direction to strike.

The Mississippi folklore version is that the rattler is the male, and the pilot (which is not a rattler) is the female of the same species. In Alabama folklore, copperheads are the male rattlesnakes. Another phase of the myth is that pilot snakes are crosses between rattlers and bull snakes.

To summarize the present views of herpetologists: there is no known relationship, either genetical or in habits, that would justify any of these reports of rattlesnake pilots.

The Vengeful Mate

That snakes travel in pairs, and wreak vengeance on anyone who may injure their mates, is a very ancient myth. It is often attributed to rattlesnakes. The basic facts of natural history involved have been discussed before, with the conclusion that the sexes are together only during a brief mating season.

There is a tall tale about a rattler pair, related as an item of family history. A party crossing Kansas in a prairie schooner in 1853 killed a large female rattler and tied the body to the back of the wagon with the head dragging. This was early in the morning. That day they made fifty miles. The next morning a child of the party was fatally bitten as she lay rolled in her blankets. It was the supposition that the male had kept pace with the party for the entire distance.

Then there is the account of Old Betsy, a Florida diamondback kept by an old hunter as a decoy. Betsy was blind and had had her fangs removed for safety, but was reported to have lured many a male rattler to his doom.

Swallowing Young for Their Protection

The most controversial of all reptile myths, the one whose denial brings out the greatest flood of indignant eyewitness protestations, is that of the snake mother that swallows her young upon any threat of danger, and then disgorges them when the menace has passed. This is an ancient and world-wide story, one not attributed to rattlesnakes alone; it was told of vipers long before rattlesnakes had been heard of by Europeans. But it is occasionally given an original rattlesnake touch by having the mother call her offspring to her unique refuge by sounding a warning rattle.

As the story is generally told, an intruder comes upon a mother snake surrounded by her numerous progeny. Alarmed, she sounds a warning, usually by hissing, whereupon the little snakes rush to her, and one by one enter her mouth and glide quickly down her throat. When the danger has passed they emerge and resume their previous diversions.

One writer adds the interesting variant of seeing a mother with the head of a young rattler sticking out of a corner of her mouth like a cigar. Another claims that in one instance there were two mothers and two broods, and the young knew which mother to dart into for refuge. A typical early eyewitness account is that of the man who once saw seventy young go into a mother's mouth at his approach.

The question arises: how can the story be denied in the face of so many confirmatory accounts? Various explanations have been given: unborn snakes, of species that bear living young (in contrast with those that lay eggs), are found in the body of the mother, and are assumed to have been found in the stomach; snakes are seen while eating young snakes or lizards; young snakes actually disappear in the grass or into a hole under the mother. But beyond all these explanations, the most appropriate bears upon the mental confusion existing between things seen, and those read or heard of in times past. Every telling strengthens the picture and the memory, and makes it more certainly a personal occurrence. Several investigators have commented that, in nearly every eyewitness account, a considerable time, usually from twenty to fifty years, has elapsed between the date of the supposed event and its telling. In one newspaper report, a man 105 years old described a swallowing-of-the-young episode he had witnessed at the age of eighteen. Certainly some mother, of all the hundreds under observation in zoos and laboratories would by now have temporarily forgotten her cage and opened her mouth to protect her young. But none has done so.

A San Antonio dealer, who handled from forty thousand to fifty thousand snakes per year, has said that two thousand young had been born to his rattlers one year. The mothers paid no attention to the young. The dealer experimented and found that an adult rattler's gastric juices would kill a young snake in about twenty minutes.

Mother Nourishing Young

A peculiar myth, accessory to the young-swallowing idea, is that mother rattlesnakes nourish their young in their stomachs when they swallow them. This has even been seriously suggested by the great naturalist John Burroughs. There is not the slightest foundation for this story, originally proposed because of difficulty of conceiving how young rattlers could eat the same food as the adults. It is now known that, although they can and do eat young mice, they depend on lizards more than do the adults.

The Fatal Boot

One of the best-defined rattlesnake legends is that which may be called the story of the fatal boot. Briefly it is as follows: a man, while plowing in a field, is bitten by a rattler and succumbs. Some time later—as measured in months or years—his son, now grown to man's estate, mysteriously sickens and dies. After another interval the same sad fate overwhelms a second son; and it is only then that the family physician, seeking the cause of these successive tragedies, finds that each son had, before his death, worn a pair of boots inherited from the unfortunate parent. In the inside of one boot there is found protruding the fang of the rattler that had so foully slain the father, and which, undiscovered through the years, has lain awaiting the day when it should strike down his offspring.

In this fable there is a considerable variation in the relationships of the triads involved in the fatality, some of them being: a father and two sons; a grandfather, father, and son; or three successive husbands of the widow (serves them right).

The weakness of the story lies in the belief that the point of a fang could contain enough dried venom to do serious damage. For the fang aperture, which conceivably might hold enough dried venom to cause at least a slight pain, lies well back to the point; and any venom on the point would be scraped off by passage through the leather of the boot.

One poor man was said to have scratched his forefinger with the fang of a dried snake skull and almost lost an arm. But was the trouble caused by venom? I had a somewhat similar experience: I shoved my hand into a new glove into which a pin holding the price tag protruded. The result was serious indeed, but it was an ordinary infection; there was neither snake nor venom involved. Experiments with venom on a dried fang discount all stories based on a danger from this source.

There is a tale of a man who killed a rattler and left it in a field. Seven years later, while grubbing for potatoes he stuck himself with a fang and suffered severe poisoning. Another account makes the interval eighteen years. In any case, assuming the accident to have happened, it was clearly an ordinary infection. No tiny quantity of a protein poison, unprotected from the elements for all those intervening years, could have produced such a result.

Here is a modern tall-story variant of the boot myth: a cowboy runs over a rattler and two successive garage men who repair the resulting puncture die from being punctured by the fang. The original boot story is now being embroidered in Texas: after the third human death, the fang taken from the boot is buried, only to be dug up by a dog with fatal results to itself.

The Roadrunner's Cactus Corral

One of the most colorful, as well as widespread, rattler myths of the Southwest concerns the roadrunner, a sprightly inhabitant of this arid land. According to the story, the roadrunner, upon finding a rattler asleep in the open, builds a wall about it with cactus stems or lobes. When the rattler awakes, it endeavors to escape. Prevented by the impenetrable wall and stuck by its spines, it becomes enraged, bites itself, and dies. In some versions the roadrunner eats its victim; in others, it merely gives a characteristic cry of victory and goes elsewhere to seek another customer.

As given, the story contains two essential weaknesses: first, rattlesnakes readily glide through or over cactus; they often seek cactus beds as places of refuge and seem undismayed by the spines. Second, rattlesnakes are virtually immune to their own venoms.

The Cabin Built on a Den

One of the early myths no longer heard is of the family of emigrants who were so unfortunate as to erect their home (presumably in winter) backed against a rock ledge. The ledge proved to be a rattlesnake den, and when the family was celebrating its housewarming with a glorious fire on the hearth, the warmed-up snakes issued forth in great numbers to the discomfort of the occupants. Probably this folklore is the outgrowth of the finding of a rattler or two in some settler's cabin, certainly not an unusual event.

The Rattler and the Wagon Wheel

A snake myth current in my youth, when I drove many a mile in a fringed surrey, has now almost died out because of transportation changes. The story was that a rattler, upon being run over by the wheel of a buckboard or such-like vehicle, coiled around the felloe of the wheel and then, at the top of the travel, was thrown into the lap of one of the occupants, usually the girl of the story. This yarn had a sufficient vogue in its day, so that many a driver was glad to have the horses turn out to avoid a crossing rattler. The automobile almost ruined this story, especially since the advent of the modern fender.

The Child Feeds a Rattlesnake

An ancient legend, which has crossed the Atlantic to be applied to rattlesnakes, is that of the child who befriends a snake. In the usual version a child, generally a girl, repeatedly carries her lunch out into the yard to eat. Finally the curiosity of her parents is aroused; they follow her and watch her sit down to enjoy her bowl of bread and milk. At once a huge snake—often a rattlesnake, if told in the United States—issues from the grass and coils beside her, to be fed milk and bread with a spoon. The horrified parents destroy the snake, with the result that the child, pining for her playmate, goes into a decline and dies.

There are many stories, some no doubt true, some legendary, about children found playing innocently with rattlesnakes. Doubtless this has often happened, since young children are usually not afraid of snakes. Although serious and sad accidents have resulted from occurrences of this kind, many have ended without harm if the rattler has not been frightened or roughly handled. From such has grown the dangerous myth that rattlesnakes will not bite children. In 1931 a boy of nine came to the reptile house of the San Diego Zoo, carrying a young rattler that he thought was a harmless snake.

The Thankful Rattlesnakes

A tall tale of early days was of the miner who kept at large in his cabin a tame rattler to which he fed the fattest mice available. Finally he was rewarded when his pet gave a warning rattle upon the entrance of an intruder who sought his cache of gold. This yarn, highly embroidered, eventually reached the lower strata of vaudeville. In this version the kindly rattler bound the burglar to a convenient bedpost with his coils and then hung his tail out of the window to rattle for the police. In one version of this story, a mother rattler held the burglar at bay while her infants rattled for the police. There was also the tale of the kindly hunter who saved the life of a big rattler, and was himself saved when he fell on a track in front of a train, whereupon the rattler pulled the hunter's red bandana out of his pocket and flagged the train.

Other stories of this type include the thankful rattler that brought his fellows to form a rope long enough to rescue the man who had befriended him from a mine shaft into which he had fallen.

The Deep Freeze

There are a number of tall tales involving frozen rattlesnakes, such as that of the man in winter who built a fence with five hundred posts, four to fourteen feet long, driven into the ground. But when it warmed up, the posts proved to be rattlers; they crawled off dragging two miles of barbed wire. Then there was the frozen rattler that was used first as a cane, and later as a poker. Every time it thawed out it was put outside to freeze. And the tale of the old nurse who went outside on a cold night to gather firewood; when she put the load down before the fire the sticks thawed out and crawled away. Another horrid tale concerns the man, who, to sleep beyond the reach of the numerous rattlers round about, built himself a hammock. Awakened by a jerking in the morning he found that what he had thought were ropes, proved to be rattlers now thawed out.

Tribal Heroes

Some tall tales apply to particular persons. Among these are rattle-snake stories attributed to Pecos Bill, the west Texas cowpuncher's Paul Bunyan, and Febold Feboldson, a man of similar stature recently arisen in the Great Plains. Pecos Bill, knowing that rattlers eat moth balls, but not chili powder, prepared some special balls with centers of mixed chili and nitroglycerine, but with exteriors of the usual naphthalene. When the coating melted off and the chili burned their insides, the rattler struck their tails on the rocks and exploded the nitroglycerine. Pecos Bill also rode a mountain lion, using a ten-foot rattler as a quirt. In capturing the rattler, to be fair, he gave it the first three bites.

Febold Feboldson owned a fourteen-foot rattler named Arabella. Once, when Febold was tied up by Indians, Arabella squirted venom on the rope, disintegrating it. Again, on a stroll, Arabella rattled the snakes' national anthem, and she and Febold were immediately surrounded by a huge assemblage of her compatriots. Febold gave himself up for lost, but Arabella rattled Brahms's "Lullaby" and put her friends to sleep. To carry Arabella, Febold had a basket back of his saddle called the "rattle seat" later corrupted to "rumble seat." Febold invented the booby trap by hol-lowing out every fourth watermelon in his patch and putting a rattler inside. This soon caused the Indians to quit stealing the melons, but as Febold was unable to remember which melons were salted, he couldn't sell them without violating the Pure Food and Drugs Act and so had to give them away.

MISCELLANEOUS BELIEFS AND FOLKLORE

Incipient Myths

One myth growing out of World War II is certainly destined for im-mortality; this is the one that tells of the soldier in training, who is advancing at a crawl under a curtain of live machine-gun fire just over-head. He meets a rattler face to face—usually a "huge sidewinder"—and raises up into the stream of bullets with disastrous results. I think I have heard this story as happening in every one of the Southwestern camps where troops were trained for desert warfare. I suppose it may have actu-ally occurred once.

A story rapidly gaining the status of folklore tells of a prospector who, while searching for fluorescent minerals with an ultraviolet lamp, almost picks up a glowing rattler by mistake. (Actually, I have found by test that rattlesnake bodies do not fluoresce in ultraviolet light, although the rattles glow slightly and the fangs strongly).

Dreams

There are many superstitions concerning dreams about snakes. Generally they foretell the existence of an enemy; if you kill the snake, you will overcome the enemy. An Alabama belief has it that if you dream of a rattler a conjure man has "put something down for you." If the rattler tried to bite you and failed, you have escaped the conjure doctor's tricks. A small rattler indicates a weak doctor, a large one a strong.

Superstitions about Meeting Rattlers

There are many superstitions of a world-wide character having to do with encountering snakes, such as the disaster to be expected if one sees a snake at the start of a journey or fails to kill the first snake seen in the spring. As these beliefs apply to all snakes, rattlesnakes are rarely mentioned specifically. It is reported that in Madras people postpone their journeys if they see a cobra or rattlesnake. I suppose that anyone seeing a rattlesnake at large anywhere in Asia, Africa, or Europe, would be justified in taking even stronger measures.

Rattlesnakes and Weather

There are many myths, world-wide in derivation and spread, concerning the relationships of snakes to weather.

Some specific applications of weather lore to rattlesnakes are the following: drape a rattler on its back over a log and it will rain in three days. If rattler tracks are many and directed to high ground, rain will be abundant; or, if rattlers are unusually vicious, rain will follow. Turning a live or a dead rattler on its back will bring rain.

Miscellaneous Myths, Legends, and Stories

Here are a number of myths and stories about rattlesnakes that have appeared in print, or have been heard voiced by visitors at the zoo, that do not lend themselves to classification.

A shirt spotted with rattlesnake venom cannot be cleaned. A Nebraska folk belief is that a sidewinder will kill any living thing it touches. If the sidewinder sticks you with a horn it will be fatal. (Actually, the horns are soft and flexible.)

A man was badly poisoned by a rope burn where a rattler had bitten the rope. A rattler struck a boy's leggings; his dog licked the leggings and died (West Virginia folk tale). A person bitten by a rattler over which birds are circling is not likely to die (Texas superstition). The tale is told of the use of rattlers to cure the bites of stinging snakes, for only bites by rattlers will do this. What a stinging snake may be remains undetermined. A band of hogs ate so many rattlesnakes they grew rattles on their tails.

Remarks made by people in front of rattlesnake cages at the zoo: all snakes shed their skins except rattlesnakes. Rattlers dislike certain colors, especially blue. Rattlesnakes won't rattle at night. The rattle is used as an anchor from which a striking snake lunges; also as a brake when sliding downhill. The reason rattlesnakes are not found in the vicinity of Puget Sound — it is true, they are not — is because the available water is so pure that rattlesnakes cannot generate venom from it and therefore starve to death. Television antennas attract rattlesnakes. (This one, at least, we may be sure did not come down from Pliny.) As animals without gall bladders are immune to rattlesnake bite (so the legend runs), a woman was heard to inquire whether her recent operation put her in that category.

Conclusion

Of all these myths, that which has most deeply affected human impressions and attitudes toward rattlesnakes is the one that pictures these snakes as malignant, vindictive, and crafty, with an especial hatred for mankind.

But a rattlesnake is only a primitive creature with rudimentary perceptions and reactions. Dangerous it surely is, and I hold no brief for its survival except in remote areas where its capacity to destroy harmful rodents may be exercised without danger to man or his domestic animals. But that the rattlesnake bears an especial enmity toward man is mythical. It seeks only to defend itself from injury by intruders of superior size, of which man is one. It could not, through the ages, have developed any especial enmity for man, since the first human being any rattlesnake may encounter is usually the last.

Bibliography

Bellaris, A. D'A
  1957. Reptiles. London.
  1970. The Life of Reptiles, Vols. 1 & 2. London.
Benedict, F. G.
  1932. The Physiology of Large Reptiles with Special Reference to the Heat Production of Snakes, Tortoises, Lizards, and Alligators. Carn. Inst., Publ. 425, pp. x + 539.
Bullock, T. H., and R. B. Cowles
  1952. Physiology of an Infrared Receptor: The Facial Pit of Pit Vipers. Science, vol. 115, no. 2994, pp. 541–543.
Bullock, T. H., and F. P. J. Diecke
  1956. Properties of an Infra-red Receptor. Jour. Physiol., vol. 134, no. 1, pp. 47–87.
Bullock, T. H., and W. Fox
  1957. The Anatomy of the Infra-red Sense Organ in the Facial Pit of Pit Vipers. Quart. Jour. Microscop. Sci., vol. 98, part 2, pp. 219–234.
Carr, A.
  1963. The Reptiles. New York.
Cochran, Doris M.
  1944. Dangerous Reptiles. Smithsonian Rept. for 1943, pp. 275–324.
Conant, R.
  1975. A Field Guide to Reptiles and Amphibians of the U.S. and Canada East of the 100th Meridian, 2nd edition. Boston.
Copeia
  Journal of the American Society of Ichthyologists and Herpetologists.
Cott, H. B.
  1940. Adaptive Coloration in Animals. London.
Cowles, R. B.
  1953. The Sidewinder: Master of Desert Travel. Pac. Discovery, vol. 6, no. 2, pp. 12–15.

Cowles, R. B., and C. M. Bogert
1944. A Preliminary Study of the Thermal Requirements of Desert Reptiles. Bull. Am. Mus. Nat. Hist., vol. 83, art. 5, pp. 261–296.
Cowles, R. B., and R. L. Phelan
1958. Olefaction in Rattlesnakes. Copeia, no. 2, pp. 77–83.
Curran, C. H., and C. F. Kauffeld
1937. Snakes and Their Ways. New York.
Dowling, H., editor
1974. HISS Yearbook of Herpetology. New York.
Fitch, H. S., and B. Glading
1947. A Field Study of a Rattlesnake Population. Calif. Fish and Game, vol. 33, no. 2, pp. 103–123.
Gadow, H.
1901. Amphibia and Reptiles. Cambridge Nat. Hist. (London), vol. 8.
Gans, C.
1966. The Biting Behavior of Solenoglyph Snakes — Its Bearing on the Pattern of Envenomation. Intern. Symp. Animal Venoms, pp. xii–xiii. Abstract.
Gloyd, H. K.
1938. Methods of Preserving and Labeling Amphibians and Reptiles for Scientific Study. Turtox News, vol. 16, no. 3, pp. 49–53, 66–67.
Goin, C., and L. Goin.
1976. Introduction to Herpetology, 3rd edition. San Francisco.
Herpetologica
Journal of the Herpetologist's League.
Keegan, H. L., and W. V. Macfarlane (editors)
1963. Venomous and Poisonous Animals and Noxious Plants of the Pacific Region. New York.
Minton, S. A., and M. R. Minton
1969. Venomous Reptiles. New York.
Moore, G. M., et al.
1968. Poisonous Snakes of the World. U. S. Govt. Printing Office, Washington, D.C.
Morris, R., and D. Morris
1965. Snakes and Men. New York.
1957. Surf Rattlers. Tex. Game and Fish, vol. 15, no. 3, p. 17.
Oliver, J. A.
1955a. The Natural History of North American Amphibians and Reptiles. Princeton, N. J.
1955b. The Natural History of North American Amphibians and Reptiles. Princeton, N. J.
1958. Snakes in Fact and Fiction. New York.
1963. Snakes in Fact and Fiction. New York.
Parker, H. W.
1965. Natural History of Snakes. Brit. Mus.
Pope, C. H.
1955. The Reptile World. New York.
1961. The Giant Snakes. New York.

Porges, N.
  1953. Snake Venoms, Their Biochemistry and Mode of Action. Science, vol. 117, no. 3029, pp. 47–51.
Russell, F. E.
  1965. Venomous Animals and Their Toxins. Smithson. Rept. (1964), pp. 477–487.
Schmidt, K. P., and R. F. Inger
  1957. Living Reptiles of the World. Garden City, New York.
Shannon, F. A.
  1953. Comments on the Treatment of Reptile Poisoning in the Southwest. Southwest. Med., vol. 34, no. 10, pp. 367–373. [See also pp. 634–635, in Current Therapy, H. F. Conn, editor, Philadelphia, 1955.]
Shaw, C. E.
  1951. Male Combat in American Colubrid Snakes with Remarks on Combat in Other Colubrid and Elapid Snakes. Herpetologica, vol. 7, part 4, pp. 149–168.
Shaw, C. E., and S. Campbell.
  1974. Snakes of the American West. New York.
Smith, H. M.
  1946. Handbook of Lizards of the United States and Canada. New York.
Stebbins, R. C.
  1954. Amphibians and Reptiles of Western North America. New York, pp. xxii + 528.
  1966. A Field Guide to Western Reptiles and Amphibians. Boston.
  1972. Amphibians and Reptiles of California. Berkeley.
Zimmerman, A. A., and C. H. Pope
  1948. Development and Growth of the Rattle of Rattlesnakes. Fieldiana: Zoöl, vol 32, no. 6, pp. 357–413.

Index

NOTE: For rattlesnake species and subspecies, see list pp. 9–10, and under scientific name in the index (common names are cross-referenced). For other snakes and animals mentioned, see under common name.